Experimental Fluid Mechanics

R. J. Adrian · M. Gharib · W. Merzkirch
D. Rockwell · J. H. Whitelaw

Springer

Berlin
Heidelberg
New York
Barcelona
Hong Kong
London
Milan
Paris
Singapore
Tokyo

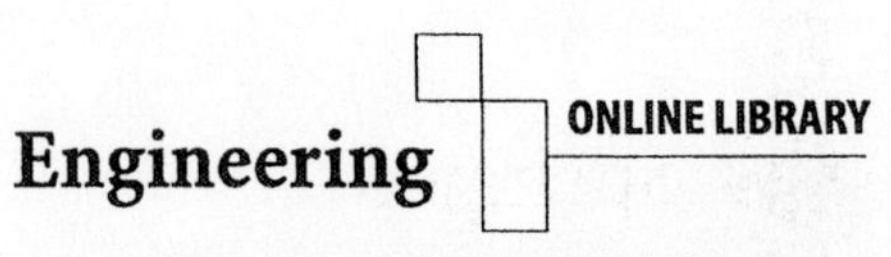

M. Raffel · C. E. Willert · J. Kompenhans

Particle Image Velocimetry

A Practical Guide

Corrected 3rd Printing

With 165 Figures and 24 Tables

Authors

Dr.-Ing. Markus Raffel
Dr. Christian E. Willert
Dr. Jürgen Kompenhans

Deutsche Forschungsanstalt
für Luft- und Raumfahrt e. V
Bunsenstraße 10
37073 Göttingen
Germany

ISBN 3-540-63683-8 Springer-Verlag Berlin Heidelberg New York

Cataloging-in-Publishing Data applied for

Die Deutsche Bibliothek - CIP-Einheitsaufnahme
Raffel, Markus: Particle image velocimetry: a practical guide / Markus Raffel; Christian E. Willert; Jürgen Kompenhans. - Berlin; Heidelberg; New York; Barcelona; Budapest; Hongkong; London; Mailand; Paris; Singapur; Tokio: Springer, 1998
ISBN 3-540-63683-8

Springer-Verlag Berlin Heidelberg New York
a member of BertelsmannSpringer Science+Business Media GmbH

http://www.springer.de

Typesetting: Camera-ready copy from author
Cover-design: design & production, Heidelberg
Printed on acid-free paper SPIN: 10793354 61/3020 hu - 5 4 3 2 -

Series Editors

Preface

That moment, when after many months of work, the endeavored result turns into shape as printed pages, is a good opportunity to lean back and to reflect on the history of the subject of this book: particle image velocimetry (PIV). The rapid development, which this measuring technique that allows us to capture the flow velocity of whole flow fields in a fraction of a second, has undergone in the past decade can best be characterized by the experience of the authors gained during their research work in the Institute of Fluid Mechanics of the Deutsches Zentrum für Luft- und Raumfahrt (DLR). The first applications of particle image velocimetry outside of the laboratory in wind tunnels as performed in the mid-eighties were characterized by the following time scales: time required to set up the PIV system and to obtain well focused photographical PIV recordings was 2 to 3 days, time required to process the film was 0.5 to 1 day, time required to evaluate a single photographical PIV recording by means of optical evaluation methods was 24 to 48 hours. Today, with modern video cameras and fast computers it is possible to focus on line, to capture up to several hundred recordings per minute, and to evaluate a digital recording within a few seconds.

Even more important than this remarkably improved performance of the PIV technique, is its unique ability to capture instantaneous flow fields and thus to allow the detection of spatial structures in unsteady flows quantitatively, which is not possible with other experimental techniques. A number of investigations in very different areas from aerodynamics to biology, from turbulence research to applications in the space shuttle, from fluid mechanics to two phase flows have proven this. Due to this wide range of possible applications of PIV the number of research groups employing the PIV technique world wide has increased from a handful at the beginning of the eighties to far beyond a thousand today.

The third reason for the increasing interest in PIV is the demand for experimental flow field data for the validation of numerical codes. For this purpose the development of the PIV technique must go on in order to achieve higher accuracy, higher spatial and temporal resolution, and larger observation areas and volumes of experimental data.

Due to the number of different applications of PIV and due to the number of different possibilities to illuminate, to record and to evaluate, the many different technical modifications of the PIV technique have been developed in the past. Even for experts it was not always easy to decide which implementation of the technique would be best suited for a given application. During the last few years this has changed. Due to the development of modern cross correlation video cameras and appropriate fast software algorithms, the digital implementation of PIV seems to be the first choice for most applications. Therefore, the authors of this book felt that it was timely to compile the knowledge about the basic principles of PIV and the main guidelines for its implementation in practice. Most of the material covered in this book has already been published in conference proceedings or in scientific journals. However, this information is widespread and cannot be easily found by someone who wants to start employing the PIV technique for his special problems. Moreover, most publications illuminate the problems only from a specific point of view.

Organization of the book

The intention of this book is to present in a more general context mainly those aspects of the PIV technique relevant to applications. This strategy is supported by the experience of the authors which is based on their own work in the development of PIV for nearly 15 years and more than 30 different applications of DLR's mobile PIV system in aerodynamics and related areas ranging from subsonic to transonic flows, from turbulence research to the investigation of sprays, and from small test facilities to large industrial wind tunnels. The authors have also considerably contributed to the Courses on Particle Image Velocimetry held during the last 5 years in their laboratory. The presentation of the material in this book takes into account the feedback from the participants of these courses as well.

This practical guide to particle image velocimetry provides in a condensed form all that information relevant for the planning, performance and understanding of experiments employing the PIV technique. It is mainly intended for engineers, scientists and students, who have already some basic knowledge of fluid mechanics and nonintrusive optical measurement techniques. For many researchers and engineers, planning to utilize PIV for their special industrial or scientific applications, PIV is just an attractive tool with unique features which may help them to gain new insights in problems of fluid mechanics. These people are usually not interested in becoming specialists in this field first before starting their investigations. On the hand side some of the basic properties of particle image velocimetry must be well understood before a correct interpretation of the results is possible. Our hope is that this practical guide on particle image velocimetry will serve this purpose by providing an easy transfer of the know how gathered by the authors during many years to the readers of this book. It will help the readers to avoid beginners' errors and bring them to a position to obtain high quality results when employing PIV right from the beginning of their work. For those, already working in the field of PIV, this book may serve as a reference to further publications containing more details, which may be first consulted in case of open questions. As with all publications, also in this book the amount of information which can be presented must be limited. Nevertheless, it is our hope that we have been able to collect all information relevant to practical work with PIV.

About the authors

Markus Raffel received his degree in mechanical engineering in 1990 from the Technical University of Karlsruhe and his doctorate in 1993 from the University of Hannover, Germany. He started working on particle image velocimetry at DLR Göttingen in 1991 with emphasis on the development of PIV recording techniques in high-speed flows. In this process he applied the method to a number of aerodynamic problems mainly in the context of rotorcraft investigations.

Christian Willert received his BS degree in Applied Science from the University of California at San Diego (UCSD) in 1987. Subsequent graduate work in experimental fluid mechanics at UCSD lead to the development of several nonintrusive measurement techniques for application in water (particle tracing, 3-D particle tracking, digital PIV). After receiving his Ph.D. in Engineering Sciences in 1992, he assumed post-doctoral positions first at the Institute for Nonlinear Science (INLS) at UCSD, then at the Graduate Aeronautical Laboratories at the California Institute of Technology (Caltech). In April 1994 he joined DLR Göttingen's measurement sciences group as part of an exchange program between Caltech and DLR (i.e. Center for Quantitative Visualization, CQV). There he continues to work in the development and application of PIV techniques with special emphasis on wind tunnel applications.

Jürgen Kompenhans received his doctorate in physics in 1976 from the University of Göttingen. Since 1977 he has been employed by DLR, the German Aerospace Center. First, he performed experimental research work on problems of aero-acoustics. For nearly 15 years now he has been involved in the development of nonintrusive measurement techniques for aerodynamic research in wind tunnels (mainly particle image velocimetry). Presently he is working in the Institute of Fluid Mechanics at DLR's research center Göttingen. Since 1985 the PIV-group, headed by J. Kompenhans, has developed and continuously improved a PIV system dedicated to the application of PIV in the rough environmental conditions of large, industrial-scale wind tunnels. This system has been successfully applied to a number of different investigations within national and international projects.

Acknowledgments

During the past decade a number of colleagues, some of them being members of our group only for a limited time, have contributed to the progress of our work: technicians, students and scientists. Among these we especially want to acknowledge the contributions of Hans. R. Höfer, Aymeric Derville, Matthew Gaydon, Markus Fischer, Markus Wiegel, Christian Kähler, Andreas Schröder, Olaf Ronneberger, Hannes Reichmuth, Rainer Höcker, Andreas Vogt, Bernward Bretthauer, Heinrich Vollmers, and Boleslaw Stasicki. Our work has been financially supported by DLR, as well as by other national and international institutions. We always found a strong interest in our work and support from our colleagues from the Measurement Methods and Flow Analysis Section of the Institute of Fluid Mechanics of DLR and its head, Karl-Aloys Bütefisch.

We especially acknowledge the support of
- Christian Kähler (typesetting, figures, layout, comments on lasers,
- and many valuable suggestions for the improvement of text and figures)
- Andreas Schröder (typesetting, figures)
- Olaf Ronneberger (comments on correlation techniques)
- Hugues Richard (typesetting, figures)
in the final phase of the preparation of this book. The section on optical evaluation techniques is based on the work of Andreas Vogt. The fluidmechanical sections on "Boundary layer instabilities" and "Turbulent boundary layers" are based on the work of Christian Kähler.

Stereo: The projection equations and the necessary nonlinear least squares fitting algorithms for stereoscopic PIV were brought to the attention of the authors by Heinrich Vollmers.

Dual-plane: For the first time dual-plane PIV was implemented during Markus Raffel's and Olaf Ronneberger's visit to Caltech. Alexander Weigand's generous offer of his experimental set-up and stimulating discussions with Jerry Westerweel and Thomas Roesgen are greatly appreciated. In fact, the cooperative work carried out in the Center for Quantitative Visualisation, jointly operated by GALCIT of Caltech and DLR, was quite successful. Special thanks to Mory Gharib.

Flow field investigations by means of PIV, discussed in this book, have been performed at the facilities of different research organizations such as DLR, the German–Dutch Wind Tunnel (DNW) with the special support of its former director, Hans Ulrich Meier, Zentrum für Angewandte Raumfahrttechnologie und Mikrogravitation (ZARM), Institut Franco–Allemand de Recherches de Saint-Louis (ISL), and others.

The comments made by Klaus Hinsch, Dirk Ronneberger and Andreas Vogt on the draft version of this book are greatly appreciated as well.

We are deeply indebted to all friends and colleagues of the worldwide PIV community who helped us to better understand the different aspects of the PIV technique during recent years by their work, their publications and conference contributions, and by personal discussions.

The authors, Göttingen, September 2000

Markus Raffel
Chris Willert
Jürgen Kompenhans

Table of Contents

1. Introduction

1.1 Historical background

Human beings are extremely interested in the observation of nature, as this was and still is of utmost importance for their survival. Human senses are especially well adapted to recognize moving objects as in many cases they mean eventual danger. One can easily imagine how the observation of moving objects has stimulated first simple experiments with set-ups and tools easily available in nature. Today the same primitive behavior becomes obvious, when small children throw little pieces of wood down from a bridge in a river and observe them floating downstream. Even this simple experimental arrangement allows them to make a rough estimate of the velocity of the running water and to detect structures in the flow such as swirls, wakes behind obstacles in the river, water shoots, etc.

However, with such experimental tools the description of the properties of the flow is restricted to qualitative statements. Nevertheless, being at the same time an artist with excellent skills and an educated observer of nature, LEONARDO DA VINCI, was able to prepare very detailed drawings of the structures within a water flow by mere observation.

Fig. 1.1. LUDWIG PRANDTL in front of his water tunnel for flow visualization in 1904

A great step forward in the investigation of flows was made after it was possible to replace such passive observations of nature by experiments carefully planned to extract information about the flow utilizing visualization techniques. A well known promoter of such a procedure was LUDWIG PRANDTL, one of the most prominent representatives of fluid mechanics, who designed and utilized flow visualization techniques in a water tunnel to study aspects of unsteady separated flows behind wings and other objects.

Figure 1.1 shows LUDWIG PRANDTL in 1904 in front of his tunnel, driving the flow manually by rotating a blade wheel [179]. The tunnel comprises an upper and lower section separated by a horizontal wall. The water recirculates from the upper open channel, where the flow may be observed, back through the lower closed duct. Two-dimensional models like cylinders, prisms, and wings can be easily mounted vertically in the upper channel, thereby extending above the level of the surface of the water.

The flow is visualized by distributing a suspension of mica particles on the surface of the water. LUDWIG PRANDTL studied the structures of the flow in steady as well as in unsteady flow (at the onset of flow) with this arrangement [166].

Being able to change a number of parameters of the experiment (model, angle of incidence, flow velocity, steady–unsteady flow) PRANDTL gained insight into many basic features of unsteady flow phenomena. However, at that time only a qualitative description of the flow field was possible. No quantitative data about flow velocity, etc., could be achieved.

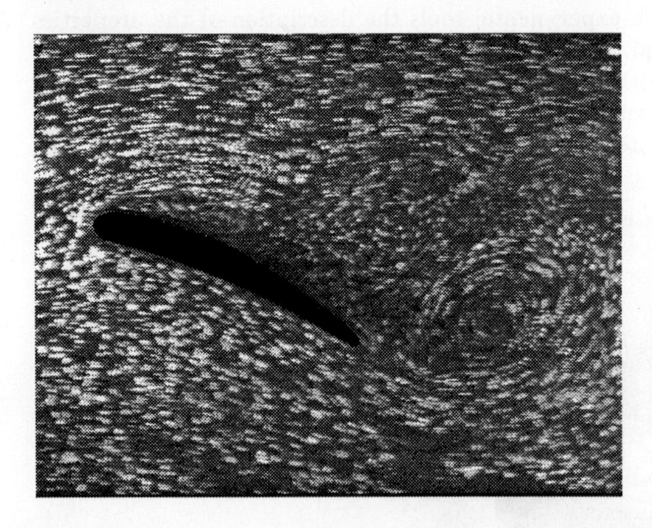

Fig. 1.2. Separated flow behind wing, visualized with modern equipment in a replica of LUDWIG PRANDTL's tunnel

Today, 90 years after LUDWIG PRANDTL's experiments, it is easily possible to also extract quantitative information about the instantaneous flow velocity field exactly from the same kind of images as were available to PRANDTL. A proof for this is given in figure 1.2. A replica of LUDWIG PRANDTL's water tunnel together with a flash lamp for illumination and a video camera have been employed to obtain a visualization of the flow by means of aluminum particles distributed on the water surface.

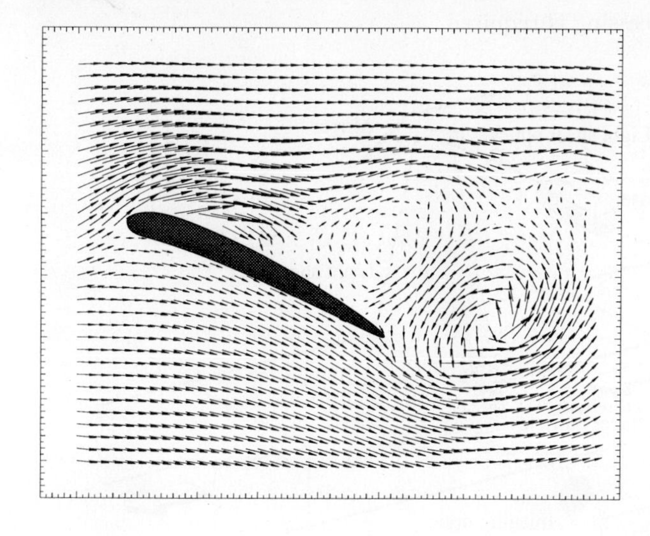

Fig. 1.3. Vector map of instantaneous velocity field corresponding to figure 1.2

Evaluation of this recording by methods which will be described later resulted in a vector map of the instantaneous velocity field shown in figure 1.3. This means that the basic principles underlying the quantitative visualization technique which is the subject of this book have already been known for a long time.

However, the scientific and technical progress achieved in the last 15 years in optics, lasers, electronics, video and computer techniques was necessary to further develop a technique for *qualitative* flow visualization to such a stage that it can be employed for *quantitative* measurement of complex instantaneous velocity fields.

1.2 Principle of particle image velocimetry (PIV)

In the following the basic features of this measurement technique, most widely named "particle image velocimetry" or "PIV", will be described briefly[1].

The experimental set-up of a PIV system typically consists of several subsystems. In most applications tracer particles have to be added to the flow. These particles have to be illuminated in a plane of the flow at least twice within a short time interval. The light scattered by the particles has to be

[1] In earlier years other names such as *speckle velocimetry*, *particle image displacement velocimetry* etc. have been used as well.

recorded either on a single frame or on a sequence of frames. The displacement of the particle images between the light pulses has to be determined through evaluation of the PIV recordings. In order to be able to handle the great amount of data which can be collected employing the PIV technique, sophisticated post-processing is required.

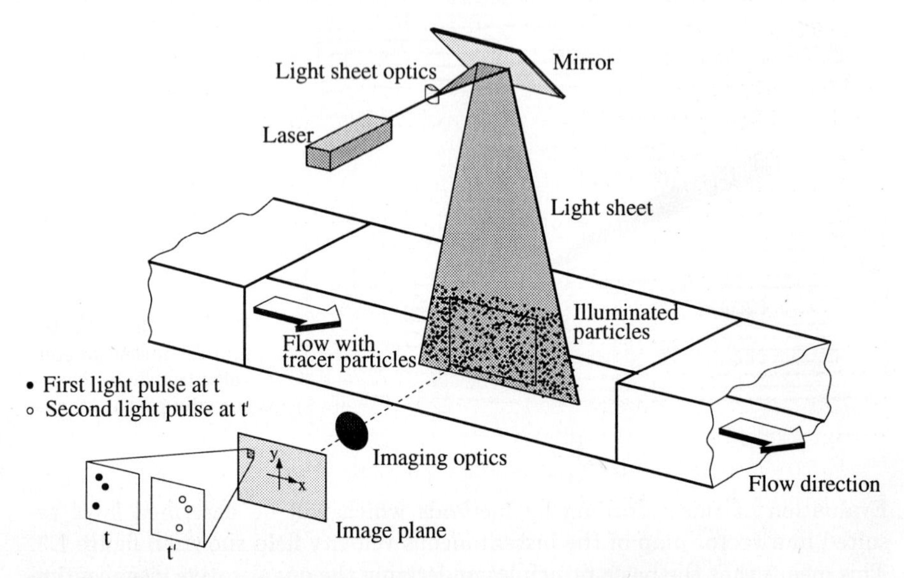

Fig. 1.4. Experimental arrangement for particle image velocimetry in a wind tunnel

Figure 1.4 briefly explains a typical set-up for PIV recording in a wind tunnel. Small tracer particles are added to the flow. A plane (light sheet) within the flow is illuminated twice by means of a laser (the time delay between pulses depending on the mean flow velocity and the magnification at imaging). It is assumed that the tracer particles move with local flow velocity between the two illuminations. The light scattered by the tracer particles is recorded via a high quality lens either on a single photographical negative or on two separate frames on a special cross correlation CCD sensor. After development the photographical PIV recording is digitized by means of a scanner. The output of the CCD sensor is stored in real time in the memory of a computer directly.

For evaluation the digital PIV recording is divided in small subareas called "interrogation areas". The local displacement vector for the images of the tracer particles of the first and second illumination is determined for each interrogation area by means of statistical methods (auto- and cross- correlation). It is assumed that all particles within one interrogation area have moved homogeneously between the two illuminations. The projection of the vector of the local flow velocity into the plane of the light sheet (2-component velocity vector) is calculated taking into account the time delay between the two illuminations and the magnification at imaging.

The process of interrogation is repeated for all interrogation areas of the PIV recording. With modern video cameras (1000×1000 sensor elements) it is possible to capture more than 100 PIV recordings per minute. The evaluation of one video PIV recording with 3600 instantaneous velocity vectors (depending on the size of the recording and of the interrogation area) is of the order of a few seconds with standard computers. If even faster availability of the data is required for on line monitoring of the flow, dedicated hardware processors are commercially available which perform evaluations of similar quality within fractions of a second.

Before going into the details of the PIV technique, some general aspects have to be discussed in order to facilitate the understanding of certain technical solutions later on.

Nonintrusive velocity measurement. In contrast to techniques for the measurement of flow velocities employing probes as pressure tubes or hot wires, the PIV technique being an optical technique works nonintrusively. This allows the application of PIV even in high speed flows with shocks or in boundary layers close to the wall, where the flow may be disturbed by the presence of probes.

Indirect velocity measurement. In the same way as with laser Doppler velocimetry the PIV technique measures the velocity of a fluid element indirectly by means of the measurement of the velocity of tracer particles within the flow, which – in most applications – have been added to the flow before the experiment started. In two phase flows particles are already present in the flow. In such a case it will be possible to measure the velocity of the particles themselves as well as the velocity of the fluid (to be additionally seeded with small tracer particles).

Whole field technique. PIV is a technique which allows one to record images of large parts of flow fields in a variety of applications in gaseous and liquid media and to extract the velocity information out of these images. This feature is unique to the PIV technique. Except Doppler global velocimetry (DGV) [177, 178], which is a new technique particularly appropriate for high speed air flows, all other techniques for velocity measurements only allow the measurement of the velocity of the flow at a single point, however in most cases with a high temporal resolution. With PIV the spatial resolution is large, whereas the temporal resolution (frame rate of recording PIV images) is limited due to technical restrictions. These features must be observed if comparing results obtained by PIV with those obtained with traditional techniques. Instantaneous image capture and high spatial resolution at PIV allow the detection of spatial structures even in unsteady flow fields.

Velocity lag. The need to employ tracer particles for the measurement of the flow velocity requires us to check carefully for each experiment whether the particles will faithfully follow the motion of the fluid elements, at least to that extent required by the objectives of the investigations. Small particles will follow the flow better.

Illumination. For applications in gas flows a high power light source for illumination is required in order that the light scattered by the tiny tracer particles will expose the photographic film or the video sensor. However, the need to utilize larger particles because of their better light scattering efficiency is in contradiction to the demand to have as small particles as possible in order that they follow the flow faithfully. In most applications a compromise has to be found. In liquid flows larger particles can usually be accepted which scatter much more light. Thus, light sources of considerably lower peak power can be used here.

Duration of illumination pulse. The duration of the illumination light pulse must be short enough that the motion of the particles is "frozen" during the pulse exposure in order to avoid blurring of the image ("no streaks").

Time delay between illumination pulses. The time delay between the illumination pulses must be long enough to be able to determine the displacement between the images of the tracer particles with sufficient resolution and short enough to avoid particles with an out-of-plane velocity component leaving the light sheet between subsequent illuminations.

Distribution of tracer particles in the flow. At qualitative flow visualization certain areas of the flow are made visible by marking a stream tube in the flow with tracer particles (smoke, dye). According to the location of the seeding device the tracers will be entrained in specific areas of the flow (boundary layers, wakes behind models, etc.). The structure and the temporal evolution of theses structures can be studied by means of qualitative flow visualization. For PIV the situation is different: a homogeneous distribution of medium density is desired for high quality PIV recordings in order to obtain optimal evaluation. No structures of the flow field can be detected on a PIV recording of high quality.

Density of images of tracer particles on the PIV recording. Qualitatively three different types of image density can be distinguished [29], which is illustrated in figure 1.5. In the case of low image density (figure 1.5 a), the images of individual particles can be detected and images corresponding to the same particle originating from different illuminations can be identified. Low image density requires tracking methods for evaluation. Therefore, this situation is referred to as "particle tracking velocimetry", abbreviated "PTV". In the case of medium image density (figure 1.5 b) the images of individual particles can be detected as well. However, it is no longer possible to identify image pairs by visual inspection of the recording. Medium image density is required to apply the standard statistical PIV evaluation techniques. In the case of high image density (figure 1.5 c) it is not even possible to detect individual images as they overlap in most cases and form speckles. This situation is called "laser speckle velocimetry" (LSV), a term which has been used at the beginning of the eighties for the medium image density case as well, as the (optical) evaluation techniques were quite similar for both situations.

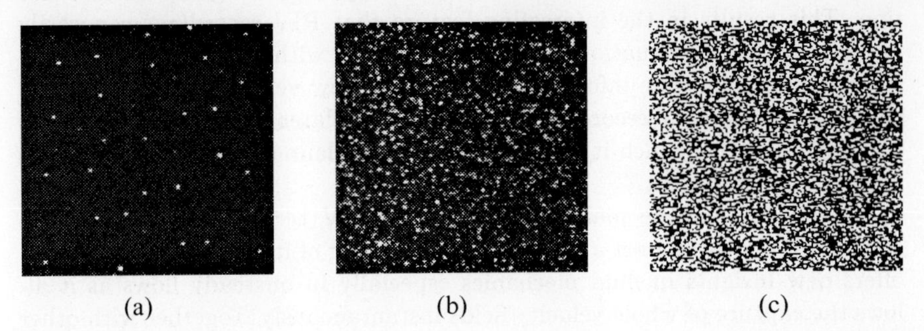

Fig. 1.5. The three modes of particle image density: (a) low (PTV), (b) medium (PIV), and (c) high image density (LSV).

Number of illuminations per recording. For both photographic and video techniques, we have to distinguish whether it is possible to store images of the tracer particles on different frames for each illumination or whether all particle images due to the different illuminations are stored on a single frame.

Number of components of the velocity vector. Due to the planar illumination of the flow field only two (in plane) components of the velocity vector can be determined in standard PIV (2C-PIV). Methods are already available to extract the third component of the velocity vector as well (stereo techniques, dual-plane PIV, holographic recording [32]). This would be labeled 3C-PIV. Both methods work in planar domains of the flow field (2D-PIV).

Extension of observation volume. In the most general way an extension of the observation volume is possible by means of holographic techniques (3D-PIV) [116]. Other methods such as establishing several parallel light sheets in a volume [32] or scanning a volume in a temporal sequence [103, 104] would be referred to as 2+1D-PIV.

Extension in time. By means of repetitively working cameras it is already possible to record temporal sequences of PIV recordings. However, as the repetition rate of pulse lasers and cameras is limited, it is not possible to record fast enough as would be required due to the temporal scales of most flows.

Size of interrogation area. The size of the interrogation area at evaluation must be small enough that velocity gradients have no significant influence on the results. Furthermore, it determines the number of independent velocity vectors and therefore the maximum spatial resolution of the velocity map which can be obtained at a given spatial resolution of the sensor employed for recording.

Repeatability of evaluation. In PIV full information about the flow velocity field (except the time delay between pulses and magnification at imaging) is stored at recording time at a very early stage of data reduction. This results in the interesting feature that PIV recordings can easily be exchanged for evaluation and post processing with others employing different techniques. The information about the flow velocity field completely contained in the PIV recording can be exploited later on in quite a different way from that for which it had originally been planned without the need to repeat the experiment.

In this section the main features of the PIV technique have been described briefly to support a general understanding of its unique features. PIV offers new insights in fluid mechanics especially in unsteady flows as it allows the capture of whole velocity fields instantaneously. Together with other quantitative flow visualization techniques [5] giving information about other important physical quantities of a fluid such as density, temperature, concentration, etc., which are already well known and widely used, and new optical methods for the measurement of quantities on the surface of a model such as pressure or deformation, a more complete experimental description of a complex flow field will be possible and will be available for comparison with the results of numerical calculations in future.

1.3 Development of PIV during the last two decades

The development of particle image velocimetry during the past 15 years is characterized by the fact that analog recording and evaluation techniques have been replaced by digital techniques. Though these analog methods have widely contributed to the fast initial success of the PIV technique the discussion of these techniques will not be one of the main objectives of this handbook on PIV. We will rather concentrate on the description of the present state of the art of PIV.

A number of sources describing the basic principles of PIV in the context of its historical development are readily available. Thus, for further information the reader is referred to the SPIE Milestone Series 99, edited by I. GRANT in 1994 [4]. This volume comprises more than 70 original papers, first published between 1932 and 1993. The majority of them originates from the eighties, including contributions about the roots of modern PIV (i.e. speckle interferometry), the early work of R. MEYNART [77], the development of low and high image density PIV, optical correlation techniques, etc. Review articles by W. LAUTERBORN and A. VOGEL (1984) [35] and by R. ADRIAN (1991) [29], which are also reprinted in the MS 99 on PIV, demonstrate the fast development and compilation of know-how about PIV within a decade.

The state of the art of PIV seen from the side of optics is described in the chapter "Particle Image Velocimetry" written by K. HINSCH in 1993 [31], included in a book on "Speckle metrology". This contribution is especially useful for the understanding of the optical aspects of PIV. It includes 104 references to other literature on PIV.

At that time strong competition with respect to the better performance of optical and digital methods in the evaluation of PIV recordings took place. Details of the theoretical fundaments of digital particle velocimetry can be found in the book *Digital particle image velocimetry – Theory and practice* published also in 1993 by J. WESTERWEEL [7]. This book includes more than 100 references.

A review paper "Particle image velocimetry: a review" by I. GRANT appeared in 1997 [30]. It gives a summary of different modifications of PIV illumination, recording and evaluation techniques, many of them not covered in this book. The paper includes 188 references.

As indicated all four publications mentioned above include a detailed bibliography of the literature on PIV, which the reader should use if more details are required on special aspects of PIV than it was possible to describe in this book. A further bibliography on PIV with nearly 1200 references was compiled by R. ADRIAN [1] and is available commercially.

The large number of references listed in the review articles demonstrate that particle image velocimetry is nowadays a well accepted tool for the investigation of velocity fields in many different areas. This also means that a number of special implementations of the PIV technique had to be developed for such different applications as, for example, in biology or in turbomachinery.

At present the widest use of PIV is made in fluid mechanics in the investigation of air and water flows. The progress made in the last years has brought PIV to such a state that it is close to being routinely applied in aerodynamic research. This is not yet the case for more complex flows (turbomachinery, two-phase flows, flames, etc.). This means that today a nearly complete and nearly stable picture of the technical aspects of PIV can best be given if looking at the demands of applications in aerodynamics or in water flows, where the technical problems are similar but usually much less severe than in air flows. Most of the technical problems in the application of PIV encountered in this special field appear in other PIV applications as well. Many of the basic considerations can easily be transferred to other applications.

1.3.1 PIV in aerodynamics

The use of the PIV technique is very attractive in modern aerodynamics, because it helps to understand unsteady flow phenomena as, for example, in separated flows above models at high angle of attack. PIV enables spatially resolved measurements of the instantaneous flow velocity field within a very short time and allows the detection of large and small scale spatial structures in the flow velocity field. Another need of modern aerodynamics is that the increasing number and increasing quality of numerical calculations of flow fields require adequate experimental data for validation of the numerical codes in order to decide whether the physics of the problem has been modeled correctly. For this purpose carefully designed experiments have to be performed in close cooperation with those scientists doing the numerical calculations. The experimental data of the flow field must possess high resolution in time and space in order to be able to compare them with high density numerical data fields. The PIV technique is an appropriate experimental tool for this task, especially if information about the instantaneous velocity field is required.

A PIV system for the investigation of air flows in wind tunnels must be operated as well in low speed flows (e.g. flow velocities of less than 1 m/s in boundary layers) as in high speed flows (flow velocities up to 600 m/s in supersonic flows with shocks). Flow fields above solid, moving, or deforming models have to be investigated. The application of the PIV technique in large, industrial wind tunnels poses a number of special problems: large observation area, long distances between the observation area and the light source and the recording camera, restricted time for the measurement, and high operational costs of the wind tunnel.

The description of the problems as given above leads to the definition of requirements which should be fulfilled when PIV is applied in aerodynamics. First of all, a high spatial resolution of the data field is necessary in order to resolve large scale as well as small scale structures in the flow. This condition directly influences the choice of the recording medium (video or photographical recording). A second important condition is that a high density of experimental data is required for a meaningful comparison with the results of numerical calculations. Thus, the image density (i.e. number of particle images per interrogation area) must be high. A powerful seeding generator (high concentration of tracer particles in the measuring volume in the flow even at high flow velocities) is needed for this purpose. As the flow velocity is measured indirectly by means of the measurement of the velocity of tracer particles added to the flow, the tracer particles must follow the flow faithfully. This requires the use of very small tracer particles. However, small particles scatter little light. This fact results in a third important condition for the application of PIV in aerodynamics: a powerful pulse laser is required for the illumination of the flow field.

1.3.2 Major technical milestones of PIV

Earlier in this section some references to papers describing the general historical development of PIV have been given. In a handbook more devoted to the technical aspects of PIV it might be of even greater interest to outline the development of PIV towards its applicability in complex flows in terms of the achievement of major technical milestones.

The understanding of some of the technical restrictions in the application of PIV in the past and their conquest may be useful for new users of the PIV technique, in order to assess the discussion in some older publications sometimes dealing with – nowadays – "strange" looking efforts to solve technical problems which no longer exist today.

The selection of these milestones was done according to the technical progress in the past as experienced by the authors in their own work. Thus, the choice is a subjective one.

Feasibility of modern PIV. The feasibility of employing the particle image velocimetry technique for the measurement of flow velocity fields in water and even in air was demonstrated in the early eighties at the VON KÁRMÁN institute in Brussels, mainly by R. MEYNART [77]. At that time the evaluation methods were based on the work done in the field of speckle interferometry (see references in [4]).

Reliable high power light sources for application in air. The use of double oscillator Nd:YAG lasers (two resonators; frequency doubled, to achieve a wavelength of $\lambda = 532\,\mathrm{nm}$ in visible light) allowed for the first time the illumination of a plane in the flow with laser pulses of the same, constant energy at any time delay between the two pulses as required by the experiment at frame rates of the order of 10 Hz [67]. Alignment of the light sheet optics and image acquisition was thus facilitated considerably.

Ambiguity removal. Especially with photographic recordings it was not possible in most cases to store the images of the tracer particles due to first and second illumination on two different recordings. Thus, the temporal sequence of the images of the tracer particles could not be distinguished. Methods to remove the ambiguity of the sign of the velocity vector had to be developed (see references in [4]). The most widely used technique was image shifting, which could be successfully applied later on even in high speed flows. By enabling the investigation of complex, unsteady 3D flow fields, this development contributed considerably to the increasing interest in PIV from the side of wind tunnel users and industry.

Generation and distribution of tracer particles in the flow. The development of powerful aerosol generators and the know how to distribute the tracer particles within the flow homogeneously improved the image density and thus the quality of the PIV recordings considerably.

Computer hardware. The improvement of computer hardware with respect to processor speed and larger memory still continues. Memory size of 16 MB and 32 bit processors, which is today's standard, allow the handling of complete digital PIV recordings (even of temporal sequences of recordings) by a personal computer, which was not possible in the eighties with 8 bit processors and the 640 KB restrictions on memory.

Improved peak finders. The spread of digital particle image velocimetry was affected by the limited size and resolution of the video sensors and hence of digital PIV recordings as compared to that of photographic recordings. The development of Gaussian peak finders allowed the determination of the location of the displacement peak with further improved accuracy. Thus, smaller interrogation windows could be utilized, leading to an increase of spatial resolution (number of vectors) in digital particle image velocimetry.

Cross correlation video camera. Today progressive scan video cameras allow users to store the images of the tracer particles on separate frames for each illumination [150]. This feature immediately solves the problem of ambiguity removal. A sensor size of 1000×1000 pixels together with the application of cross correlation methods with superior signal-to-noise ratio at evaluation yield velocity vector fields of nearly the same quality as was possible only with 35 mm photographic film in the past.

Theoretical understanding of PIV. At the beginning of the development of particle image velocimetry the understanding of the technique was a more intuitive one. Progress was often made just by trial and error. In the past few years the theoretical understanding of the basic principles of the PIV technique has been improved considerably. Such theoretical considerations as well as simulations of the recording and evaluation process give useful information on many parameters important for the planning of an experiment utilizing PIV.

In this chapter a brief introduction to the basic principles of PIV and to some of its problems and technical constraints to be kept in mind has been given.

Next, the different topics will be described in more detail. We will start with providing the background of the most important physical principles. In the following the mathematical background of PIV evaluation will be discussed. With this knowledge the path has been prepared for the understanding of the recording, evaluation, and post processing methods applied in PIV. Furthermore the present state of the technical development of stereo and dual-plane PIV will be described – methods allowing access to the third component of the velocity vector in planar domains. In the final chapter examples of the application of PIV will be presented, thereby explaining the specific problems experienced at each measurement due to the different properties of the flow under investigation.

2. Physical and technical background

2.1 Tracer particles

It is clear from the principle of PIV as described that PIV – in contrast to hotwire or pressure probe techniques – is based on the direct determination of the two fundamental dimensions of the velocity: length and time. On the other hand, the technique measures indirectly, because it is the particle velocity which is determined instead of fluid velocity. Therefore, fluid mechanical properties of the particles have to be checked in order to avoid significant discrepancies between fluid and particle motion.

2.1.1 Fluid mechanical properties

A primary source of error is the influence of gravitational forces if the densitys of the fluid ρ and the tracer particles ρ_p do not match. Even if it can be neglected in many practical situations, we will derive the gravitationally induced velocity $\boldsymbol{U}_g$ from STOKES drag law in order to introduce the particle's behavior under acceleration. Therefore, we assume spherical particles in a viscous fluid at a very low Reynolds number. This yields:

$$\boldsymbol{U}_g = d_p^2 \frac{(\rho_p - \rho)}{18\mu} \boldsymbol{g} \tag{2.1}$$

where $\boldsymbol{g}$ is the acceleration due to gravity, μ the dynamic viscosity of the fluid, and d_p is the diameter of the particle.

In analogy to equation (2.1), we can derive an estimate for the velocity lag of a particle in a continously accelerating fluid:

$$\boldsymbol{U}_s = \boldsymbol{U}_p - \boldsymbol{U} = d_p^2 \frac{(\rho_p - \rho)}{18\mu} \boldsymbol{a} \tag{2.2}$$

where $\boldsymbol{U}_p$ is the particle velocity. The step response of $\boldsymbol{U}_p$ typically follows an exponential law if the density of the particle is much greater than the fluid density:

$$\boldsymbol{U}_p(t) = \boldsymbol{U}\left[1 - \exp\left(-\frac{t}{\tau_s}\right)\right] \tag{2.3}$$

with the relaxation time τ_s given by:

$$\tau_s = d_p^2 \frac{\rho_p}{18\mu} .$$

If the fluid acceleration is not constant or STOKES drag does not apply (e.g. at higher flow velocities) the equations of the particle motion become more difficult to solve, and the solution is no longer a simple exponential decay of the velocity. Nevertheless, τ_s remains a convenient measure for the tendency of particles to attain velocity equilibrium with the fluid. The result of equation (2.3) is illustrated in figure 2.1 where the time response of particles with different diameters is shown for a strong deceleration in an air flow.

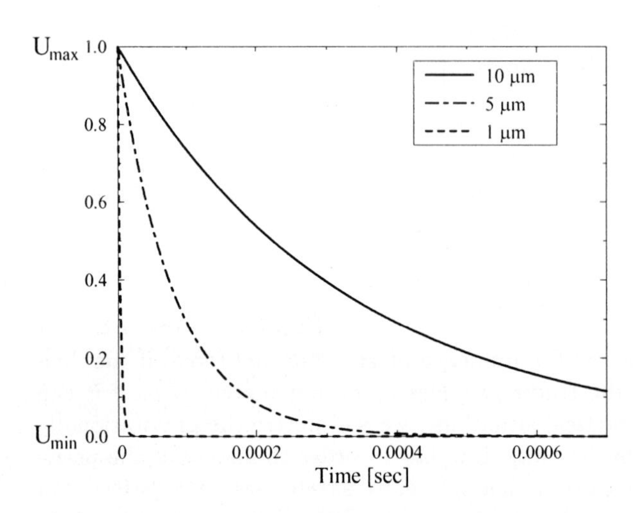

Fig. 2.1. Time response of oil particles with different diameters in a decelerating air flow

When applying PIV to liquid flows the problems of finding particles with matching densities are usually not severe, and solid particles with adequate fluid mechanical properties can often be found. Usually their size can easily be determined before suspension into the liquid and will not change afterwards. Some of the most often used tracers are listed in section 2.1.3. However, early applications of PIV have already shown that difficulties arise in providing high quality seeding in gas flows compared to applications in liquid flows [42, 59, 62, 93, 142]. The problems are similar to those, which are faced when applying laser Doppler velocimetry. From equation (2.2) it can be seen that due to the difference in density between the fluid and the tracer particles, the diameter of the particles should be very small in order to ensure good tracking of the fluid motion. On the other hand, the particle diameter should not be too small as light scattering properties have also to be taken into account as will be shown in the following section. Therefore, it is clear that a compromise has to be found. This problem is discussed in the literature extensively [63, 76, 78]. The most commonly used seeding particles for PIV investigations of gaseous flows are listed in table 2.2 on page 20. For most of our applications we used oil particles which were generated by means of a Laskin nozzle (see section 2.1.3), with a mean diameter of the oil particles being about 1 μm. It is well known from LDV measurements in gas flows that the size and the distribution of the tracer particles may change during the travel from the aerosol generator to the test section, where the measurements take place. It is therefore advisable to gain information about the particles and especially about the velocity lag directly from the observation area [42, 62, 132]. The result of one of our examinations of this problem is presented in figure 2.2.

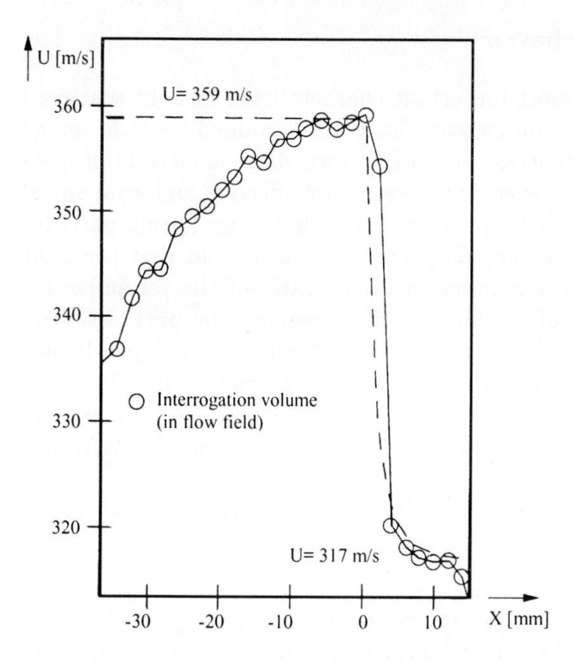

Fig. 2.2. Comparison of the experimental (PIV, particle diameter: 1.7 μm) and theoretical (dashed line) result for the change of the U-component of the instantaneous velocity vector along a line in the flow field about a bluff cylinder when crossing a shock

The U-component of the instantaneous flow velocity along one line of a PIV recording of a transonic flow field is printed. The measured flow velocity U drops from 359 m/s in front of a shock to 317 m/s within a distance of 8 mm. The real extent of the shock ($\approx 10^{-3}$ mm) cannot be resolved due to the finite size of the interrogation areas ($\approx 2-3$ mm diameter, when projected back into the flow field, indicated as circles in figure 2.2). However, a calculation of the velocity lag of particles with a diameter of 1.7 μm, carried out according to a theoretical approach, considering effects of compressibility, deformation of droplets, and high Reynolds numbers [92] yields a similar relation between velocity and distance as measured with PIV (compare the dashed line in figure 2.2). This shows that it does not make much sense to utilize tracer particles with a diameter much smaller than 1 μm for this experiment, because of the fact that during the evaluation of the PIV recording the velocities are averaged within each interrogation area. Smaller tracer particles would be only necessary if a higher spatial resolution in the vicinity of the shock was required.

Besides the bias error which might be introduced, large particle sizes can cause data drop-out in critical areas of the flow field as for example in vortex cores, shear flows or boundary layers. In some cases the velocity lag due to centrifugal forces in a vortex leads to only small errors of measurement but due to its integration from vortex generation to the light sheet the particle density becomes too low for adequate measurements.

2.1.2 Light scattering behavior

In this section some of the most important characteristics of light scattered by tracer particles will be summarized. Since the obtained particle image intensity and therefore the contrast of the PIV recordings is directly proportional to the scattered light power, it is often more effective and economical to increase the image intensity by properly choosing the scattering particles than by increasing the laser power. In general it can be said that the light scattered by small particles is a function of the ratio of the refractive index of the particles to that of the surrounding medium, the particles' size, their shape and orientation. Furthermore, the scattered light depends also on polarization and observation angle. For spherical particles with diameters larger than the wavelength of the incident light, MIE's scattering theory can be applied. A detailed description and discussion is given in the literature [15].

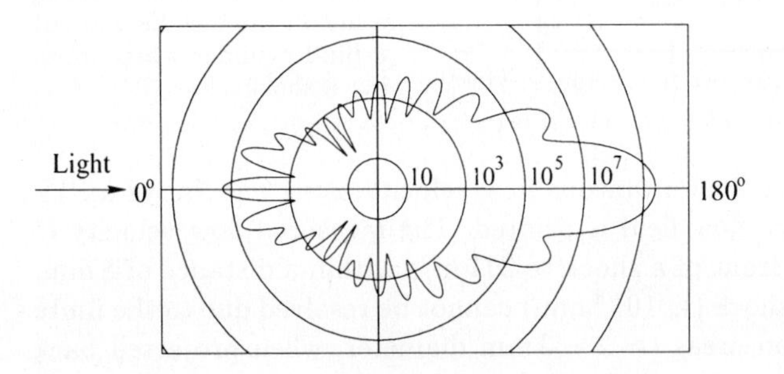

Fig. 2.3. Light scattering by a 1 μm oil particle in air

Figures 2.3 and 2.4, show the polar distribution of the scattered light intensity for oil particles of different diameters in air with a wavelength λ of 532 nm, according to Mie's theory. The intensity scales are in logarithmic scale and are plotted so that the intensity for neighboring circles differ by a factor of 100. The Mie scattering can be characterized by the normalized diameter, q, defined by:

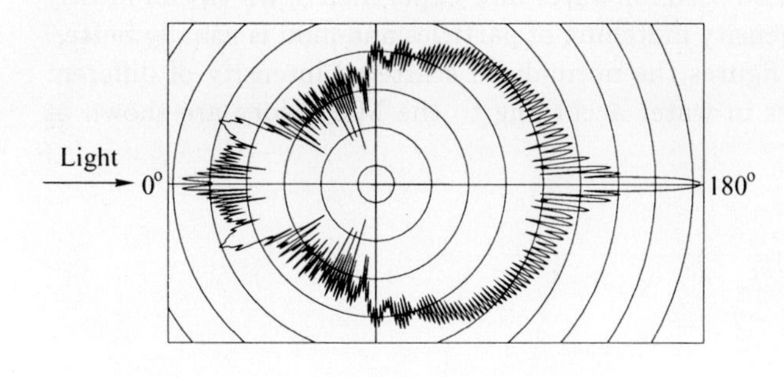

Fig. 2.4. Light scattering by a 10 μm oil particle in air

$$q = \frac{\pi d_\mathrm{p}}{\lambda} .$$

If q is larger than unity, approximately q local maxima appear in the angular distribution over the range from 0° to 180°. For increasing q the ratio of forward to backward scatter intensity will increase rapidly. Hence, it would be advantageous to record in forward scatter, but, due to the limited depth of field, recording at 90° has most often to be used. In general the light scattered paraxially (i.e. at 0° or 180°) from a linearly polarized incident wave is linearly polarized in the same direction and the scattering efficiency is independent of polarization. In contrast, the scattering efficiency for most other observation angles strongly depends on the polarization of the incident light. Furthermore, for observation angles in the range from 0° to 180° the polarization direction can be partially turned. This is particularly important if image separation or image shifting depending on polarization of the scattered light has to be applied. Therefore, such a technique works well only for certain particles, for example 1 μm diameter oil particles in air.

There is a clear tendency for the scattered light intensity to increase with increasing particle diameter. However, if we recall that the number of local maxima and minima is proportional to q, it becomes clear, that the function of the light intensity versus particle diameter is characterized by rapid oscillations if only one certain observation angle is taken into account. One implication is that particle images of high intensity do not always imply that the particle crossed the center of the measurement volume. Hence, a determination of the out-of-plane particle displacement by analysing particle positions in a light sheet with known intensity profile by the image intensity is usually not feasible. When averaging over a range of observation angles, which is determined by the observation distance and the recording lens aperture, the intensity curve is smoothed considerably. The average intensity roughly increases with q^2, and as already mentioned above, the scattering efficiency strongly depends on the ratio of the refractive index of the particles to that of the fluid. Since the refractive index of water is considerably larger than that of air, the scattering of particles in air is at least one order of magnitude more powerful compared to particles of the same size in water. Therefore, much larger particles have to be used for water flow experiments, which can mostly be accepted since the density matching of particles and fluid is usually better. In the following three figures the normalized scattered intensity of different diameter glass particles in water according to the Mie theory are shown at $\lambda = 532\,\mathrm{nm}$.

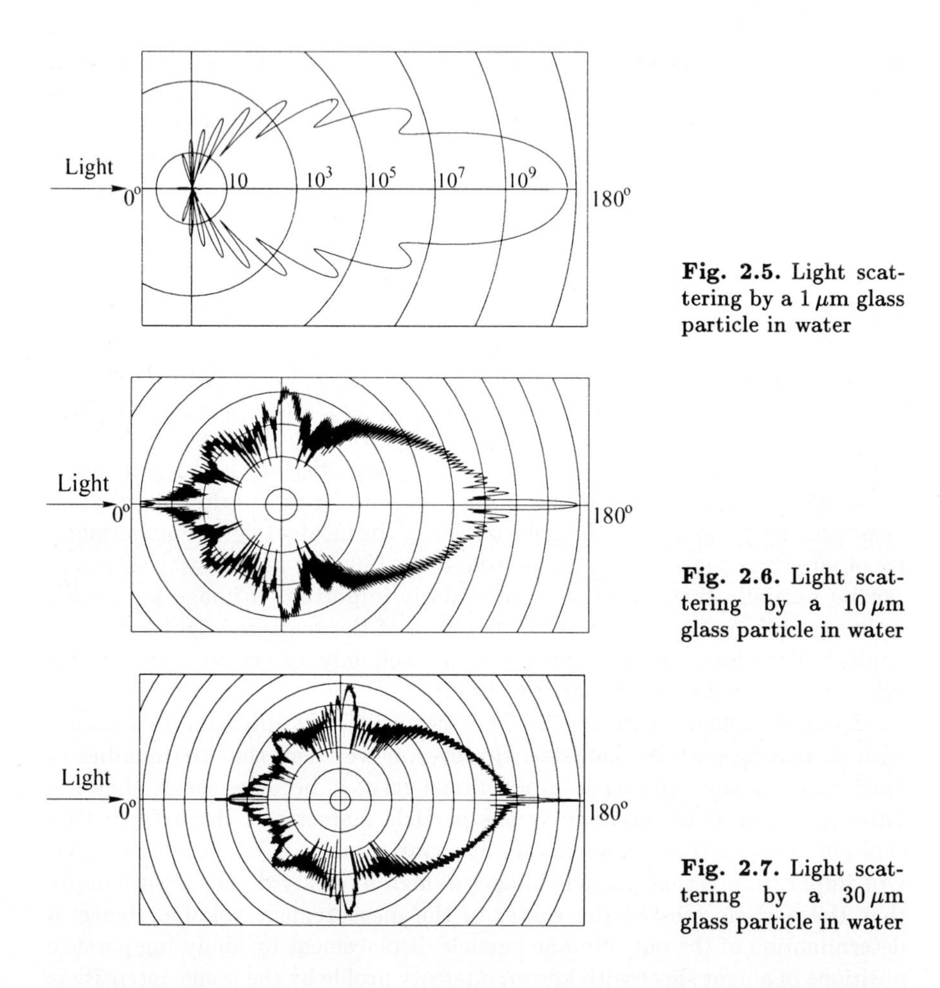

Fig. 2.5. Light scattering by a 1 μm glass particle in water

Fig. 2.6. Light scattering by a 10 μm glass particle in water

Fig. 2.7. Light scattering by a 30 μm glass particle in water

As can be seen from all Mie scattering diagrams, the light intensity is not blocked by the particles but spread in all directions. Therefore, for a large number of particles inside the light sheet massive multiscattering appears. Then the light which is focused by the recording lens is not only due to direct illumination but also due to fractions of light, which have been scattered by more than one particle. In the case of heavily seeded flows this considerably increases the intensity of individual particle images, because the intensity of directly – at 90° to the incident illumination – recorded light is orders of magnitude smaller than that scattered in the forward scatter range.

One interesting implication is that not only can larger particles be used to increase the scattering efficiency but also the number density of the particles. However, two problems limit this effect from being intensively used. First, the background noise and therefore the noise on the recordings will increase significantly. Second, if – as is usually the case – polydisperse particles (i.e. particles of different sizes) are used, it is finally not sure whether the number of visible particles has been increased by simply increasing the number of very large particles. Since images of larger particles clearly dominate PIV evaluation, it would be difficult to give sure estimates on the effective particle size and the corresponding velocity lag.

2.1.3 Particle generation and supply

Descriptions of seeding particles and their characteristics have been given in many scientific publications. In contrast to that, little information can be found in the literature on how to practically supply the particles into the flow under investigation. Sometimes seeding can be done very easily or does not even have to be done. The use of natural seeding is sometimes acceptable, if enough visible particles are naturally present to act as markers for PIV. In almost all other work it is desirable to add tracers in order to achieve sufficient image contrast and to control particle size. For most liquid flows this can easily be done by suspending solid particles into the fluid and mixing them in order to get a homogeneous distribution.

Table 2.1. Seeding materials for liquid flows

Type	Material	Mean diameter in μm
Solid	Polystyrene	10 – 100
	Aluminum	2 – 7
	Glass spheres	10 – 100
	Granules for synthetic coatings	10 – 500
Liquid	Different oils	50 – 500
Gaseous	Oxygen bubbles	50 – 1000

A number of different particles which can be used for flow visualization, LDV, and PIV are listed in table 2.1 for liquid and in 2.2 for gas flows. For our experiments in oil and water flows we used coated glass spheres of approximately 10 μm diameter as is shown in figure 2.8 for two different magnifications. They offer good scattering efficiency and a sufficiently small velocity lag.

In gas flows the supply of tracers is very often more critical for the quality and feasibility of the PIV measurement and the health of the experimentalists

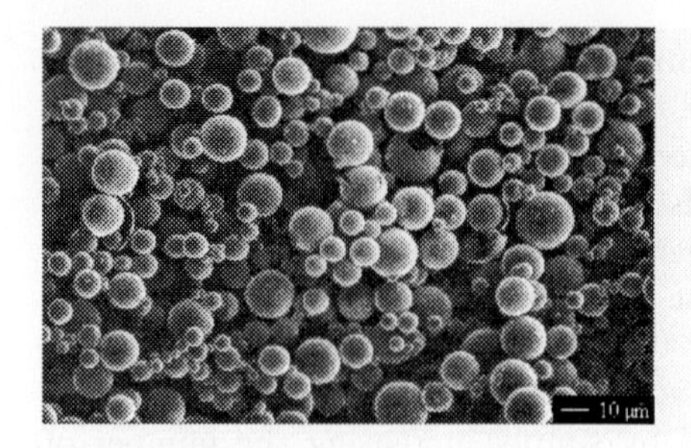

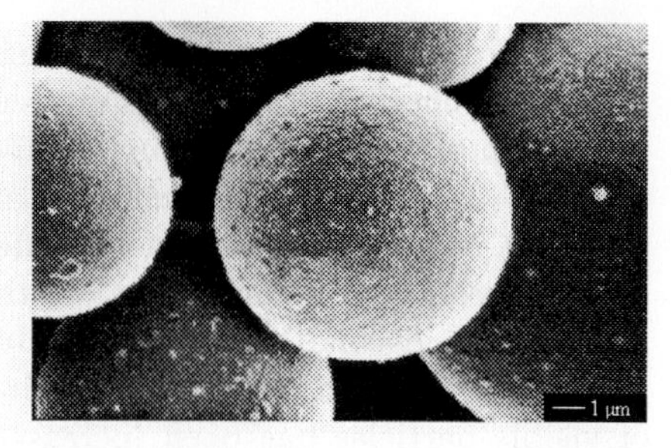

Fig. 2.8. Glass particles: ×500 and ×5000

if they have to breath seeded air for example in wind tunnels in an open test section. The particles which are often used are not easy to handle because many liquid droplets tend to evaporate rather quickly and solid particles are difficult to disperse and very often agglomerate. The particles can therefore not simply be supplied a long time before the measurement, but must be injected into the flow shortly before the gaseous medium enters the test section. The injection has to be done without significantly disturbing the flow, but in a way and at a location that ensures homogeneous distribution of the tracers. Since the existing turbulence in many test set-ups is not strong enough to mix the fluid and particles sufficiently, the particles have to be supplied from a large number of openings. Distributors, like rakes consisting of many small pipes with a large number of tiny holes, are often used. Therefore, particles which can be transported inside small pipes are required.

Table 2.2. Seeding materials for gas flows

Type	Material	Mean diameter in μm
Solid	Polystyrene	0.5 – 10
	Aluminum	2 – 7
	Magnesium	2 – 5
	Glass micro-balloons	30 – 100
	Granules for synthetic coatings	10 – 50
	Dioctylphathalate	1 – 10
Smoke		< 1
Liquid	Different oils	0.5 – 10

A number of techniques are used to generate and supply particles for seeding gas flows [63, 76, 78]: dry powders can be dispersed in fluidized beds or by air jets. Liquids can be evaporated and afterward condensed in so-called condensation generators, or liquid droplets can directly be generated in atomizers. Atomizers can also be used to disperse solid particles suspended in evaporating liquids, or to generate tiny droplets of high vapor pressure liquids (e.g. oil) that have been mixed with low vapor pressure liquids (e.g. alcohol) which evaporate before the test section. For seeding wind tunnel flows condensation generators, smoke generators and monodisperse polystyrene or latex particles injected with water-ethanol are most often used for flow visualization and LDV. For most of the PIV measurements in air flows Laskin nozzle generators and oil have been used. These particles offer the advantage of not being toxic, they stay in air at rest for hours, and do not change in size significantly under various conditions. In recirculating wind tunnels they can be used for a global seeding of the complete tunnel volume or for a local seeding of a stream tube by a seeding rake with a few hundred tiny holes. A technical description of such an atomizer is given below.

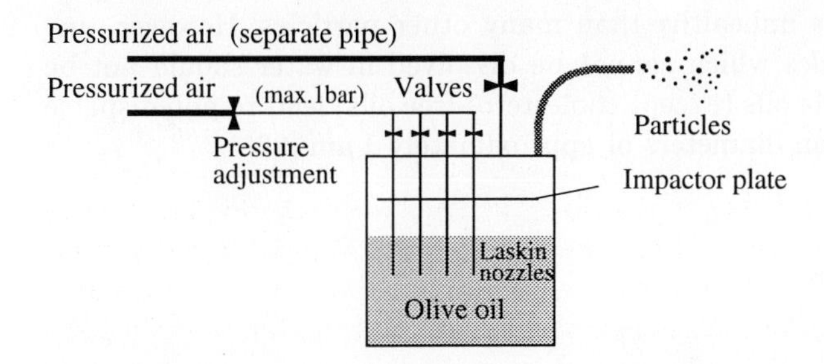

Fig. 2.9. Oil seeding generator

The aerosol generator consists of a closed cylindrical container with two air inlets and one aerosol outlet. Four air supply pipes – mounted at the top – dip into vegetable oil inside the container. They are connected to one air inlet by a tube and each has a valve. The pipes are closed at their lower ends (see figure 2.10). Four Laskin nozzles, 1 mm in diameter, are equally spaced in each pipe [46].

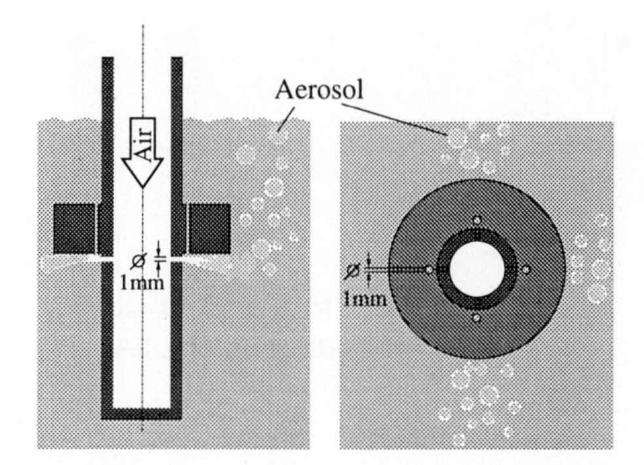

Fig. 2.10. Sketch of a Laskin nozzle

A horizontal circular impactor plate is placed inside the container, so that a small gap of about 2 mm is formed by the plate and the inner wall of the container. The second air inlet and the aerosol outlet are connected directly to the top. Two gauges measure the pressure on the inlet of the nozzles and inside the container, respectively. Compressed air with 0.5 to 1.5 bar pressure difference with respect to the outlet pressure is applied to the Laskin nozzles and creates air bubbles within the liquid. Due to the shear stress induced by the tiny sonic jets small droplets are generated and carried inside the bubbles towards the oil surface. Big particles are retained by the impactor plate; small particles escape through the gap and reach the aerosol outlet. The number of particles can be controlled by the four valves at the nozzle inlets. The particle concentration can be decreased by an additional air supply via the second air inlet. The mean size of the particles generally depends on the type of liquids being atomized, but is only slightly dependent on the operating pressure of the nozzles. Vegetable oil is the most commonly used liquid since oil droplets are believed to be less unhealthy than many other particles. However, any kind of seeding particles which cannot be dissolved in water should not be inhaled. Most vegetable oils (except cholesterol-free oils) lead to polydisperse distributions with mean diameters of approximately 1 μm [92].

2.2 Light sources

2.2.1 Lasers

Lasers are widely used in PIV, because of their ability to emit monochromatic light with high energy density, which can easily be bundled into thin light sheets for illuminating and recording the tracer particles without chromatic aberrations. In figure 2.11 a typical configuration of a laser is shown. Generally speaking, as shown in figure 2.11, every laser consists of three main components.

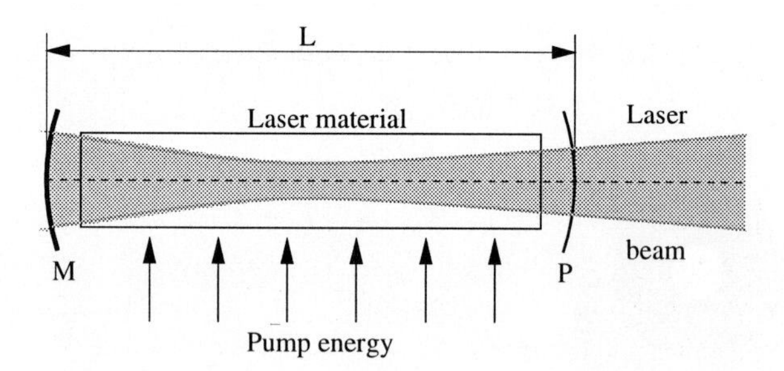

Fig. 2.11. Schematic diagram of a laser

The laser material consists of an atomic or molecular gas, semiconductor or solid material.

The pump source excites the laser material by the introduction of electromagnetical or chemical energy.

The mirror arrangement allows an oscillation within the laser material.

In the following we will describe the principle of gas lasers and give an overview of the lasers used in PIV.

It is well known from quantum mechanics that each atom can be brought into various energy states by three elementary kinds of interaction with electromagnetic radiation. This can be illustrated in an energy level diagram, as shown in figure 2.12 for a hypothetical atom with only two possible energy states. An *excited* atom at level E_2 usually drops back to the state E_1 after a very short, but not exactly defined period of time and emits the energy $E_2 - E_1 = h\nu$ in the form of a randomly directed photon. This process is called *spontaneous emission.*

However, if, on the other hand, a photon with "appropriate" frequency ν impinges on an atom, then two effects are possible: either – in the case of *absorption* – an atom in the state E_1 can receive the energy $h\nu$, i.e. it becomes *raised* to E_2 and the photon is absorbed; or the incident photon can stimulate an atom in the excited E_2 state into a specific, nonspontaneous, transition to E_1. Then, in addition to the incident photon, a second photon in phase with the first occurs, i.e. the impinging wave was coherently amplified (*stimulated emission*).

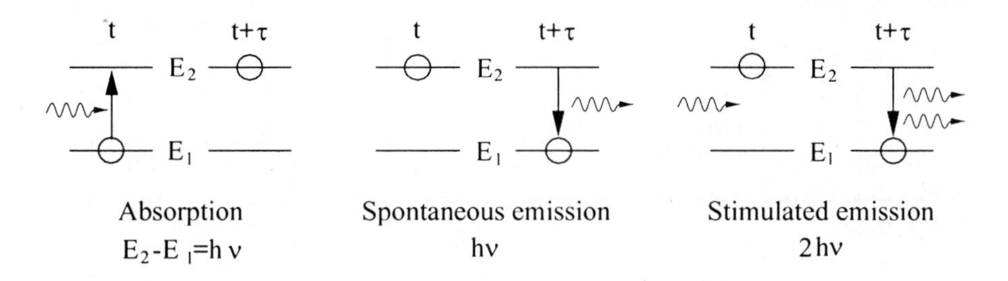

Fig. 2.12. Elementary kinds of interactions between atoms and electromagnetic radiation

When there are large numbers of atoms, one of the two processes – absorption or stimulated emission – predominate: if there are more atoms in the E_2 state than in the E_1 state (i.e. the population density $N_2 > N_1$ [atoms/m^3]), then stimulated emission predominates, and, in the case of $N_1 > N_2$, absorption.

Since the laser can only operate if a *population inversion* is forced to take place ($N_2 > N_1$) , external energy has to be transferred to the laser material because atoms usually exist in their ground state. This is achieved by a different pump mechanism depending on the kind of laser material. Solid laser materials are generally pumped by electromagnetic radiation, semiconductor lasers by electronic current, and gas lasers by collision of the atoms or molecules with electrons and ions.

It should be noted that in a system which consists of only two energy states, as described so far, no population inversion can be achieved, because when the number of atoms N_2 in level E_2 equals the number N_1 in level E_1, absorption and stimulated emission are equally likely and the material will become transparent at the frequency $\nu = (E_2 - E_1)/h$. In other words: the number of transitions from the upper level E_2 to the lower level E_1 and vice versa are on average the same. Hence, at least three energy levels of the laser medium are essential to achieve population inversion. But a three level system is not very efficient because a fraction of more than 50% of the atoms of the system has to be exited in order to amplify an impinging photon. This means that the energy needed for the excitation of this fraction is lost for the amplification. In the case of a four level laser the lower laser level E_2 does not coincide with the basic level E_1 and therefore remains unoccupied at room temperature. In this way it is easier to achieve the population inversion and a four level laser requires substantially less pumping power. This is illustrated in figure 2.13. If for instance state E_4 is achieved by optical pumping at frequency ν according to $h\nu = E_4 - E_1$, then a rapid nonradiative transition to the upper laser level E_3 occurs. The atoms remain in this so-called metastable state E_3 for a relatively long interim period before they drop down to the unoccupied lower laser level E_2.

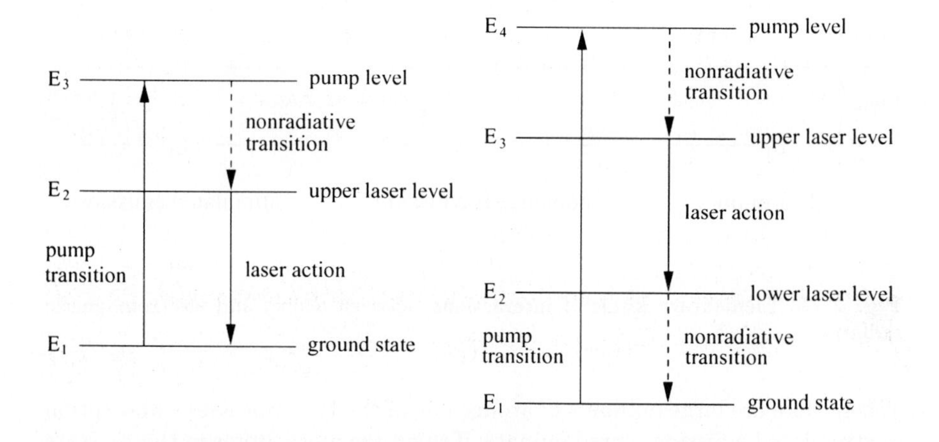

Fig. 2.13. Level diagrams of three (left) and four (right) level lasers

As a consequence of population inversion through energy transfer by the pump mechanism spontaneous emission occurs in all directions which causes excitation of further neighboring atoms. This initiates a rapid increase of stimulated emission and therefore of radiation in a chain reaction.

In the case of a cylindrical shape of laser material the rapid increase of radiation occurs in a defined direction, because the amplification increases with increasing length of the laser medium. With an optical resonator (mirror arrangement) the laser material can be extended to form an oscillator. The simplest way to achieve this is to place the material between two exactly aligned mirrors. In this case, a photon which impinges randomly on one of the mirror surfaces is reflected and amplified in the laser material again. This process will be repeated and generates an avalanche of light which increases exponentially with the number of reflections, finally resulting in a stationary process. In other words, standing waves are produced on the resonator length with the condition

$$L = \frac{m\lambda}{2n} \tag{2.4}$$

where n is the refractive index, m an integer number, and L the resonator length. Since the frequency ν according to the transition $\nu\, h = E_2 - E_1$ does not correspond to exactly one wavelength, but rather to a spectrum of a certain band width $\Delta\nu$ depending on the transition time τ of the process these conditions can be fulfilled by different wavelengths λ or frequencies ν and the resonator can oscillate in many axial modes with distinct frequencies.

Consecutive modes are separated by a constant difference $\Delta\nu = c/(2\, L\, n)$, wherein c is the speed of light. Moreover, the cross-section of the laser beam can be divided into several ranges oscillating in antiphase with intermediate node lines, i.e. different transverse modes can be sustained as well (see figure 2.14). Their occurrence depends on resonator design and alignment. The lowest order transverse mode TEM_{00} (TEM = transverse electric mode; index = node in X- and Y-direction) is most commonly used, because it produces a beam with uniform phase and a Gaussian intensity distribution versus the beam cross-section.

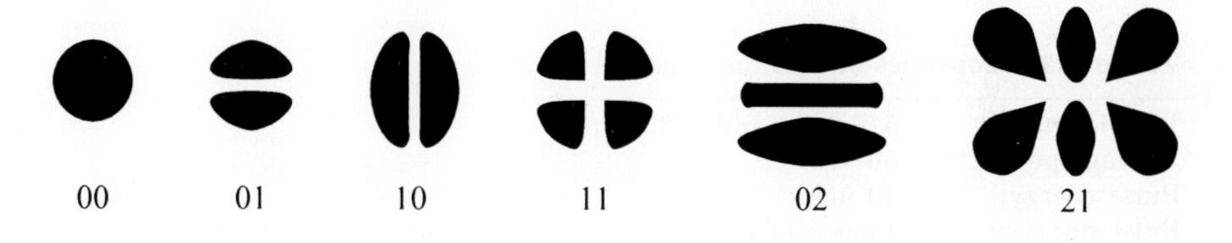

Fig. 2.14. Examples of different transverse modes

There are various types of resonators with different mirror curvatures. The confocal resonator, shown in figure 2.11, is particularly stable and easy to adjust. Hemispherical resonators use one planar and one concave mirror, and critical resonators use two planar mirrors. Critical resonators offer the advantage of having no beam waist inside the laser rod and therefore using its whole volume. However, they are sensitive for thermal lens effects and misalignment.

In gas lasers, which are usually used for continuous operations (CW = continuous wave), free electrons are accelerated by an electrical field resulting in an excitation of the gaseous medium. The plasma tube in which the excitation takes place is closed by Brewster windows (plates tilted at the polarization angle). Therefore, these lasers emit linearly polarized light. In the case of optically pumped solid-state lasers, the light of the rod-shaped flash lamp is concentrated on the laser rod by a cylindrical mirror with an elliptical cross-section. In lasers, which basically consist of luminescence diodes, two polished, parallel surfaces work as resonator mirrors. The laser light perpendicular to the p-n junction is more divergent due to the lower aperture width.

Helium-neon lasers (He-Ne lasers $\lambda = 633\,\text{nm}$) are the most common as well as the most efficient lasers in the visible range. He-Ne lasers have rather been used for the optical evaluation of photographic images in PIV than for illuminating the flow. The power of commercial models ranges from less than 1 mW to more than 10 mW. The laser transitions take place in the neon atom. During the electrical discharge within a gaseous mixture of helium and neon, the helium atoms are excited by collisions with electrons. In the second step the upper laser levels of the Ne atoms are populated through collisions with metastable helium atoms. This is the main pump mechanism. The major properties of the He-Ne laser that are important for the evaluation of PIV images are the coherence of the laser beam and the Gaussian distribution of the laser beam intensity (TEM_{00}). If necessary, beam quality can be further improved by means of spatial filters (see section 5.3.2).

Copper-vapor lasers (Cu lasers $\lambda = 510\,\text{nm}, 578\,\text{nm}$) are the most important neutral metal vapor lasers for PIV. The wavelengths of Cu lasers are within the yellow and green spectrum (table 2.3).

Table 2.3. Properties of a commercial Cu laser

Wavelength:	510.6 nm and 578.2 nm
Average power:	50 W
Pulse energy:	10 mJ
Pulse duration:	15 ns – 60 ns
Peak power:	$<$ 300 kW
Pulse frequency:	5 kHz – 15 kHz
Beam diameter:	40 mm
Beam divergence:	$0.6 \cdot 10^{-3}$ rad

These lasers are characterized by their high average power (typically 1–30 W) and an efficiency of up to one percent. Continuous wave operations are not possible due to the long life-span of the lower laser level. During pulse operations, repetition rates within the kHz range can be achieved.

This type of laser has intensively been developed during the past decade because the copper laser is an important pump source for dye lasers. Whereas most lasers are cooled, metal vapor lasers need thermal insulation so that the operating temperature for vaporizing the metal can typically reach 1500°C. Two electrodes are located at the ends of a thermally insulated ceramic tube with a pulsed charge burning in between them. For improving discharge quality, neon is added as the buffer gas at a pressure of around 3000 Pa. Table 2.3 lists some of the key properties of a typical Cu laser as an example.

Argon-ion lasers (Ar^+ lasers $\lambda = 514\,\text{nm}, 488\,\text{nm}$) are gas lasers, similar to the He-Ne lasers described above. In argon lasers, very high currents have to be achieved for ionization and excitation. This is technologically much more complicated compared to He-Ne lasers. Typically the efficiency of these lasers is on the order of a tenth of a percent. These lasers can supply over 100 W in the blue-green range and 60 W in the near ultraviolet range. Emission is produced at several wavelengths through the use of broadband laser mirrors. Individual wavelengths can be selected by means of Brewster prisms in the laser resonator. The individual wavelengths can be adjusted by turning the prism. The most important wavelengths are 514.5 and 488.0 nm. Nearly all conventional inert gas ion lasers supply TEM_{00}. Despite the extreme load on the tubes resulting from the high currents, product lives of several thousand operating hours can be achieved. Since argon lasers are frequently used for LDV measurements, they are often found in fluid mechanics laboratories. In PIV they can easily be used for low speed water investigations.

Semiconductor lasers offer the advantage to be very compact. The laser material is typically 1 cm long and has a diameter of 0.5 mm. The total efficiency of a commercial diode laser pumped Nd:YAG system is around 7%. Since heating is considerably reduced, these types of pumped lasers supply a very good beam quality of over 100 mW in the TEM_{00} mode during continuous operations. The diode laser is interesting for PIV because it can be used as a seed laser to improve the coherence length of flash lamp pumped Nd:YAG lasers for use in holographic PIV. A particularly interesting variant is the combination of a diode-pumped laser oscillator and a flash lamp pumped amplifier. Together with other optical components, like vacuum-pinholes and phase-conjugated mirrors, this concept offers very good beam properties, but at the same time, its initial purchase costs are also high.

Ruby lasers (Cr^{3+} lasers $\lambda = 694\,\text{nm}$), historically the very first lasers, use ruby crystal rods containing Cr^{3+} ions as the active medium. They are pumped optically by means of flash lamps. As already mentioned above, the ruby laser is a three level system which has the disadvantage that approximately 50% of the atoms must be excited before population inversion takes place. The high pumping energy needed can usually only be achieved during pulse mode operations. The wavelength of the ruby laser is 694.3 nm. Like other solid-state lasers, the ruby laser can also be operated normally or in Q-switched mode. (For details about Q-switched mode see the next paragraph.)

The ruby laser is particularly interesting for PIV because it delivers very high pulse energies and its beam is well suited for holographic imaging because of its good coherence. Its disadvantage is that the low repetitive rates hamper the optical alignment and its light is emitted at the edge of the visible spectrum. Photographic films are usually not sensitive for red light and also modern CCD cameras are usually optimized for smaller wavelengths.

Neodym-YAG laser (Nd:YAG lasers $\lambda = 532\,\text{nm}$) are the most important solid-state laser for PIV in which the beam is generated by Nd^{3+} ions. The Nd^{3+} ion can be incorporated into various host materials. For laser applications, YAG crystals (yttrium-aluminum-garnet) are commonly used. Nd:YAG lasers have a high amplification and good mechanical and thermal properties. Excitation is achieved by optical pumping in broad energy bands and nonradiative transitions into the upper laser level.

The fact that solid-state lasers can be pumped with white light results from the arrangement of the atoms which form a lattice. The periodic arrangement leads to energy bands formed by the upper energy levels of the single atoms. Therefore, the upper energy levels of the system are not discrete as in the case of single atoms, but are continuous.

As already mentioned the Nd:YAG laser is a four-level system which has the advantage of a comparably low laser threshold. At conventional operating temperatures, the Nd:YAG laser only emits the strongest wavelength, 1064 nm. In the relaxation mode the population inversion takes place as soon as the threshold is reached, with this threshold value depending on the design of the laser cavity. In this way, many successive laser pulses can be obtained during the pump pulse of the flash lamp. By including a quality switch (Q-switch) inside the cavity the laser can be operated in a triggered mode. The Q-switch has the effect of altering the resonance characteristics of the optical cavity. If the Q-switch is operated, allowing the cavity to resonate at the most energetic point during the flashlamp cycle, a very powerful laser pulse, the so-called giant pulse, can be achieved. Q-switches normally consist of a polarizer and a Pockels cell, which change the quality of the optical resonator depending on the Pockels cell voltage. The Q-switched mode is in general more interesting and is usually used in PIV. Even if the Q-switches can be used to generate more than one giant pulse out of one resonator, PIV lasers are mostly designed as double-oscillator systems. This enables the user to adjust the separation time between the two illuminations of the tracer particles independently of the pulse strength. The beam of Q-switch lasers is linearly polarized. For PIV, and many other applications, the fundamental wavelength of 1064 nm is frequency-doubled using special crystals. (For details about these so called KDP crystals see the next section.) After separation of the frequency-doubled portion, approximately one-third of the original light energy is available at 532 nm. Nd:YAG lasers are usually driven in a repetitive mode. Since the optical properties of the laser cavity change with changing temperature, good and constant beam properties will only be obtained at a nominal repetition rate and flashlamp voltage. Due to thermal lensing the beam quality, which is very often poor compared to other laser types, decreases significantly when, for example, single pulses are used. This is not that critical for telescopic resonator arrangements but very important for modern critical resonator systems. The coherence length of pulsed Nd:YAG lasers is normally on the order of a few centimeters. For holographic recording, lasers with a narrow spectral bandwidth have to be used. This is usually done by injection from a smaller semiconductor laser into the cavity by a partially reflecting mirror. Then, the laser pulse builds up from this small seeding pulse of narrow bandwidth, resulting in coherence lengths of 1 or 2 meters. However, very precise laser timing and temperature control for the primary cooling circuit are required for this purpose.

2.2.2 Features and components of Nd:YAG lasers for PIV

Commercially available Nd:YAG laser rods are up to 150 mm long and have diameters of up to 10 mm. Typically pulse energies of 400 mJ or more can be achieved out of one oscillator. In this case, more than one flashlamp and critical resonators with plane mirror surfaces have to be used. The price paid for the high output power that can be achieved with those resonators is that the beam profile tends to be very poor: hot spots and different ring modes can often be found. In order to improve the beam profile, output mirrors with a reflectivity that varies with the radius are frequently used. However, even with these mirrors the beam profile is sometimes very poor, even if it is specified to be 80% Gaussian in the near and 95% in the far field. Two laser systems of the same manufacturer often have different beam properties depending on the laser rod properties, and the alignment of the laser. Since a good beam profile is absolutely essential for PIV (see chapter 3) it must be specified not only in the near and in the far field – as most manufacturers do – but also in the mid-field in a distance of 2–10 m from the laser. The description of the beam intensity distribution should not only be based on a good fit to a Gaussian distribution, but also on the minimum and maximum energy in order to ensure a hole-free intensity distribution without hot spots.

In figure 2.15 the intensity profiles versus the light sheet thickness measured at four different distances from the laser are shown. The light sheet optics used for this experiment are shown in figure 2.20. The peak value of the distribution has been adjusted close to full scale for each position (1.8 m, 3.3 m, 4.3 m, 5.8 m). It can be seen that the thickness of the light sheet increases slowly with the distance from the laser. A small side peak is visible at every position but seems to vanish at 5.8 meters. The fluctuations in the distribution are minimized at position three (4.3 m). When assessing these light sheet profiles it has to be taken into account that the loss of correlation during the evaluation of PIV recordings is mainly influenced by the light sheet intensity distribution at recording (see chapter 3). The light energy contained in the side peak will be lost in most situations, because a very small flow component in the lateral direction would displace particles from bright towards dark areas and would therefore lead to only one illumination of the tracer particles. For flow fields without any significant out-of-plane velocity component the light sheet can be focused more precisely and a better, more Gaussian-like, intensity profile can be obtained in the out-of-plane direction.

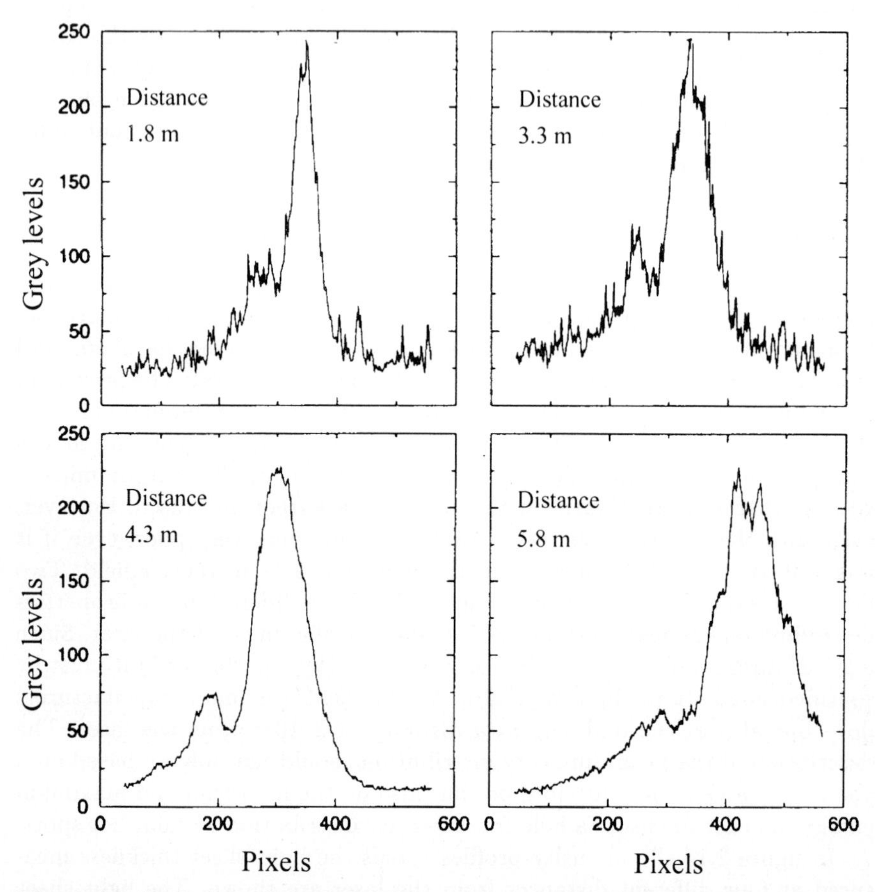

Fig. 2.15. Evolution of the light sheet profile with increasing distance from the laser

However, even in those cases the intensity profile versus the light sheet height strongly depends on the laser beam properties. If data is lost in some regions of the observation area due to insufficient illumination, the result of the whole measurement may become questionable. In other words, pulse energy is one value required for the specification of light sheet illumination, but definitely not the most important one.

In table 2.4 some critical parameters are listed which should be specified when assessing a double oscillator Nd:YAG laser. The laser system and all specifications are made with respect to a wavelength of 532 nm and a repetition rate of 10 Hz of the two pulses, unless otherwise stated. All trigger signals of the laser are TTL compatible.

Table 2.4. Properties and specifications of modern Nd:YAG PIV-laser systems

Repetition rate[a]	10 Hz
Pulse energy for each of two pulses	320 mJ
Roundness at 8 m from laser output[b]	75%
Roundness at 0.5 m from laser output[c]	75%
Spatial intensity distribution at 8 m from laser output[d]	< 0.2
Spatial intensity distribution at 0.5 m from laser output[d]	< 0.2
Linewidth	1.4 cm^{-1}
Power drift over 8 hours[e]	< 5%
Energy stability[f]	< 5%
Beam pointing stability[g]	100 μrad
Deviation from colinearity of laser beams	< 0.1 mm/m
Beam diameter at laser output	9 mm
Divergence[h]	0.5 mrad
Jitter between two following laser pulses	2 ns
Delay between two laser pulses	0 to 10 ms
Resolution	5 ps
Working temperatures	15°–35°C
Cooling water[i]	10° C –25°C
Power requirements	220–240 V, 50 Hz

[a] And integral fractions of 10 Hz, eg. 5 Hz, 2.5 Hz etc.
[b] Ratio between two perpendicular axis (major and minor axis).
[c] If laser beam is elliptical, major axis of both oscillators is parallel.
[d] $|(I_{max} - I_{min})|/|(I_{max} + I_{min})|$, with I being the peak intensity in the spatial distribution limited by the diameter at half maximum for both oscillators.
[e] Without readjustment of phase-matching for ambient temperatures of 18°C< T < 35°C.
[f] Shot to shot, peak to peak, 100% of shots.
[g] RMS, on 200 alternating pulses at the focal plane of a 2 m lens.
[h] Full angle on 200 pulses at e^{-2} of the peak, 85% of total energy.
[i] Secondary circuit, 10 l / min pressure, 1.5 to 3 bar.

In figure 2.16 a laser system with telescopic resonators is shown. It offers only 2 × 70 mJ pulse energy but has the advantage of a good and stable beam profile. Based on the authors' experience it can be said that the beam profile of this laser has been stable during more than ten years of application.

In figure 2.17 a laser system with critical resonators is shown. These systems typically have 150–450 mJ per pulse. The disadvantages of critical resonators have already been described above. In general it has to be mentioned that beam properties very much depend on manufacturers' know-how and the tuning of each individual laser.

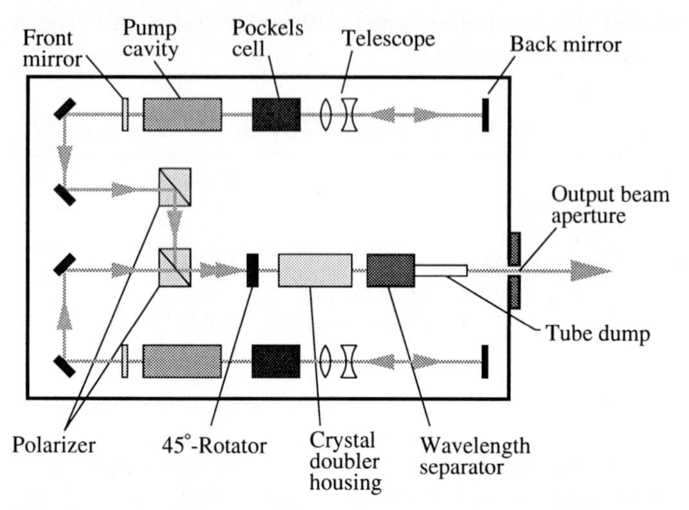

Fig. 2.16. Double oscillator laser system with telescopic resonators

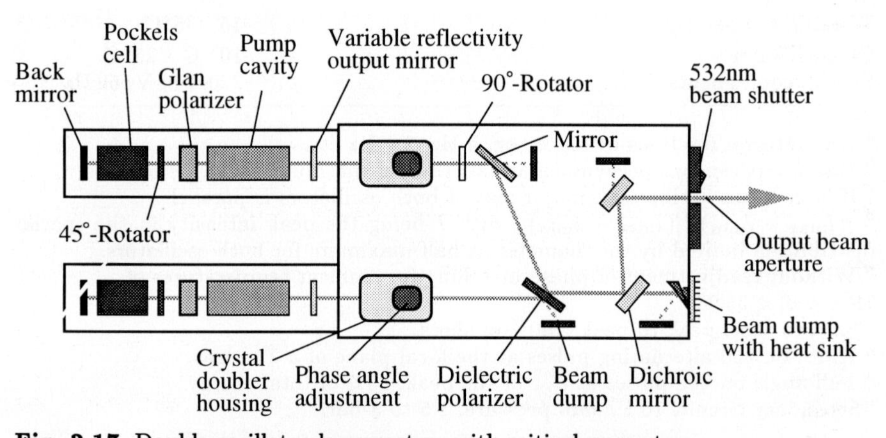

Fig. 2.17. Double oscillator laser system with critical resonators

In the following section a short description of Nd:YAG laser components is given.

The *pumping chamber* contains the Nd:YAG rod and a linear flashlamp which are sealed into their respective mountings with O-rings. These two components are surrounded by ceramic reflectors which provide efficient optical pumping of the laser rod. Glass filter plates absorb the ultraviolet radiation from the flashlamp.

The *output mirror* has a plane surface in most cases with a partially reflecting coating facing into the cavity. The opposite plane surface has an antireflection coating. In some cases the output mirror has a curved surface with a variable reflectivity coating decreasing from the center to the edges.

The *back mirror* has a highly reflective surface facing towards the cavity. Usually, this mirror has a slightly curved surface.

Describing the decay of energy inside a resonator can be done by introducing a quality factor or Q-factor. This factor can be changed with the help of the *Q-Switch.* The Q-switch normally consists of a polarizer plate, a temperature stabilized Pockels cell crystal and a beam path correcting prism. It is driven by high voltages and is used to release the energy stored in the laser rod in a giant pulse by rapidly changing the resonance conditions. Its principle is as follows: during the beginning of the flashlamp pulse the (voltage dependent) birefringence of the Pockels cell makes it act as a quarterwave plate. The polarization of the light that passed the polarizer is rotated by 90° on its way through the crystal towards the mirror and backwards towards the polarizer. The reflected light is then rejected by the polarizer. Therefore no laser oscillation and no light amplification will take place. When the energy stored in the laser rod reaches a maximum the Pockels cell voltage is altered and the polarization of passing light will no longer be changed. As a result the laser oscillation begins immediately and the energy stored is extracted in a pulse of only a few nanoseconds duration. Since the birefringence of the Pockels cell is also temperature dependent, the Pockels cell is generally temperature stabilized.

An intracavity *telescope* is used to avoid high order modes within the cavity. It also compensates for thermal lensing effects inside the laser rod.

A *second harmonic generator* is a nonlinear crystal used for the frequency doubling of the Nd:YAG laser emission. Simply speaking it converts infrared light of a wavelength of 1064 nm into visible green light of 532 nm. The process of frequency doubling only takes place when the crystal is oriented such that the direction of propagation of the pump beam is at a specific angle to the crystal axis. This condition is known as phase matching. Therefore the crystal can usually be angle tuned by the user. Since the refractive index and therefore also the actual phase matching changes with temperature the crystal has to be temperature stabilized to ensure stable conversion efficiencies. As most crystals used are hygroscopic the heating of the crystals should not be switched off in order to protect its surface from moisture. The crystal most commonly used is called KD*P. It can be cut in two different orientations (Type I or Type II) and has to be chosen depending on the final configuration of the laser. For Type I crystals the incident laser light has to be linearly, typically vertically or horizontally, polarized. The frequency doubled light emerges with a polarization which is orthogonal to that of the pump radiation. This type of doubler is used for PIV laser systems with two polarization directions as, for example, shown in figure 2.17. In order to generate green light of identical polarization one Type II crystal is generally used. Therefore, the infrared laser light must have two polarization components. The second harmonic will then have one polarization direction parallel to one of both original components depending on the orientation of the crystal. In order to provide two components of the incident laser light its linear polarization is turned by an angle of 45° using a polarization rotator.

A *polarization rotator* is a crystal which continuously rotates the polarization angle of linearly polarized light when it propagates through it. The rate of rotation is dependent on the material, its thickness, and the wavelength. A 45° rotator will be used when a Type II doubling crystal is used. A 90° rotator might be used in front of the beam combination optics, if two oscillators of identical orientation are used (see figure 2.17).

A *prism harmonic separator* can be used to separate the second harmonic wave by deflecting it into an energy dump. Two energy dumps are provided, one for the fundamental and one for the third harmonic wave. These separators are most efficient when used with only one polarization direction, as the reflection losses at the prism surfaces are lower for one polarization (see figure 2.16).

A *dichroic mirror* has maximum reflectivity for one given wavelength. The fundamental and any unwanted harmonic waves pass through such a mirror and can therefore be steered into an external energy dump.

2.2.3 White light sources

Even if most PIV investigations are performed using laser light sheets, white light sources might also be used. Due to the finite extension of these sources and since white light cannot be collimated as well as monochromatic light, they clearly have some disadvantages. On the other hand, the spectral output of sources like Xenon lamps is well suited for use with CCD cameras because of their spectral sensitivity. Systems are commercially available which can easily be triggered and offer a repetition rate that matches the video rate. Two flashlamps can be linked by optical fiber bundles in order to achieve short pulse separation times. If the outputs of the fibers are arranged in line, the generation of a light sheet is considerably simplified. The main advantage of these white light sources is – besides costs – that applications are not hampered by laser safety rules.

2.3 Light sheet optics

This section treats the optics for the illumination of the particles by a thin light sheet. Therefore we describe three different lens configurations, which have been used during various experiments. Rules for the calculation of the light sheet intensity distribution are not given herein. The reason for that is that geometric optics rules are already sufficient for a general layout of the chosen lens configuration. They do not require a special description and can easily be found in every book on optics [13]. On the other hand, more sophisticated calculations based on Gaussian optics usually require some assumptions, which are valid only for exceptional cases. Computer programs can be used in order to predict further parameters such as the light sheet thickness at the beam waist where the theoretical (geometrical) thickness is zero, but their description is beyond the scope of this book.

Optical fibers for beam delivery can be used to improve the handling of the system or for experimental situations where mirror systems would not be feasible. For CW lasers a variety of systems are available and for their use we refer to the manufacturers. Since pulsed lasers and white light sources have only limited repetition rates fiber bundles can be used for the combination of two sources for shorter pulse separation times. New developments can already deliver more than 10 mJ per pulse and further improvements can be expected [40]. However, the use of optical fibers will always be associated with a certain loss in intensity.

The essential element for the generation of a light sheet is a cylindrical lens. When using lasers with a sufficiently small beam diameter and divergence – like e.g. Argon-ion lasers – one cylindrical lens can be sufficient to generate a light sheet of appropriate shape. For other light sources – like e.g. Nd:YAG lasers – a combination of different lenses is usually required in order to generate thin light sheets of high intensity. At least one additional lens has then to be used for focusing the light to an appropriate thickness. Such a configuration is shown in figure 2.18, where also a third cylindrical lens has been added in order to generate a light sheet of constant height.

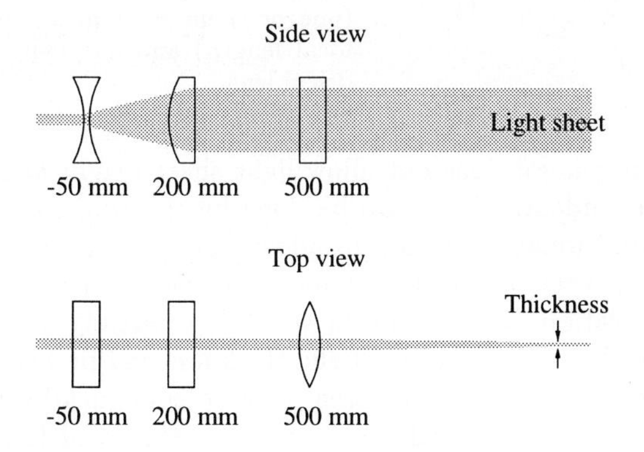

Fig. 2.18. Light sheet optics using three cylindrical lenses (one of them with negative focal length)

The reason why a diverging lens has been used first is that focal lines should be avoided. In high power pulse lasers focal points have to be avoided, as otherwise the air close to the focal point will be ionized. Focal lines usually do not ionize the air but dust particles might be burned if the area in the vicinity of the line is not covered or evacuated. In both cases acoustic radiation will occur and the beam properties will change significantly. For the light sheet shown in figure 2.18, the position of minimum thickness is given by the beam divergence of the light source and the focal length of the cylindrical lens on the right hand side, e.g. at a distance of 500 mm from the last lens for the conditions illustrated in figure 2.18.

The combination of a cylindrical lens together with two telescope lenses makes the system more versatile. This is shown in figure 2.19 where spherical lenses have been used, because they are in general easier to manufacture especially if short focal length lenses are required. The height of the light sheet shown in figure 2.19 is mainly given by the focal length of the cylindrical lens in the middle. A diverging lens – negative focal length – could also be used, however, since the focal line has a relatively large extension this configuration can be used also for pulsed lasers. The adaptation of the light sheet height has to be done by changing the cylindrical lens. The adjustment of the thickness can be easily done by shifting the spherical lenses with respect to each other.

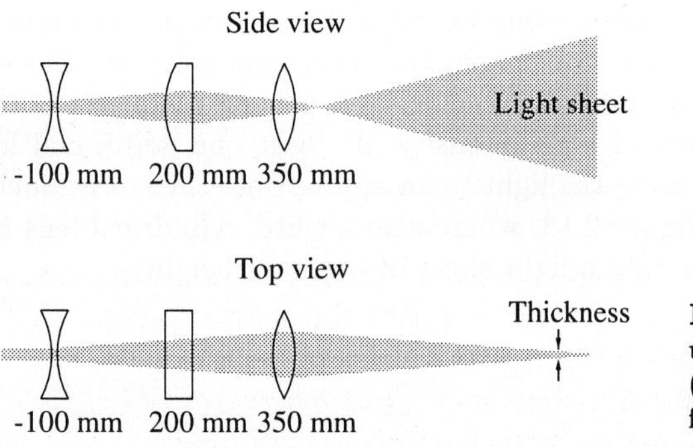

Fig. 2.19. Light sheet optics using two spherical lenses (one of them with negative focal length) and one cylindrical lens

The use of spherical lenses in general does not allow light sheet height and thickness to be changed independently. This can be done by the configuration shown in figure 2.20. Additionally this set-up allows the generation of light sheets which are thinner than the beam diameter at every location. It therefore enables the generation of light sheets which are already thin shortly after the last lens. With this arrangement the thickness can be held constantly small. However, the energy per unit area of these configurations is high. When using pulse lasers the critical region close to the focal line has to be covered in order to avoid dust or seeding particles disturbing the generation of a defined light sheet. Using a diverging cylindrical lens first would solve those problems, but the combination shown in figure 2.20 has the advantage of imaging of the beam profile from a certain position in front of the lens to the observation area while keeping its properties constant.

Simple geometric considerations can be used for these lens combinations to determine from which position the laser beam has been imaged and, if the development of the beam profile of the laser is known, this information can be used to optimize the light sheet intensity distribution. For lasers with a critical beam profile this can improve the valid data yield, because the light sheet intensity distribution especially in the out-of-plane direction is essential for the quality of the measurement (see section 3). In figure 2.15 on page 30 the evolution of a light sheet profile generated by a lens configuration similar to that shown in figure 2.20 has been shown as a function of the distance from the laser.

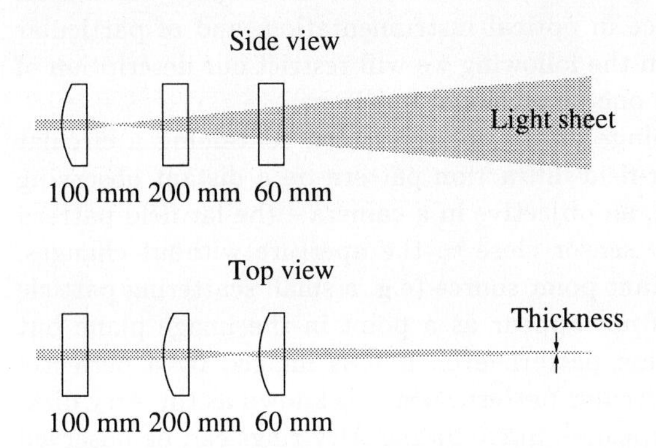

Fig. 2.20. Light sheet optics using three cylindrical lenses

A few general rules should be given here also. Uncoated lens surfaces in air exhibit a slight reflectivity of $[(n-1)/(n+1)]^2$. Since this value is on the order of 4% for common lenses the losses due to the reflection can usually be accepted. However, these reflections can cause damage if they are focused close to other optical components. In most cases this can easily be avoided by the right orientation of the lenses as demonstrated in figure 2.21.

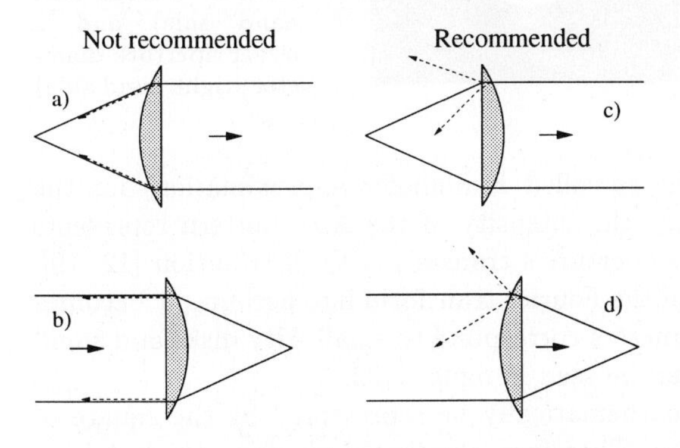

Fig. 2.21. General considerations on the orientation of lenses inside the light sheet optics

Furthermore case c and d in figure 2.21 should be used in order to minimize aberrations. For other situations it might be possible to tilt the lens slightly in order to avoid reflections on to other lenses or towards the laser or even into the resonator.

2.4 Imaging of small particles

2.4.1 Diffraction limited imaging

This section provides a description of diffraction limited imaging, which is an effect of practical significance in optical instrumentation, and of particular interest for PIV recording. In the following we will restrict our description of imaging by considering only one-dimensional functions.

If plane light waves impinge on an opaque screen containing a circular aperture they generate a far-field diffraction pattern on a distant observing screen. By using a lens – e.g. an objective in a camera – the far field pattern can be imaged on an image sensor close to the aperture without changes. However, the image of a distant point source (e.g. a small scattering particle inside the light sheet), does not appear as a point in the image plane but forms a Fraunhofer diffraction pattern even if it is imaged by a perfectly aberration-free lens [13]. A circular pattern, which is known as the Airy disk, will be obtained for a low exposure. Surrounding Airy rings can be observed for a very high exposure.

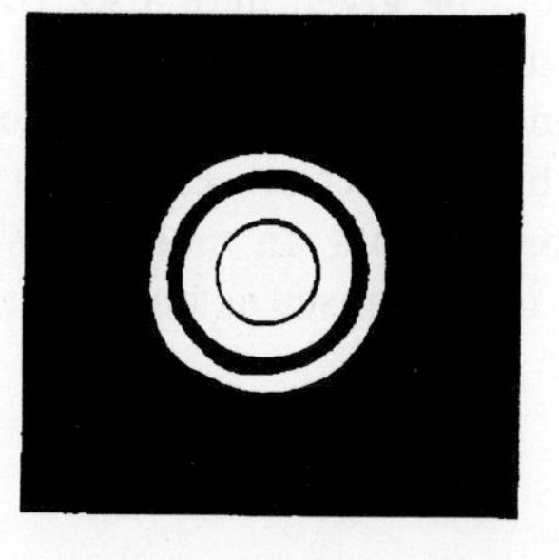

Fig. 2.22. Airy patterns for a small (left hand side) and a larger aperture diameter (right hand side)

Using an approximation (the so-called Fraunhofer approximation) for the far field it can be shown that the intensity of the Airy pattern represents the Fourier transform of the aperture's transmissivity distribution [12, 19]. Taking the scaling theorem of the Fourier transform into account, it becomes clear that large aperture diameters correspond to small Airy disks and small apertures to large disks as can be seen in figure 2.22.

The Airy function can mathematically be represented by the square of the first order Bessel function. Therefore, the first dark ring, which defines the extension of the Airy disk, corresponds to the first zero of the first order Bessel function shown in figure 2.23. The Airy function represents the impulse reponse – the so-called point spread function – of an aberration-free lens. We will now determine the diameter of the Airy disk d_{diff}, because it represents the smallest particle image that can be obtained for a given imaging configuration.

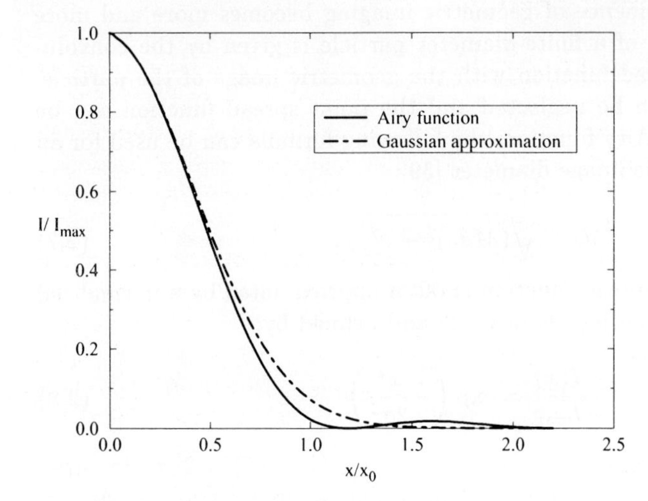

Fig. 2.23. Normalized intensity distribution of the Airy pattern and its approximation by a Gaussian bell curve

In figure 2.23 the value of the radius of the ring and therefore of the Airy disk can be found for a given aperature diameter D_{a} and wavelength λ:

$$\frac{I(x)}{I_{\mathrm{max}}} \stackrel{!}{=} 0 \quad \Rightarrow \quad \frac{d_{\mathrm{diff}}}{2x_0} = 1.22$$

with

$$x_0 = \frac{\lambda}{D_{\mathrm{a}}}\,.$$

If we consider imaging of objects in air – the same media on both sides of the imaging lens – the focus criterion is given by (see figure 2.24):

$$\frac{1}{z_0} + \frac{1}{Z_0} = \frac{1}{f} \tag{2.5}$$

where z_0 is the distance between the image plane and lens and Z_0 the distance between the lens and the object plane. Together with the definition of the magnification factor

$$M = \frac{z_0}{Z_0}$$

the following formula for the diffraction limited minimum image diameter can be obtained:

$$d_{\text{diff}} = 2.44\, f_{\#}(M+1)\,\lambda \tag{2.6}$$

where $f_{\#}$ is the f-number defined as the ratio between the focal length f and the aperture diameter D_{a} [12]. In PIV, this minimum image diameter d_{diff} will only be obtained when recording small particles – on the order of a few microns – at small magnifications. For larger particles and/or larger magnifications the influence of geometric imaging becomes more and more dominant. The image of a finite-diameter particle is given by the convolution of the point spread function with the geometric image of the particle. If lens aberrations can be neglected and the point spread function can be approximated by the Airy function, the following formula can be used for an estimate of the particle image diameter [39]:

$$d_\tau = \sqrt{(M d_{\text{p}})^2 + d_{\text{diff}}^2}\,. \tag{2.7}$$

In practice the point spread function is often approximated by a normalized Gaussian curve also shown in figure 2.23 and defined by:

$$\frac{I(x)}{I_{\max}} = \exp\left(-\frac{x^2}{2\sigma^2}\right) \tag{2.8}$$

where the parameter σ must be set to $\sigma = f_{\#}(1+M)\lambda\sqrt{2}/\pi$, in order to approximate diffraction limited imaging. This approximation is particularly useful because it allows a considerable simplification of the mathematics encountered in the derivation of modulation transfer functions, which also includes other kinds of optical aberrations of the imaging lens as will be described later.

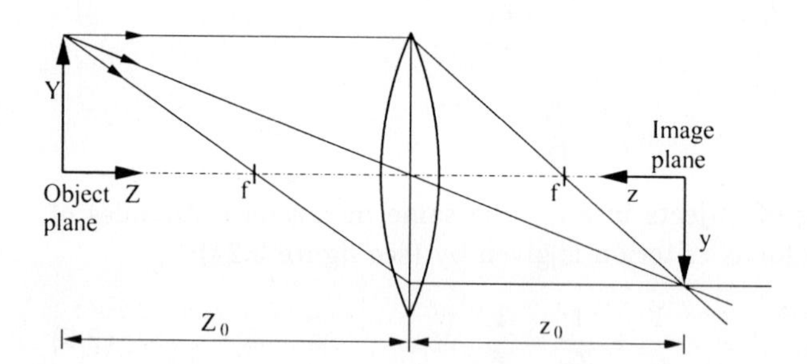

Fig. 2.24. Geometric image reconstruction

In practice there are two good reasons for optimizing the particle image diameter:

First, an analysis of PIV evaluation shows that the error in velocity measurements strongly depends on the particle image diameter (see e.g. section

5.5.2). For most practical situations, the error is minimized by minimizing both the image diameter d_τ and the uncertainty in locating the image centroid or correlation peak centroid respectively.

Second, sharp and small particle images are particulary essential in order to obtain a high particle image intensity $I_{\max}$, since at constant light energy scattered by the tracer particle the light energy per unit area increases quadratically with decreasing image areas ($I_{\max} \sim 1/d_\tau^2$). This fact also explains why increasing the particle diameter not always compensates for insufficient laser power.

Equation (2.7) shows that for a range of particle diameters greater than the wavelength of the scattered light ($d_\tau \gg \lambda$), the diffraction limit becomes less important and the image diameter increases nearly linearly with increasing particle diameter. Since the average energy of the scattered light increases with $(d_\mathrm{p}/\lambda)^2$ for particles with a diameter greater than the wavelength (Mie's theory), the image intensity becomes independent of the particle diameter, as both the scattered light and the image area increase with d_p^2.

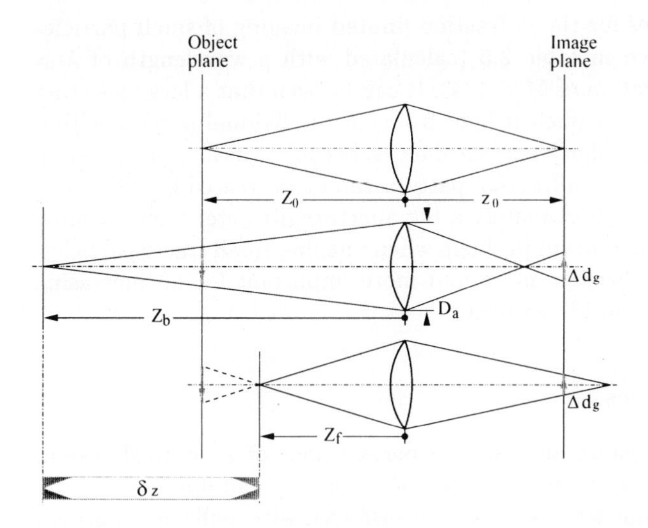

Fig. 2.25. Out-of-focus imaging

As can be seen in figure 2.25 a point in the object plane generates a sharp image at only one defined position in the image space, where the rays transmitted by different parts of the lens intersect. This point of intersection and therefore of best focusing has been determined in equation (2.5). If the distance between the lens and the image sensor is not perfectly adjusted, the geometric image is blurred and its diameter can also be determined by geometric optics. This blur of images due to misalignment of the lens does not depend on diffraction or lens aberrations. However, the minimum image diameter d_{diff} which can be obtained due to diffraction is commonly used also as the acceptable diameter of the geometric image (d_g in figure 2.25). Therefore, the particle image diameter obtained by equation (2.7) can be used to estimate the depth of field δ_Z using the following formula [26]:

$$\delta_Z = 2 f_{\#} d_{\text{diff}} (M+1)/M^2 \,. \tag{2.9}$$

Table 2.5. Theoretical values for diffraction limited imaging of small particles ($\lambda = 532\,\text{nm},\ M = 1/4,\ d_\text{p} = 1\,\mu\text{m}$)

$f_{\#} = f/D_a$	$d_\tau [\mu\text{m}]$	$\delta_Z [\text{mm}]$
2.8	4.7	0.5
4.0	6.6	1.1
5.6	9.1	2.0
8.0	13.0	4.2
11	17.8	7.8
16	26.0	16.6
22	35.7	31.4

Some theoretical values for the diffraction limited imaging of small particles ($d_\text{p} \approx 1\,\mu\text{m}$) are shown in table 2.5 (calculated with a wavelength of $\lambda = 532\,\text{nm}$ and a magnification of $M = 1/4$). It can be seen that a large aperture diameter is needed to get sufficient light from each individual particle within the light sheet, *and* to get sharp particle images, because – as already shown in figure 2.22 – the size of the diffraction pattern can be decreased by increasing the aperture diameter. Unfortunately a big aperture diameter yields a small focal depth which is a significant problem when imaging small tracer particles. Since lens aberrations become more and more important for an increasing aperture, they will be considered next.

2.4.2 Lens aberrations

In analogy to linear system analysis the performance of an optical system can be described by its impulse response – the point spread function – or by the highest spatial frequency that can be transferred with sufficient contrast. This upper frequency – the so-called resolution limit – can be obtained by the reciprocal value of the characteristic width of the impulse response. According to this, the physical dimension is the reciprocal of a length; typically it is interpreted as the number of line pairs per millimeter (lps/mm) that can be resolved. The traditional means of determining the quality of a lens was to evaluate its limit of resolution according to the so-called RAYLEIGH criterion: two point sources were said to be "barely resolved" when the center of one Airy disk falls on the first minimum of the Airy pattern of the other point source. That means that the theoretical resolution limit ρ_m is the reciprocal value of the radius of the Airy disk [13]:

$$\rho_{\mathrm{m}} = \frac{2}{d_{\mathrm{diff}}} = \frac{1}{1.22\, f_{\#}(M+1)\lambda} \,. \tag{2.10}$$

Another useful parameter in evaluating the performance of an optical system is the contrast or image modulation defined by the following equation:

$$\mathrm{Mod} = \frac{I_{\max} - I_{\min}}{I_{\max} + I_{\min}} \,. \tag{2.11}$$

The measurement of the ratio of image modulation Mod for varying spatial frequencies yields the modulation transfer function (MTF). The modulation transfer function has become a widely used means of specifying the performance of lens systems and photographic films.

In practice an approximation of the modulation transfer function can be obtained by an inverse Fourier transformation of the point spread function. Using the Gaussian approximation given by equation (2.8) and shown in figure 2.23 greatly simplifies this transformation. In the following we continue to simplify the description of imaging by considering only one-dimensional functions. The Fourier transformation FT of a one-dimensional Gaussian function is given by:

$$\sigma\sqrt{2\pi}\,\exp\left(-2\pi^2\sigma^2 r^2\right) \quad \overset{\mathrm{FT}}{\Longleftrightarrow} \quad \frac{1}{\sigma\sqrt{2\pi}}\,\exp\left(-\frac{x^2}{2\sigma}\right) \tag{2.12}$$

where σ determines the width of the Gaussian curve and r represents the variable for the spatial frequency.

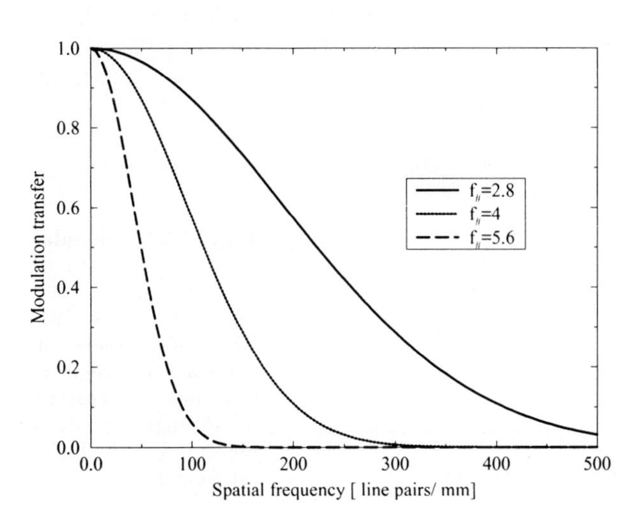

Fig. 2.26. Image modulation versus spatial frequency for a hypothetical lens systems without spherical aberrations at three different f-numbers (Gaussian approximations)

Figure 2.26 shows a plot of image modulation versus spatial frequency for three apertures of a hypothetical lens system without spherical aberration as obtained by the inverse transformation of the Gaussian approximation of the

Airy function. As shown in table 2.5, the minimum image diameter decreases with decreasing f-numbers. As a consequence, high spatial frequencies can only be recorded at small f-numbers and for a given spatial frequency r, small f-numbers yield better contrast compared to large f-numbers. Although the shape of these curves only roughly approximates the shape of MTFs of real lens systems, the qualitative behavior can clearly be seen.

However, taking lens aberrations into account results in major changes of the MTFs especially when using small f-numbers. This can be seen in figure 2.27 where measured values of a high quality 100 mm lens are presented together with a Gaussian curve fitted to the value measured at the highest frequency. These values of the modulation transfer are often given in the data sheets of a lens system for different f-numbers and magnifications or can easily be estimated when studying diagrams of the modulation versus the image height. From experience these values can be used for a rough estimation of the image diameters to be expected, regardless of the fact that they were originally measured using white light and therefore consider also chromatic aberrations, which do not have to be taken into account for monochromatic laser light. According to the previous discussion, we assume that the MTF can be described by the inverse Fourier transform of the Gaussian approximation of the normalized image intensity distribution $I(x)/I_{\mathrm{max}}$, which can be written in normalized form as:

$$\tilde{M}_{\mathrm{TF}} = \exp\left(-2\pi^2\sigma^2 r^2\right) . \tag{2.13}$$

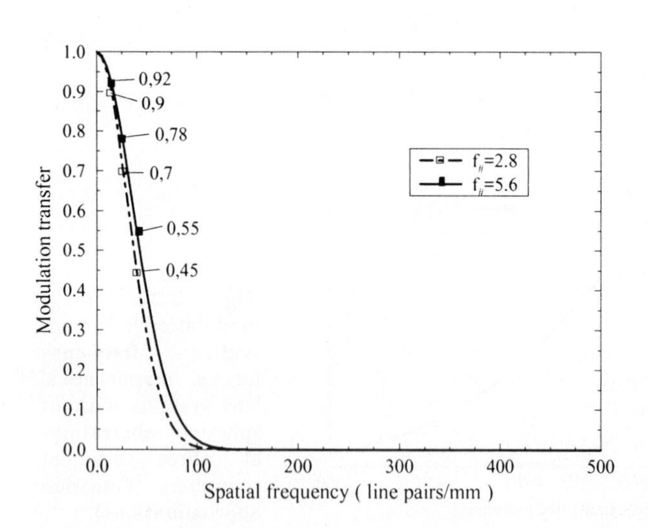

Fig. 2.27. Modulation transfer data of a high quality 100 mm lens for two different f-numbers at three spatial frequencies and a Gaussian fit through the data measured at the highest frequency

Taking a characteristic value r' for the spatial frequency ($r' \approx d_\tau$) and the corresponding MTF value $\tilde{M}_{\mathrm{TF}}$ into account, equation (2.8) can be used in order to determine σ and therefore a function through this point in the MTF (e.g. $\tilde{M}_{\mathrm{TF}}(r') = 0.55$ at $r' = 40$ line pairs/mm from figure 2.27):

$$\sigma = \sqrt{-\frac{\ln[\tilde{M}_{\mathrm{TF}}(r')]}{2\pi^2 r'^2}} \,. \tag{2.14}$$

Now, equation (2.8) can be used to approximate the normalized intensity distribution of the image. In contrast to the Bessel function the Gaussian approximation has no zero-crossings (see figure 2.23). An image diameter can therefore not be determined by taking the x-value of the first zero. This requires some kind of threshold level. In photographic PIV this threshold level represents the start of the nonlinear behavior of the film material used for recording and contact copy (see figure 2.33). As described in section 5.3.2, a two step photographic process is normally used to reduce the effects of the film fog and other background noise for subsequent optical evaluation. In digital PIV a certain threshold is normally used because of electronic background noise. In the following we assume that the lower 20% of the intensity of an image will not be used for evaluation. This assumption and equation (2.8) yield the following formula for the radius of an image of a circular object:

$$x' = \sqrt{-2\sigma^2 \ln 0.2} \tag{2.15}$$

When substituting equation (2.14) we obtain an approximation of the image diameter d' that corresponds to a circular object which has an extension $(2r')^{-1}$:

$$d' \approx 0.8\sqrt{-\frac{\ln[\tilde{M}_{\mathrm{TF}}(r')]}{r'^2}} \,. \tag{2.16}$$

For the estimation of the image diameters of smaller objects the same approximation as in equation (2.7) can be used to obtain:

$$d_\tau \approx \sqrt{-0.64\,\frac{\ln[\tilde{M}_{\mathrm{TF}}(r')]}{r'^2} - \left(\frac{M}{2r'}\right)^2} \,. \tag{2.17}$$

The definition of a MTF is a practical approach to describe the performance of optical systems. Since the underlying optical processes are much more complex, the description would be more complete when considering also the relative phase shift. However, phase shifts in optical systems occur only off axis and are of less interest than the MTF [13].

For many practical applications the MTF of a complex optical system can be assumed to be simply the product of the MTFs of the individual components. When doing so, the value of the film MTF (for example shown in figure 2.34) at a certain spatial frequency r' must be multiplied by the lens MTF value for the same spatial frequency and equation (2.17) can be used to estimate the image diameter for a given recording condition. Practical experience has shown that the resolution of a film with high sensitivity (3200 ASA) can be used to perform high quality PIV measurements even in the case of difficult recording conditions [132]. The image diameter estimated by equation (2.17) is $\approx 20\,\mu$m ($M = 1/4$, $f_{\#} = 2.8$) and is in good correspondence with image diameters found during experiments. However, image recording by video sensors cannot be described by those models adequately. The sampling due to the regular arrangement of picture elements and its influence on the PIV evaluation has to be modeled in a different manner in order to describe important effects sufficiently.

2.4.3 Perspective projection

In this section a model of the imaging geometry associated with PIV recording will be described by means of linear algebra. This model will then be used to describe and quantify the effects of perspective projection in the remainder of this section. In section 4.3.5 it serves for the determination of the image shift due to a rotating mirror and in section 7.1 the basic equations for stereoscopic PIV will be derived by means of this model.

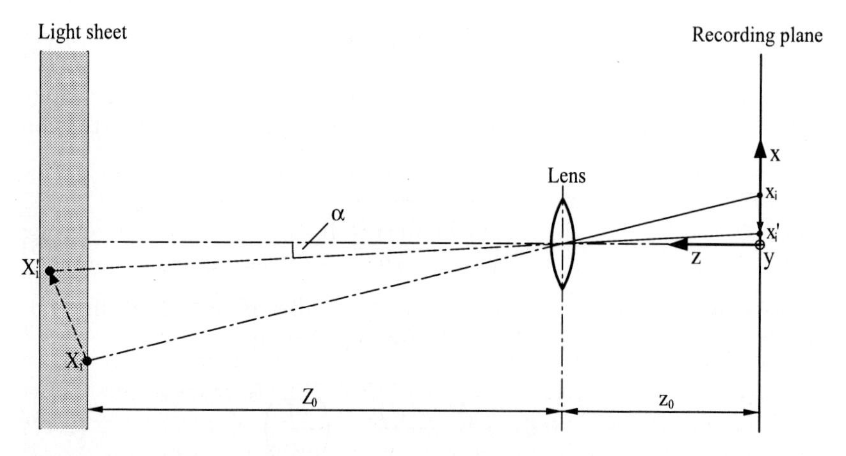

Fig. 2.28. Imaging of a particle within the light sheet on the recording plane

In order to fully explain the influence of a velocity component perpendicular to the light sheet on the location of the image points in the coordinate system x, y, z (figure 2.28), the imaging through the lens must be taken into account. Ideal imaging conditions are assumed for this calculation. Image distortions resulting from nonideal lenses can be taken into account by means of an extended model which is beyond the scope of this book. The underlying perspective projection can be modeled by defining a homogeneous coordinate system. This is common practice in digital image processing [11]. It is necessary to describe the imaging geometry in a homogeneous, four dimensional vector space, because the projection under consideration is a nonlinear transformation. The matrices used are given in appendix A.

The following operators describe the required perspective transformations: P_H for projecting a point in the light sheet on to the image plane and $\mathsf{P}_\mathrm{H}^{-1}$ for the transformation of a point from the image plane on to the light sheet plane. D_H defines the particle displacement between the light pulses by D_x, D_y, D_z. The following relation between the location of image points on the recording plane due to the imaging of a particle at position $\boldsymbol{x}_i$ and $\boldsymbol{x}'_i$ is obtained:

$$\boldsymbol{x}'_{i,\mathrm{H}} = \mathsf{P}_\mathrm{H} \circ \mathsf{D}_\mathrm{H} \circ \mathsf{P}_\mathrm{H}^{-1} \circ \boldsymbol{x}_{i,\mathrm{H}} \, . \tag{2.18}$$

After converting into the (camera-)world coordinate system the image displacement $\boldsymbol{d} = \boldsymbol{x}'_i - \boldsymbol{x}_i$ corresponding to a certain particle displacement $\boldsymbol{D}$ can be obtained:

$$x'_i - x_i = -M(D_X + D_Z\, x'_i/z_0) \tag{2.19}$$

$$y'_i - y_i = -M(D_Y + D_Z\, y'_i/z_0) \, . \tag{2.20}$$

Assuming a particle displacement only in the X, and Y directions ($D_Z \approx 0$) would simplify equation (2.19) and equation (2.20) considerably. Then, the in-plane particle displacement could easily be determined by multiplying the image displacement by $(-M)$. In this particular case, the only uncertainty of the velocity measurement would be introduced by the uncertainty in determining the image displacement and the geometric parameters. However, in practical cases a flow field is never strictly two-dimensional over the whole observation field. Moreover, conventional PIV, which was at the beginning suitable only for measurements of flow fields with weak out-of-plane components, has been adapted also for use in highly three-dimensional flows over the last decade. It can be seen in equation (2.19) and equation (2.20), that a particle displacement in the Z-direction influences the particle image displacement, especially for large magnitudes of X'_i and Y'_i at the edges of the observation field [90, 73]. This effect introduces an uncertainty in measuring the in-plane velocity components, because it cannot be separated from the in-plane components. This uncertainty will turn into a systematic error if it is assumed that PIV determines just the in-plane components even for larger viewing angles. It will be shown by a simple analysis of the recording of a hypothetical three-dimensional flow field that this systematic measurement error can increase up to more than 15% of the mean flow velocity.

2.4.4 Discussion of the perspective error

The following discussion is based on the assumption of a flow with a strong out-of-plane component. In order to describe the influence of the perspective projection qualitatively and quantitatively an example will be given: a potential vortex is assumed, similar to vortices occurring in natural flows (hurricane) and flows in technical devices (cyclone). In the case of an isoenergetic flow the Bernoulli equation yields for the velocity components V and W:

$$\sqrt{V^2 + W^2} = \frac{|\boldsymbol{U}_{\mathbf{max}}|\,\tilde{X}}{\sqrt{X^2 + Y^2}},$$

where $\boldsymbol{U}_{\mathbf{max}}$ is the maximum velocity of the vortex flow, and $\tilde{X}$ is the distance from the point where it occurs to the vortex axis.

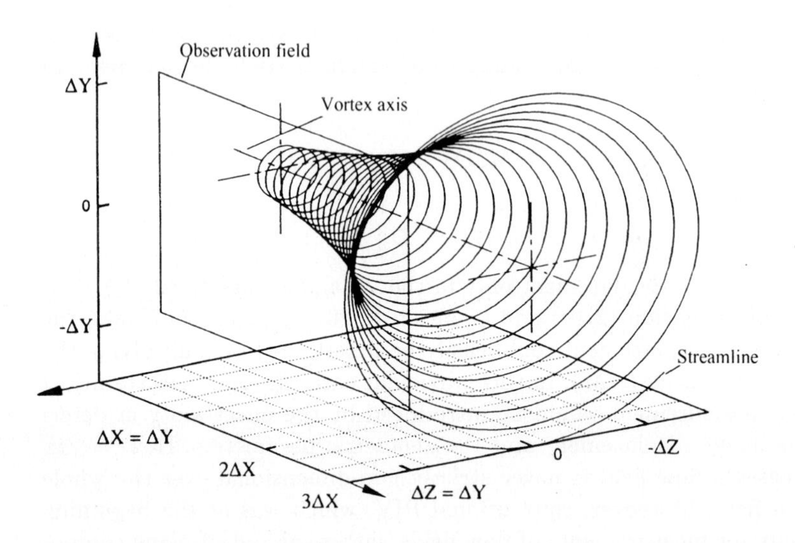

Fig. 2.29. 3D representation of the coordinate system, the observation field and a streamline ($\alpha_0 < \pi/4$ for better representation) of a potential vortex

The X-axis of the observed plane of the flow is parallel to the vortex axis, and is located outside the vortex core as illustrated in figure 2.29. The V and W components can be calculated according to the formula given above, while the U-component of the velocity is assumed to have 20% of the magnitude of the other velocity components at point $(0, 0, \Delta Z)$. This leads to the following equations for the velocity components as functions of the X, Y, Z-coordinates:

$$U = |\boldsymbol{U}_{\mathbf{max}}| \frac{1}{\sqrt{2}\,0.2}$$

$$V = -|\boldsymbol{U}_{\mathbf{max}}| \cos\left(\arctan\left(\frac{Y}{\Delta Y}\right) + \alpha_0\right) \frac{\Delta Y}{\sqrt{\Delta Y^2 + Y^2}}$$

$$W = -|\boldsymbol{U}_{\mathbf{max}}| \sin\left(\arctan\left(\frac{Y}{\Delta Y}\right) + \alpha_0\right) \frac{\Delta Y}{\sqrt{\Delta Y^2 + Y^2}}\,.$$

In the case of the following numerical simulation of the PIV recording process $\alpha_0 = \pi/4$ is assumed, so that $V = 0$ at the upper left corner of the observation area and $W = 0$ at the lower left corner. In figure 2.30 we plot the U and V components of the velocity field, which are in general obtained by a PIV measurement.

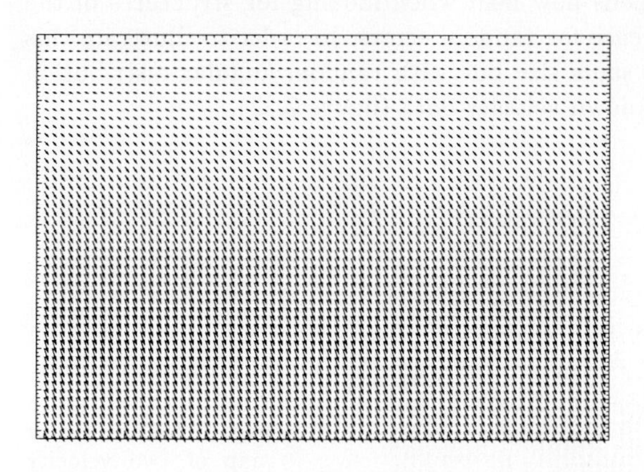

Fig. 2.30. Velocity vector map of the U and V components of a simulated potential vortex within the observation field shown in figure 2.29

However, since PIV determines the perspective projection of the three-dimensional velocities, we come to the following representation of the measured velocity components.

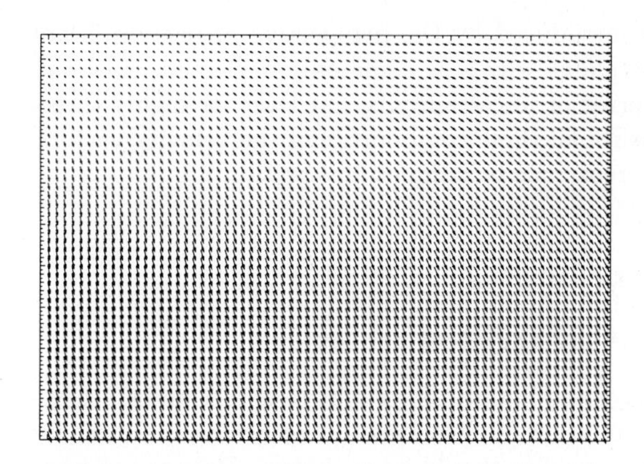

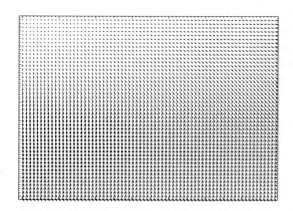

Fig. 2.31. Vector map of two velocity components after "PIV recording" ($M = 1/4$, 35 mm film, $f = 60$ mm) calculated by means of perspective transformation from the same velocity vector field as presented in figure 2.29

In figure 2.31 the values that would have been obtained by a PIV measurement of the 3D velocity vectors in the XY plane are presented, computed by means of the algorithms for the transformation described above. The computation was based on the following parameters: Magnification $M = 1/4$, film size $35 \times 24\,\text{mm}^2$, focal length of the objective lens $f = 60\,\text{mm}$. These parameters are realistic values for many practical applications. The difference between the real UV data and the calculated "PIV recording" data is up to 16.6% of the mean flow velocity in this case of idealized conditions. In order to achieve accurate data, this deviation has clearly to be assessed as a considerable error, but its systematic influence does not hamper the interpretation of the instantaneous flow field when looking for structures of the flow field as would be the case for random errors. In order to illustrate this statement, an error of the same size but with random angular distribution was superimposed on the known velocity data (figure 2.32).

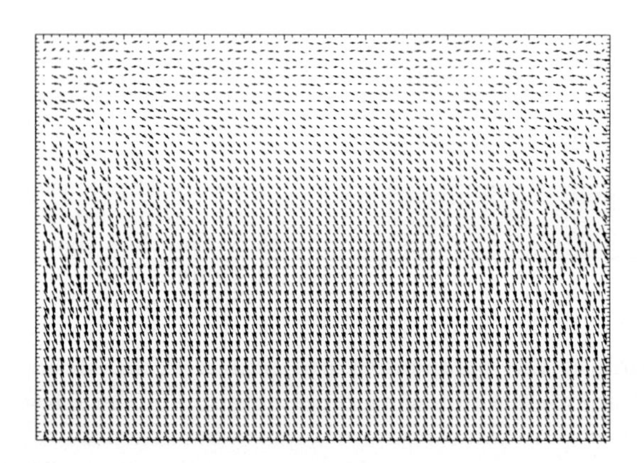

Fig. 2.32. Vector map of two velocity components with a random error of the same order as contained in the "PIV recording", shown in figure 2.31

The fact that the error introduced by the perspective projection is not that visible in the results explains why it is commonly neglected. However, the only way to avoid this error or uncertainty is – at least for highly three-dimensional flows – to measure all three components of the velocity vectors, for example by means of stereoscopic techniques (see section 7.1).

2.5 Photographic recording

2.5.1 Short description of the chemical processes

Photographic films are widely used in optical systems to detect and store optical information. In PIV as in other applications this recording is a basis for subsequent evaluation, and its quality should be optimized. Relevant properties such as spectral sensitivity, speed, resolution, photographic gamma and photographic noise have therefore to be taken into account. In the following, a short and simplified description of the chemical processes from exposure to fixing will be given, together with an introduction to the diagrams normally used to characterize film performance.

An unexposed photographic emulsion mainly consists of tiny silver halide grains in a gelatine support. The interaction of an impinging photon with the photographic emulsion consists of random events. The probability that a photon strikes a silver halide grain depends on the density of the grains. Those grains that have absorbed a sufficient number of photons are found to contain tiny patches of metallic silver. These patches are referred to as "development centers". During the developing process the chemicals used are diffusing from the surface through the gelatine. The existence of a single tiny development center precipitates the change from silver halide to silver. The unexposed grains are left unchanged and are removed during the "fixing".

2.5.2 Introduction to performance diagrams

The locally varying intensity transmittance of the emulsion after development is defined by [12]:

$$T(x,y) = \text{local average}\left(\frac{I_{\text{trans}}(x,y)}{I_{\text{inc}}}\right) . \tag{2.21}$$

I_{inc} is the intensity of a light source illuminating the recording after development, I_{trans} is the intensity of the light locally transmitted through the recording. The area of the *local average* is large with respect to the size of a film grain, but small compared with an area within which the transmitted intensity changes significantly. The local variance of $T(x,y)$ is – besides other process parameters – caused by the local variance of the prior exposure E, which is the integral of the intensity per unit area over the exposure time Δt:

$$E = \int_{\Delta t} I \, \mathrm{d}t . \tag{2.22}$$

HURTER and DRIFFIELD demonstrated that the logarithm of the intensity transmittance $\log[1/T(x,y)]$, the so-called photographic density D_{photo}, is proportional to the silver mass per unit area of a developed transparency. Plots of the photographic density versus the logarithm of exposure $\log(E)$ are therefore commonly referred to as Hurter–Driffield curves. They are often given in data sheets of photographic films. The following figures display diagrams for two films of different sensitivity, which are widely used in black and white photography. It can be seen in figure 2.33 that the density is equal to a minimum value referred to as "gross fog" when the exposure is below a certain level. The density increases with increasing exposure at the so-called "toe". For further increasing exposures it is linearly proportional to the logarithm of exposure (see figure 2.33). The slope of the linear region is referred to as the photographic gamma γ. For further increasing exposures the curve saturates at a constant level. This region is called the "shoulder" of the Hurter–Driffield curve. The linear region of the curve is the portion generally used for PIV recording. Films can be classified as high contrast film for a photographic gamma of two or more or low contrast films for $\gamma < 1$. The photographic gamma can further be varied by varying the development time. This effect and the nonlinear features of photographic emulsions described by their Hurter–Driffield curves can be used in an additional photographic process – a contact copy – which can be performed after evaluation and which reduces the noise in optical PIV evaluation as described in section 5.3.

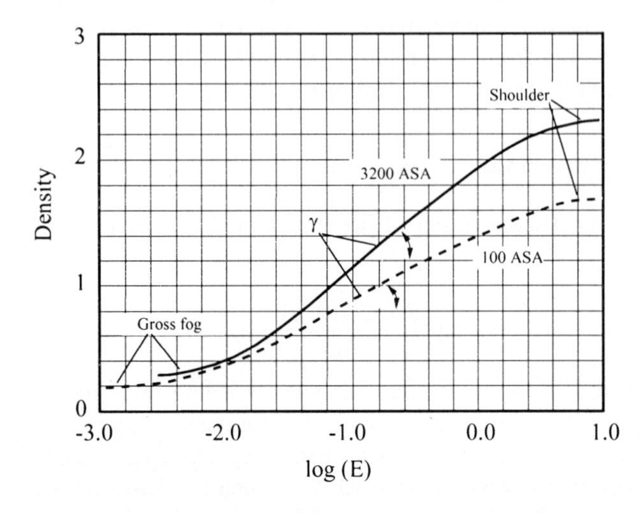

Fig. 2.33. Hurter–Driffield curves for two standard black and white films

Up to now it has been assumed that any spatial variation of the incident light during recording will be transferred into corresponding spatial density variations on the film. However, this is not valid if the spatial scale of exposure variations is too small. Generally speaking, a given photographic emulsion possesses only a limited spatial frequency response. Therefore, the highest spatial frequency that can be recorded with a sufficient contrast is a further important film specification. However, for many applications the evaluation of film resolution properties only on the basis of an upper limit of spatial frequency response is not sufficient. Especially for PIV, where high resolution is required, the response of a film over the entire operating frequency range is more useful. In figure 2.34 the spatial frequency response of black and white films with sensitivity of 100 ASA and 3200 ASA respectively is shown. It can be seen that the film with the higher sensitivity offers lower frequency resolution. Such frequency response diagrams (MTF's) have become a widely used means of specifying the performance of films and lens systems. The concept of MTFs and the analysis of their effect on the obtainable particle image diameter has been described in more detail in section 2.4. Another important diagram for selecting the proper film for PIV recording is the curve of the spectral sensitivity. The sensitivity is defined as the reciprocal value of the exposure E, which is necessary to obtain a certain photographic density. Figure 2.35 shows such curves for two commonly used films. The curves are given for a density that is the density of the film fog plus one, in other words, they represent $\log(1/E)$ which is needed to obtain a film transmittance that is a tenth of the transmittance of the film fog. It can be seen that the 3200 ASA film shows superior sensitivity especially in the green. This is one of the reasons why frequency doubled Nd:YAG lasers with a wavelength of 532 nm (vertical line in figure 2.34) are widely used in photographic PIV.

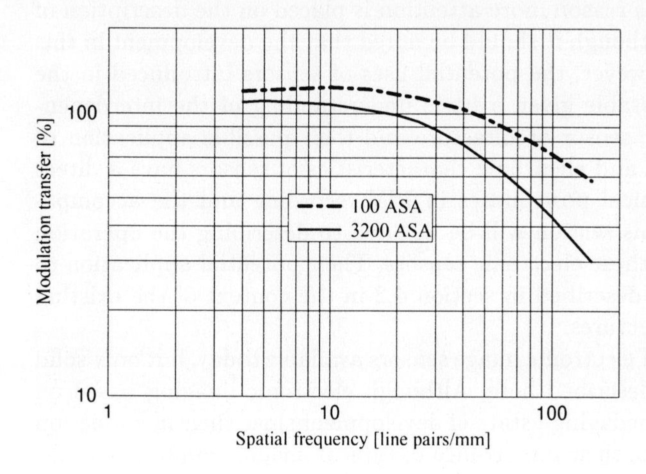

Fig. 2.34. Spatial frequency response (MTF) of two standard black and white films

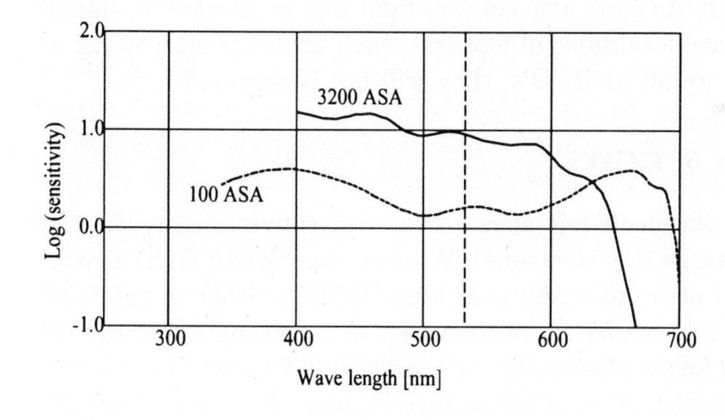

Fig. 2.35. Spectral sensitivity of two standard black and white films

2.6 Digital image recording

Recent advances in electronic imaging have provided an attractive alternative to the photographic methods of PIV recording. Immediate image availability and thus feedback during recording as well as a complete avoidance of photochemical processing are but a few of the apparent advantages brought about with electronic imaging. The present trends suggest that electronic recording will be of increased importance to PIV in the future forcing the photographic methods aside. For this reason more attention is placed on the description of electronic recording although it should be noted that the development in this area is very rapid. However, the potential uses of sensors introduced in the future should be assessable given a basic understanding of the interdependence between current sensor architecture and their possible application to PIV. Since the optical and electronic characteristics of sensors have a direct influence on the technical possibilities in PIV recording and the accompanying error sources, this section will be devoted to describing the operation and characteristics of these electronic sensors. Their potential application to PIV recording will be described in section 4.2 in the context of the existing variety of CCD architectures.

There is a variety of electronic image sensors available today, but only solid state sensors will be described here. Although electronic imaging based on vidicon tubes has reached a high state of development since their introduction more than 50 years ago, their importance to typical imaging applications has decreased dramatically in favor of solid state imagers. Of these, the charge coupled device, or CCD, has found the most widespread use. Charge injection devices (CID) and CMOS imagers also convert light into an electronic signal. Although these units have a number of features, such as direct addressing or on-chip processing, not found in CCD's, they will not be covered here.

2.6.1 Characteristics of CCD's

In general the CCD is an electronic sensor that can convert light, that is, photons, into electric charge (i.e. electrons). When we speak of a CCD sensor, we generally refer to an array of many individual CCD's, either in the form of a line (e.g. in a line scan camera), or arranged in a rectangular array (of course other specialized forms also exist). The individual CCD element in the sensor is called a pixel, which is short for picture element. Its size is generally on the order of $10 \times 10\,\mu\text{m}^2$, or 100 pixels per mm.

The operation of these pixels is best described by referring to the schematic cross-section shown in figure 2.36. The CCD is built on a semiconducting substrate, typically silicon, with metal conductors on the surface, an insulating oxide layer, an n-layer (anode) and a p-layer (cathode) below that. A small voltage applied between the metal conductors and the p-layer generates an electric field within the semiconductor. The local minimum in the electric field that is formed below the center of the pixel is associated with a lack of electrons and is known as a potential well. In essence the potential well is equivalent to a capacitor allowing it to store charge, this is, electrons. When a photon of proper wavelength enters the p-n junction of the semiconductor an electron-hole pair is generated. In physics this effect is known as the photoelectric effect. While the *hole*, considered as a carrier of positive charge, is absorbed in the p-layer, the generated electron (or charge) migrates along the gradient of the electric field toward its minimum (i.e. potential well) where it is stored. Electrons continue to accumulate for the duration of the pixel's exposure to light. However, the pixel's storage capacity is limited, described by its full-well capacity which is measured in electrons per pixel. Typical CCD sensors have a full-well capacity on the order of 10 000 to 100 000 electrons per pixel. When this number is exceeded during exposure (overexposure) the additional electrons migrate to the neighboring pixels which leads to image blooming. This effect is significantly reduced through specialized antiblooming architectures incorporated in modern CCD sensors: the overflowing charge is captured by conductors as it migrates toward the neighboring CCD cells.

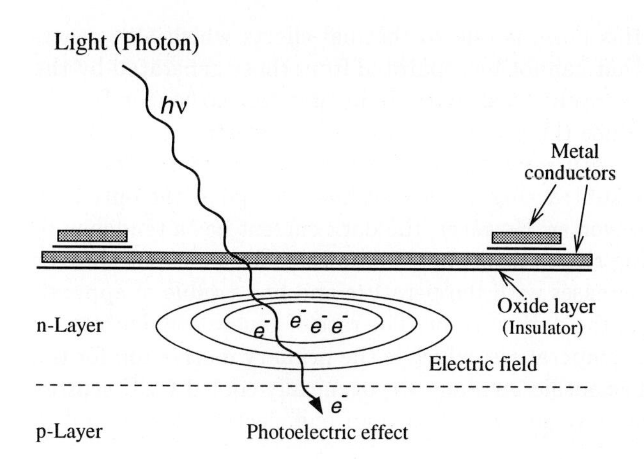

Fig. 2.36. Simplified model of a pixel

Another characteristic of a pixel is its fill factor or aperture which is defined as the ratio of its optically sensitive area and its entire area. This value can reach 100% for special, scientific-grade, back-illuminated sensors or may be as low as 15% for complex interline-transfer sensors which will be described in a later section. The primary reason for the limited aperture of most pixels is opaque areas on the surface of the sensors, either metal conductors used to form the potential wells and facilitate the transport of the accumulated charge to the readout port(s), or areas which are masked off to locally store charge before it is read out. Two methods exist for improving the fill factor: back-thinning is a costly process which removes the back of the substrate to a few tens of microns such that the sensor may be exposed from the back. Back-thinned CCDs are custom built and are frequently applied in astronomy and spectroscopy. Also the process cannot be applied to all CCD architectures because opaque regions are frequently needed to temporarily store collected charge. An alternative and more economical approach to enhance the fill factor is to deposit an array of microlenses on to the sensor allowing each pixel to collect more of the incoming light. The light sensitivity of each pixel may then be improved up to a factor of three.

2.6.2 Sources of noise

As with any electronic device, the CCD pixel is subject to electronic noise. For many electronic imaging applications the issue of noise only plays a secondary role in that it corrupts the visual perception of the image. In the case of PIV the light scattered from small particles is ideally captured on an otherwise black background. Due to the limited light scattering efficiency of the tiny tracer particles the recorded signal will sometimes only barely exceed the background noise level of the sensor as the observation area and observation distance is increased.

A major source of this noise is due to thermal effects which also generate electron-hole pairs that cannot be separated from those generated by the photoelectric effect: as a result weak particle images can no longer be distinguished from noise. Since the production rate of the electron-hole pairs is constant at a given temperature and exposure, this dark current or dark count can be accounted for by subtracting a constant bias voltage at the output of the charge-to-voltage converter. However, the dark current has a tendency to fluctuate over time giving rise to noise, better known as dark current noise or dark noise which also increases with temperature and has a value of approximately the square root of the dark current. The rate of generation doubles for every 6–7°C increase in temperature, which is the primary motivation for the use of cooled sensors in scientific imaging. Cryogenically cooled CCD sensors as applied in astronomy may generate less than one electron per second in each pixel.

Another source of noise is read noise or shot noise which is a direct consequence of the charge-to-voltage conversion during the readout sequence. In general the read noise increases with the readout frequency which is why many scientific applications require *slow-scan* cameras. Under standard operating conditions a normal CCD camera will have a noise level of several hundred electrons generated in each pixel for the period of integration (1/25 or 1/30 s). A careful optimization of the conversion electronics, a reduced readout frequency as well as cooling of the sensor may limit the read noise to a few electrons RMS per pixel. Up to now the prohibitive cost of these specialized cameras has made their use for PIV recording unfeasible. Nevertheless an increasing availability of cameras based on Peltier-cooled (i.e. electrically cooled) CCD sensors can be of interest for PIV applications.

2.6.3 Spectral characteristics

Similar to photographic film, the CCD sensor has a sensitivity and spectral response. A pixel's sensitivity or quantum efficiency, QE, is defined as the ratio between the number of collected photoelectrons and the number of incident photons per pixel and is measured in collected charge over light intensity $\mathrm{Cb}/(\mathrm{J}\cdot\mathrm{cm}^2)$. Alternatively, units of current, $I = Q/\Delta t$, over incident power, $P = E/(\Delta t \cdot \mathrm{Area})$, are used: $\mathrm{A}/(\mathrm{W}\cdot\mathrm{cm}^2)$. To a large extent this value depends on the pixel's architecture, that is, its aperture (i.e. fill factor), material and thickness of the optically sensitive area. Due to the width and position of the frequency-dependent band-gap of silicon, the CCD substrate material, photons of different frequencies will penetrate the sensor differently resulting in a wavelength dependent quantum efficiency of the CCD sensor. Examples for the spectral response of several sensors are given in figure 2.37.

To reduce the susceptability to infrared light, many commercially available CCD cameras come equipped with an infrared filter in front of the CCD sensor. Other filters may also be used to match the CCD's spectral characteristics to that of the human eye.

The responsivity of a CCD element expresses the ratio of useful signal voltage to exposure for a given illumination. This quantity depends on both the quantum efficiency and the on-chip charge-to-voltage conversion.

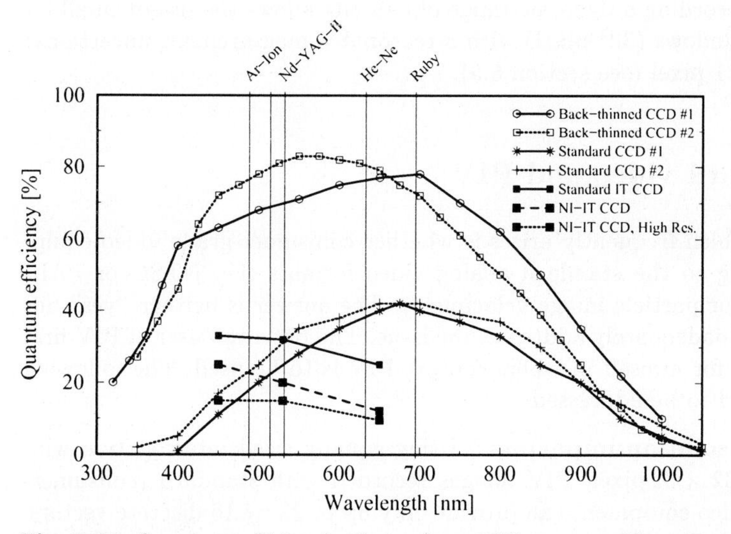

Fig. 2.37. Quantum efficiencies for various CCD sensors. The principle frequencies for the most common laser sources are shown as vertical lines (IT = interline transfer sensor architecture, NI-IT = noninterlaced, interline transfer sensor architecture).

2.6.4 Linearity and dynamic range

Since each electron captured in the potential well adds linearly to the cumulative collected charge, the output signal voltage for the individual pixel is practically directly proportional to the collected charge. Nonlinearities are usually due to overexposure or poorly designed output amplifiers. With adequate design, linearities with deviations of less than 1% are possible. Linearity is of importance in PIV recording when small particle images are to be located with accuracies below half a pixel. Any nonlinear behavior during recording jeopardizes the capability of measuring the particle image displacement in the subpixel regime. Especially if the particles themselves are to be located and tracked as in PTV, a linear dependency between recorded signal and scattered light is of importance.

The CCD's dynamic range is defined as the ratio between the full-well capacity and the dark current noise. Since the dark current noise is temperature dependent, the dynamic range of a CCD increases as the temperature is lowered.

Standard video sensors operating at room temperature typically have a dynamic range of 100–200 gray levels which exceeds that of human perception. Once digitized the useful signal is 7–8 bits in depth. With additional cooling and careful camera design a dynamic range exceeding 65 000 gray levels (16 bits/pixel) is possible. For the application of electronic imaging in digital PIV recording a dynamic range of 6–8 bits allows the use of small interrogation windows (32^2 pixel) with a reasonable measurement uncertainty of less than 0.1 pixel (see section 5.5).

2.7 Standard video and PIV

A question which frequently arises is whether comsumer-grade video equipment adhering to the standard analog video formats (i.e. NTSC or PAL) can be used for particle image velocimetry. The answer is neither "yes" nor "no" since it inadequately addresses the issue. The potential user of PIV first has to clarify for himself in which context PIV is to be used. The following questions need to be addressed:

Is spatial resolution important? Using equally sized interrogation windows of 32×32 pixel, PIV images recorded with standard (consumer-grade) video equipment can provide only up to 22 by 16 discrete vectors; a 1000×1000 pixel sensor provides more than 30 by 30 discrete vectors, while the utilization of photographic techniques can yield even more.

Is measurement accuracy important? Is the PIV system to be used as a quantitative visualization tool or is it intended to get accurate estimates of vorticity or circulation? Due to the analog nature of the standard video signal a small frame-to-frame jitter during the digitization process can cause pixels to be slightly misaligned which in turn increases the measurement uncertainty in the displacement data (i.e. velocity data). The problem typically worsens when standard (analog) video recorders are used.

Is temporal resolution important? The primary advantage of standard video equipment is that it can be used to record image sequences at 25 Hz (PAL) or 30 Hz (NTSC) using standard equipment. If the flow under investigation is slow enough that it can be resolved temporally at this frame rate, PIV at this lower resolution may be of interest. If, in addition, the PIV images can be stored digitally as they are being recorded, a good measurement accuracy in the PIV data may be achieved at the same time.

2.8 The video standard

Video is a general term describing the television type of imaging and is typically associated with the analog transmission of images whose information is coded in time, line-by-line. The video standards in use today are derivatives of standards such as NTSC-1 which was established in 1948 by the National Television Standards Committee (NTSC). Frame rates, line counts and other characteristics were determined at that time according to the then available technology (see table 2.6).

Table 2.6. Characteristics of black & white video standards.

Main area of use	Europe	North America/Japan
B& W	CCIR	RS-170 (EIA)
Color	PAL/SECAM	NTSC
Format	4:3 aspect ratio	4:3 aspect ratio
	2 interlaced fields per frame	2 interlaced fields per frame
Frame rate	25 frames/s	29.97 frames/s
	50 fields/s	59.94 fields/s
Resolution	625 lines	525 lines
	574 lines visible	484 lines visible
Row scan time	15.625 kHz = 64 μs	15.734 kHz = 63.55 μs
	52.48 μs active	52.80 μs active
	11.52 μs retrace	10.75 μs retrace
Bandwidth	5.5 MHz	4.2 MHz

To this date, the standard (television) video signal is transmitted in an interlaced format in which alternating fields of either odd or even lines of the image are transmitted at twice the video frame rate. This procedure provides the viewer with a uniform image appearance without the need for increasing the video signal's bandwidth. The transmission scheme shown in figure 2.38 applies to the image display on the monitor as well as to the readout of the recording camera or any other device employing video signal transmission. When a video image is digitized, for instance, the odd field (i.e. odd lines) is digitized, top-to-bottom before the first even line of the second, even field is sampled (figure 2.38, right). As a result adjacent vertical pixels will always be sampled with a time-separation of one field, that is 1/50th (PAL) or 1/60th (NTSC) of a second. Most common video cameras provide the signal in a similar manner such that the final digitized image actually will consist of two images at half the vertical resolution with one field of time delay between them. On an analog display monitor the image will be perceived to flicker. Even the electronic shutters on most video-format CCD sensors operate on a field-to-field basis. In essence interlaced video makes the implementation of single-exposure double-frame PIV more difficult than it needs to be. Nevertheless, there are some methods of successfully using video for PIV recording as given in the next section. The introduction of new video standards (HDTV or digital television) may alleviate many of the problems brought about by the old standards, but may well bring up others (e.g. compression artifacts).

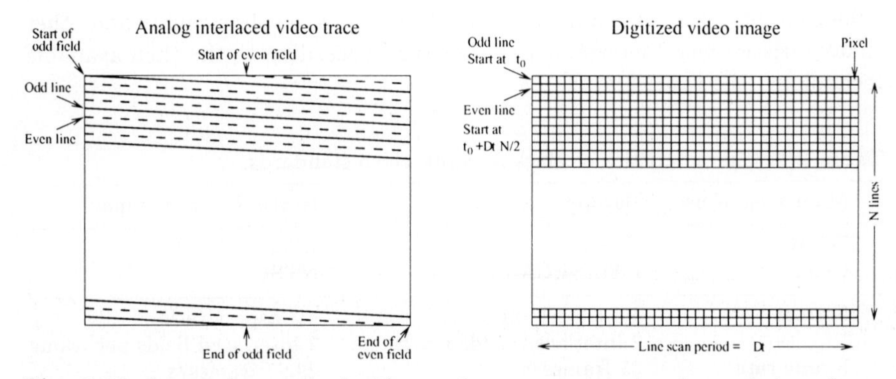

Fig. 2.38. Images in standard video equipment are interlaced, consisting of two separate fields (left). Digitization preserves the interlaced nature of the analog image (right).

3. Mathematical background of statistical PIV evaluation

A detailed mathematical description of statistical PIV evaluation has been given by ADRIAN [37]. This early work from 1988 concentrated on auto-correlation methods and was later expanded to cross-correlation analysis [66]. Most of the characteristics and limitations of the statistical PIV evaluation have been described therein. The most complete and careful mathematical description of digital PIV has been given by WESTERWEEL [7]. In this chapter a simplified mathematical model of the recording and subsequent statistical evaluation of PIV images will be presented. For this purpose the two-dimensional spatial estimator for the correlation will be referred to as the *correlation.* First, we analyze the cross-correlation of two frames of singly exposed recordings, then we expand the theory for the evaluation of doubly exposed recordings. The motivation for why auto- and cross-correlation methods are employed in PIV evaluation will be given in chapter 5.

3.1 Particle image locations

Typically, PIV recordings are subdivided into interrogation areas during evaluation. These areas are called interrogation spots – in the case of optical interrogation – or interrogation windows when digital recordings are considered. Due to reasons stated in the following, for cross-correlation analysis those interrogation areas need not necessarily be located at the same position of the PIV recording. Their geometrical backprojection into the light sheet will be referred to as interrogation volumes in the following (see figure 3.1). Two interrogation volumes used for statistical evaluation together define the measurement volume. Now, a single exposure recording is considered. It consists of a random distribution of particle images, which correspond to the following pattern of N tracer particles inside the flow:

$$\boldsymbol{\Gamma} = \begin{pmatrix} \boldsymbol{X_1} \\ \boldsymbol{X_2} \\ \cdot \\ \cdot \\ \cdot \\ \boldsymbol{X_N} \end{pmatrix} \quad \text{with} \quad \boldsymbol{X_i} = \begin{pmatrix} X_i \\ Y_i \\ Z_i \end{pmatrix}$$

being the position of a tracer particle in a $3N$-dimensional space. $\boldsymbol{\Gamma}$ describes the state of the ensemble at a given time t. $\boldsymbol{X_i}$ is the position vector of the particle i at time t. For more details about the mathematical description of the tracer ensemble, see [7]. Lower case letters refer to the coordinates in the image plane (figure 3.1) such that

$$\boldsymbol{x} = \begin{pmatrix} x \\ y \end{pmatrix}$$

is the image position vector in this plane.

In the remainder of this section we will assume that the particle position and the image position are related by a constant magnification factor M for simplicity, such that:

$$\boldsymbol{X_i} = \boldsymbol{x_i}/M \quad \text{and} \quad \boldsymbol{Y_i} = \boldsymbol{y_i}/M \ .$$

As already described in section 2.4.3, a more complex model of imaging geometry has to be used to take the effect of perspective projection into account.

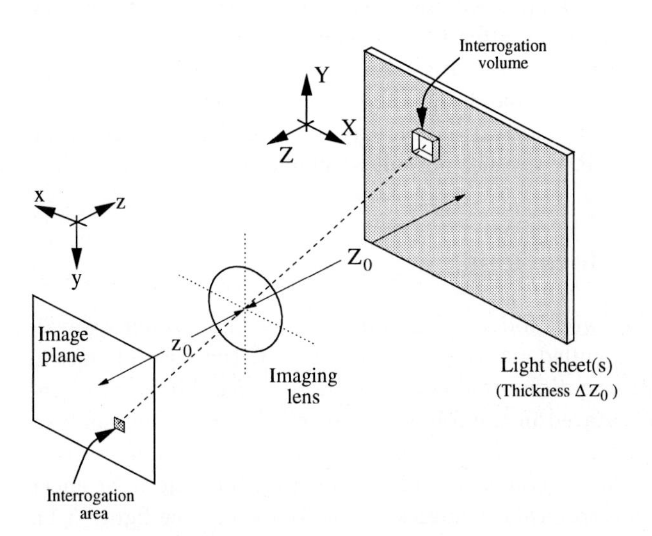

Fig. 3.1. Schematic representation of geometric imaging

3.2 Image intensity field

In this section a mathematical representation of the intensity distribution in the image plane is given. It is assumed that the image can best be described by a convolution of the geometric image and the impulse response of the imaging system, the so-called point spread function. For infinite small particles and perfectly aberration-free well focused lenses the amplitude of the point spread function can mathematically be described by the square of the first order Bessel function the so-called Airy function (see section 2.4).

A more complex model of imaging has to include imperfections of lenses and photographic films or sensors. For lenses and photographic films an estimation of the main effects besides diffraction can be obtained by analyzing their modulation transfer functions (MTF's) (see section 2.4.2). For CCD sensors a careful analysis requires more complex models, which have not yet been described in the PIV literature sufficiently. The description of digital imaging of very small objects is especially important, because the systematic arrangement of sensor elements can cause significant bias errors in statistical particle image displacement estimation (so-called peak locking; see section 5.4).

In the following we assume the point spread function of the imaging lens $\tau(\boldsymbol{x})$ to be Gaussian versus x and y (see appendix B.2), which is a common practice in literature and a good approximation for the point spread function of real lens systems [37, 7]. The convolution product of $\tau(\boldsymbol{x})$ with the geometric image of the tracer particle at the position $\boldsymbol{x}_i$ therefore describes the image of a single particle located at position $\boldsymbol{X}_i$. In the following we will assume infinitely small geometric particle images which would be the case for small particles imaged at small magnifications. Therefore, we use the Dirac delta-function shifted to position $\boldsymbol{x}_i$ to describe the geometric part of the particle image. Thus, the image intensity field of the first exposure may be expressed by:

$$I = I(\boldsymbol{x}, \boldsymbol{\Gamma}) = \tau(\boldsymbol{x}) * \sum_{i=1}^{N} V_0(\boldsymbol{X}_i)\, \delta(\boldsymbol{x} - \boldsymbol{x}_i) \qquad (3.1)$$

where $V_0(\boldsymbol{X}_i)$ is the transfer function giving the light energy of the image of an individual particle i inside the interrogation volume V_{I} and its conversion into an electronic signal or optical transmissivity[1]. $\tau(\boldsymbol{x})$ is considered to be identical for every particle position. The visibility of a particle depends on many parameters as for example the scattering properties of the particle, the light intensity at the particle position, the sensitivity of the recording optics and the sensor or film at the corresponding image position. In the following we assume that the particles at every position have the same scattering properties and the recording optics and media have a constant sensitivity over the image plane.

In many situations different weight is put on different locations inside the interrogation area. This can be done by a multiplication of the recorded image intensity with weight kernels in the case of digital evaluation or implicitly due to the spatial intensity distribution of the interrogating laser beam in the case of optical evaluation. In the following we assume that Z is the viewing

[1] Strictly speaking equation (3.1) is valid only for incoherent light. For coherent light a term considering the interference of overlapping particle images has to be included [7]. In most practical situations the particle images do not overlap. Therefore, we use equation (3.1) also for coherent illumination.

direction and that the light intensity inside the interrogation volume is only a function of Z and that the image intensity finally analyzed depends on X and Y only due to the weight function. Therefore, $V_0(\boldsymbol{X})$ just describes the shape, extension, and location of the actual interrogation volume:

$$V_0(\boldsymbol{X}) = W_0(X,Y)\, I_0(Z) \tag{3.2}$$

where $I_0(Z)$ is the intensity profile of the laser light sheet in the Z direction and $W_0(X,Y)$ is the interrogation window function geometrically back projected into the light sheet. This is mathematically not correct, because it does not consider the convolution with the point spread function. For rectangular interrogation windows this means that in our mathematical description we neglect the effects of partially cropped images at the edges of the interrogation area. However, we will use this simple model of the interrogation volumes in the flow, because it also simplifies the description of PIV evaluation:

$$I_0(Z) = I_Z \exp\left(-8\,\frac{(Z-Z_0)^2}{\Delta Z_0^2}\right)$$

might be used to describe the Gaussian intensity profile of the laser light sheet, where ΔZ_0 is the thickness of the light sheet measured at the e^{-2} points and I_Z is the maximum intensity of the light sheet. $W_0(X,Y)$ can be described in a similar way if a Gaussian window function with a maximum weighting W_{XY} at position X_0, Y_0 has to be considered:

$$W_0(X,Y) = W_{\mathrm{XY}} \exp\left(-8\,\frac{(X-X_0)^2}{\Delta X_0^2} - 8\,\frac{(Y-Y_0)^2}{\Delta Y_0^2}\right) .$$

Since many pulsed lasers used for PIV have an intensity distribution which is closer to a top-hat function than to a Gaussian function and since digitized recordings are commonly interrogated with rectangular windows, $V_0(\boldsymbol{X})$ can be also defined as a rectangular box:

$$I_0(Z) = \begin{cases} I_{\mathrm{Z}} & \text{if} \quad |Z-Z_0| \leq \Delta Z_0/2 \\ 0 & \text{elsewhere} \end{cases} \tag{3.3}$$

$$W_0(X,Y) = \begin{cases} W_{XY} & \text{if} \quad |X-X_0| \leq \Delta X_0/2 \quad \text{and} \quad |Y-Y_0| \leq \Delta Y_0/2 \\ 0 & \text{elsewhere.} \end{cases} \tag{3.4}$$

The factor $I_0(Z_i)$ represents the amount of light received from the particle i inside the flow, and located at distance $|Z_i - Z_0|$ from the center plane of the laser light sheet. ΔZ_0 is the light sheet thickness and therefore the extension of the interrogation volume in the Z direction. $\Delta X_0 = \Delta x_0/M$ and $\Delta Y_0 = \Delta y_0/M$ is the extension of the interrogation volume in the X direction and Y direction respectively. With $\tau(\boldsymbol{x}-\boldsymbol{x}_i) = \tau(\boldsymbol{x}) * \delta(\boldsymbol{x}-\boldsymbol{x}_i)$ (see appendix B.1) and the assumption that the particle images under consideration do not overlap equation (3.1) can alternatively be written as:

$$I(\boldsymbol{x},\boldsymbol{\Gamma}) = \sum_{i=1}^{N} V_0(\boldsymbol{X_i})\,\tau(\boldsymbol{x}-\boldsymbol{x_i})\,. \quad \text{(see appendix B)} \tag{3.5}$$

This expression for the image intensity field will be intensively used in the following sections. In the following we will illustrate different representations of the intensity field and their correlation by giving an example for the recording of three arbitrarily located particles.

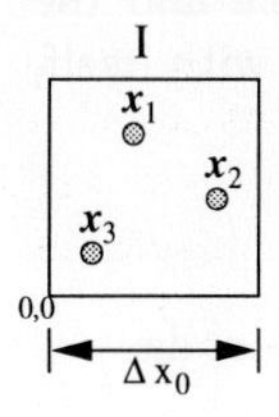

Fig. 3.2. Example of an intensity field I (single exposure)

3.3 Mean value, auto-correlation, and variance of a single exposure recording

In this section we will determine spatial estimators for the mean value and the variance of the image intensity field, because these quantities will be used for the normalization of the cross-correlation. Furthermore, auto-correlation and auto-covariance of a single exposure intensity field will be introduced. The main equations used in the following are taken from PAPOULIS [22, 23]. The spatial average is defined as:

$$\langle I(\boldsymbol{x},\boldsymbol{\Gamma})\rangle = \frac{1}{a_{\mathrm{I}}}\int_{a_{\mathrm{I}}} I(\boldsymbol{x},\boldsymbol{\Gamma})\,\mathrm{d}\boldsymbol{x}$$

where a_{I} is the interrogation area. Employing equation (3.5) yields:

$$\langle I(\boldsymbol{x},\boldsymbol{\Gamma})\rangle = \frac{1}{a_{\mathrm{I}}}\int_{a_{\mathrm{I}}} \sum_{i=1}^{N} V_0(\boldsymbol{X_i})\,\tau(\boldsymbol{x}-\boldsymbol{x_i})\,\mathrm{d}\boldsymbol{x}\,.$$

The mean value of the intensity field can be approximated by:

$$\mu_{\mathrm{I}} = \langle I(\boldsymbol{x},\boldsymbol{\Gamma})\rangle = \frac{1}{a_{\mathrm{I}}}\sum_{i=1}^{N} V_0(\boldsymbol{X_i})\int_{a_{\mathrm{I}}} \tau(\boldsymbol{x}-\boldsymbol{x_i})\,\mathrm{d}\boldsymbol{x}\,.$$

We can now derive the auto-correlation of the single exposure intensity field in a similar way:

$$\begin{aligned} R_{\mathrm{I}}(\boldsymbol{s},\boldsymbol{\Gamma}) &= \langle I(\boldsymbol{x},\boldsymbol{\Gamma}) I(\boldsymbol{x}+\boldsymbol{s},\boldsymbol{\Gamma})\rangle \\ &= \frac{1}{a_{\mathrm{I}}}\int_{a_{\mathrm{I}}} \sum_{i=1}^{N} V_0(\boldsymbol{X_i})\,\tau(\boldsymbol{x}-\boldsymbol{x_i}) \sum_{j=1}^{N} V_0(\boldsymbol{X_j})\,\tau(\boldsymbol{x}-\boldsymbol{x_j}+\boldsymbol{s})\,\mathrm{d}\boldsymbol{x} \end{aligned}$$

where $\boldsymbol{s}$ is the separation vector in the correlation plane. By distinguishing the terms $i \neq j$ which represent the correlation of different particle images and therefore randomly distributed noise in the correlation plane and the $i = j$ terms which represent the correlation of each particle image with itself, we come to the following representation:

$$\begin{aligned} R_{\mathrm{I}}(\boldsymbol{s},\boldsymbol{\Gamma}) &= \frac{1}{a_{\mathrm{I}}}\sum_{i\neq j}^{N} V_0(\boldsymbol{X_i})\,V_0(\boldsymbol{X_j}) \int_{a_{\mathrm{I}}} \tau(\boldsymbol{x}-\boldsymbol{x_i})\,\tau(\boldsymbol{x}-\boldsymbol{x_j}+\boldsymbol{s})\,\mathrm{d}\boldsymbol{x} \\ &+ \frac{1}{a_{\mathrm{I}}}\sum_{i=j}^{N} {V_0}^2(\boldsymbol{X_i}) \int_{a_{\mathrm{I}}} \tau(\boldsymbol{x}-\boldsymbol{x_i})\,\tau(\boldsymbol{x}-\boldsymbol{x_j}+\boldsymbol{s})\,\mathrm{d}\boldsymbol{x}\;. \end{aligned}$$

Following the decomposition proposed by ADRIAN, we can write:

$$R_{\mathrm{I}}(\boldsymbol{s},\boldsymbol{\Gamma}) = R_{\mathrm{C}}(\boldsymbol{s},\boldsymbol{\Gamma}) + R_{\mathrm{F}}(\boldsymbol{s},\boldsymbol{\Gamma}) + R_{\mathrm{P}}(\boldsymbol{s},\boldsymbol{\Gamma})$$

where $R_{\mathrm{C}}(\boldsymbol{s},\boldsymbol{\Gamma})$ is the convolution of the mean intensities of I and $R_{\mathrm{F}}(\boldsymbol{s},\boldsymbol{\Gamma})$ is the fluctuating noise component both resulting from the $i \neq j$ terms. $R_{\mathrm{P}}(\boldsymbol{s},\boldsymbol{\Gamma})$ finally is the self-correlation peak located at position $(0,0)$ in the correlation plane. It results from the components that correspond to the correlation of each particle image with itself ($i = j$ terms).

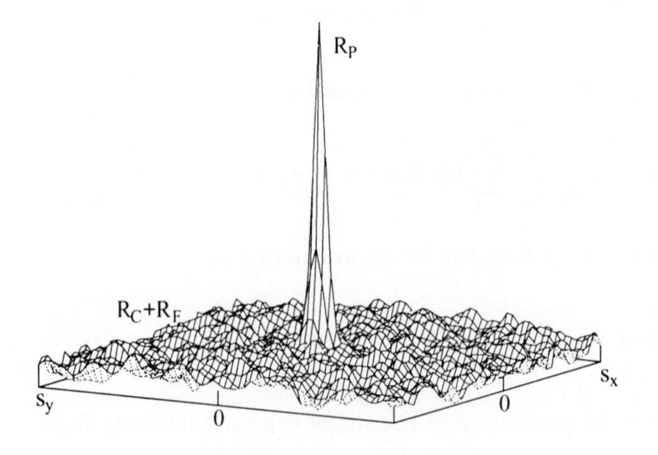

Fig. 3.3. Composition of peaks in the auto-correlation function

We will now concentrate on this central peak in order to evaluate its features. For a Gaussian particle image intensity distribution

$$\tau(\boldsymbol{x}) = K \exp\left(-\frac{8\,|\boldsymbol{x}|^2}{d_\tau^2}\right)$$

it can be shown that the auto-correlation $R_\tau(\boldsymbol{s})$ is again a Gaussian function with a width that is $\sqrt{2}\,d_\tau$ (see appendix B.3). Consequently $R_P(\boldsymbol{s}, \boldsymbol{\Gamma})$ may be rewritten as following:

$$R_P(\boldsymbol{s}, \boldsymbol{\Gamma}) = \sum_{i=1}^{N} V_0{}^2(\boldsymbol{X_i}) \exp\left(\frac{-8|\boldsymbol{s}|^2}{(\sqrt{2}\,d_\tau)^2}\right) \frac{1}{a_I} \int_{a_I} \tau^2\left(\boldsymbol{x} - \boldsymbol{x_i} + \frac{\boldsymbol{s}}{2}\right) \mathrm{d}\boldsymbol{x}\,.$$

In the remainder of this book we will always use $R_\tau(\boldsymbol{s})$ instead of:

$$\exp\left(\frac{-8|\boldsymbol{s}|^2}{(\sqrt{2}\,d_\tau)^2}\right) \frac{1}{a_I} \int_{a_I} \tau^2\left(\boldsymbol{x} - \boldsymbol{x_i} + \frac{\boldsymbol{s}}{2}\right) \mathrm{d}\boldsymbol{x}$$

taking into account that its features are mainly the same also for non-Gaussian $\tau(\boldsymbol{x})$: the maximum of $R_\tau(\boldsymbol{s})$ is located at $|\boldsymbol{s}| = 0$ and the characteristics of its shape is given by the particle images shape. Therefore, we will write R_P as[2]:

$$R_P(\boldsymbol{s}, \boldsymbol{\Gamma}) = R_\tau(\boldsymbol{s}) \sum_{i=1}^{N} V_0{}^2(\boldsymbol{X_i})\,.$$

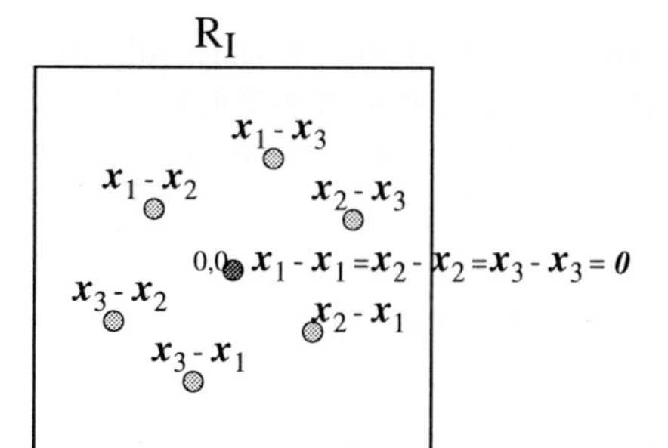

Fig. 3.4. Schematic representation of the auto-correlation of the intensity field I given in figure 3.2

In figure 3.4 the schematic of the auto-correlation of the example intensity field I is given. Correlation peaks (R_P and R_F) occur at locations which are given by the vectorial differences between particle image locations. Their strength is proportional to the number of all possible differences which result

[2] The dependency of $R_\tau(\boldsymbol{s})$ has been disregarded

in that location. For intensity fields with zero mean value the auto-correlation equals the auto-covariance. For nonzero mean values of the intensity field the auto-covariance $C_{\mathrm{I}}(\boldsymbol{s})$ can be obtained by [22]:

$$C_{\mathrm{I}}(\boldsymbol{s}) = R_{\mathrm{I}}(\boldsymbol{s}) - {\mu_{\mathrm{I}}}^2 .$$

An estimator of the variance of the intensity field can be obtained by:

$$\sigma_{\mathrm{I}}^2 = C_{\mathrm{I}}(\mathbf{0}, \boldsymbol{\Gamma}) = R_{\mathrm{I}}(\mathbf{0}, \boldsymbol{\Gamma}) - {\mu_{\mathrm{I}}}^2 = R_{\mathrm{P}}(\mathbf{0}, \boldsymbol{\Gamma}) - {\mu_{\mathrm{I}}}^2 .$$

3.4 Cross-correlation of a pair of two singly exposed recordings

As already mentioned before, PIV recordings are most often evaluated by locally cross-correlating two frames of single exposures of the tracer ensemble. The mathematical background of this technique will now be described.

In the following, a constant displacement $\boldsymbol{D}$ of all particles inside the interrogation volume is assumed, so that the particle locations during the second exposure at time $t' = t + \Delta t$ are given by:

$$\boldsymbol{X_i}' = \boldsymbol{X_i} + \boldsymbol{D} = \begin{pmatrix} X_i + D_X \\ Y_i + D_Y \\ Z_i + D_Z \end{pmatrix} .$$

We furthermore assume that the particle image displacements are given by:

$$\boldsymbol{d} = \begin{pmatrix} M D_X \\ M D_Y \end{pmatrix}$$

which is a simplification of the perspective projection that is only valid for particles located in the vicinity of the optical axis (see section 2.4.3).

We come to the following representation of the image intensity field for the time of the second exposure (see equation 3.5):

$$I'(\boldsymbol{x}, \boldsymbol{\Gamma}) = \sum_{j=1}^{N} V_0'(\boldsymbol{X_j} + \boldsymbol{D})\, \tau(\boldsymbol{x} - \boldsymbol{x_j} - \boldsymbol{d})$$

where $V_0'(\boldsymbol{X})$ defines the interrogation volume during the second exposure. If we first consider identical light sheet and windowing characteristics, the cross-correlation function of two interrogation areas can be written as:

$$R_{\mathrm{II}}(\boldsymbol{s}, \boldsymbol{\Gamma}, \boldsymbol{D}) = \frac{1}{a_{\mathrm{I}}} \sum_{i,j} V_0(\boldsymbol{X_i})\, V_0(\boldsymbol{X_j} + \boldsymbol{D}) \int_{a_{\mathrm{I}}} \tau(\boldsymbol{x} - \boldsymbol{x_i})\, \tau(\boldsymbol{x} - \boldsymbol{x_j} + \boldsymbol{s} - \boldsymbol{d})\, \mathrm{d}\boldsymbol{x}$$

where $\boldsymbol{s}$ is the separation vector in the correlation plane. Analogous to the procedure used in the previous section we come to:

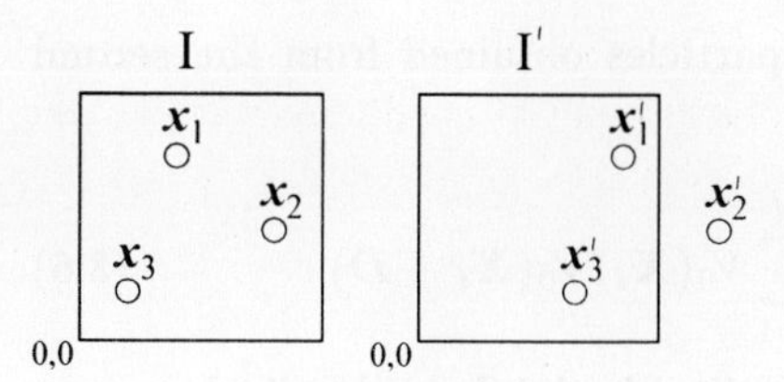

Fig. 3.5. The intensity field I recorded at t and the intensity field I' recorded after a time delay of Δt at t'

$$R_{\mathrm{II}}(\boldsymbol{s}, \boldsymbol{\Gamma}, \boldsymbol{D}) = \sum_{i,j} V_0(\boldsymbol{X_i})\, V_0(\boldsymbol{X_j} + \boldsymbol{D})\, R_\tau(\boldsymbol{x_i} - \boldsymbol{x_j} + \boldsymbol{s} - \boldsymbol{d}) \,.$$

By distinguishing the terms $i \neq j$ which represent the correlation of different randomly distributed particles and therefore mainly noise in the correlation plane and the $i = j$ terms, which contain the displacement information desired, we obtain:

$$\begin{aligned} R_{\mathrm{II}}(\boldsymbol{s}, \boldsymbol{\Gamma}, \boldsymbol{D}) &= \sum_{i \neq j} V_0(\boldsymbol{X_i})\, V_0(\boldsymbol{X_j} + \boldsymbol{D})\, R_\tau(\boldsymbol{x_i} - \boldsymbol{x_j} + \boldsymbol{s} - \boldsymbol{d}) \\ &+ \; R_\tau(\boldsymbol{s} - \boldsymbol{d}) \sum_{i=1}^{N} V_0(\boldsymbol{X_i})\, V_0(\boldsymbol{X_i} + \boldsymbol{D}) \,. \end{aligned}$$

Again, we can decompose the correlation into three parts:

$$R_{\mathrm{II}}(\boldsymbol{s}, \boldsymbol{\Gamma}, \boldsymbol{D}) = R_{\mathrm{C}}(\boldsymbol{s}, \boldsymbol{\Gamma}, \boldsymbol{D}) + R_{\mathrm{F}}(\boldsymbol{s}, \boldsymbol{\Gamma}, \boldsymbol{D}) + R_{\mathrm{D}}(\boldsymbol{s}, \boldsymbol{\Gamma}, \boldsymbol{D})$$

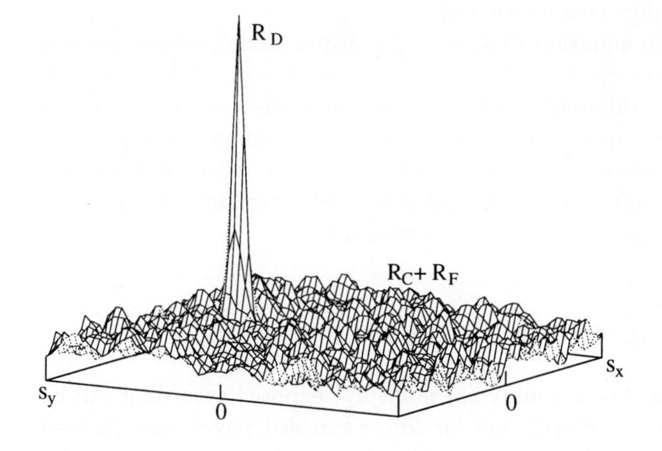

Fig. 3.6. Composition of peaks in the cross-correlation function

where $R_{\mathrm{D}}(\boldsymbol{s}, \boldsymbol{\Gamma}, \boldsymbol{D})$ represents the component of the cross-correlation function that corresponds to the correlation of images of particles obtained from the first exposure with images of identical particles obtained from the second exposure ($i = j$ terms):

$$R_D(\boldsymbol{s}, \boldsymbol{\Gamma}, \boldsymbol{D}) = R_\tau(\boldsymbol{s} - \boldsymbol{d}) \sum_{i=1}^{N} V_0(\boldsymbol{X}_i)\, V_0(\boldsymbol{X}_i + \boldsymbol{D}) \,. \tag{3.6}$$

Hence, for a given distribution of particles inside the flow, the displacement correlation peak reaches a maximum for $\boldsymbol{s} = \boldsymbol{d}$. Therefore, as already anticipated, the location of this maximum yields the average in-plane displacement, and thus the U and V components of the velocity inside the flow.

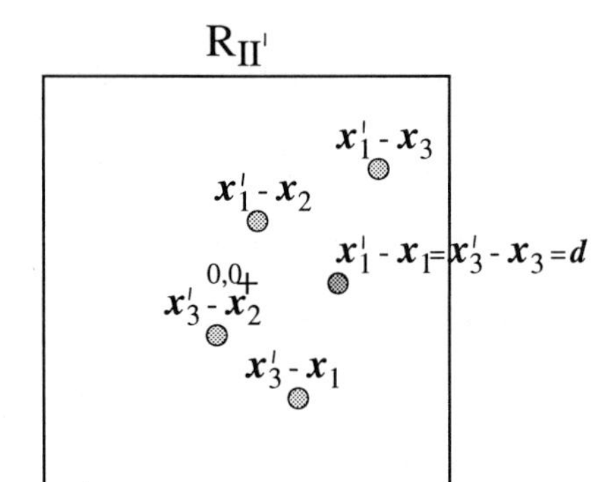

Fig. 3.7. Schematic representation of the cross-correlation of the intensity fields I and I' given in figure 3.5

In figure 3.7 the schematic of the cross-correlation of the example intensity fields I and I' is given. Nearly the same correlation peaks occur as in the auto-correlation shown in figure 3.4 but at locations which are displaced by $\boldsymbol{d}$. Correlations of $\boldsymbol{x}'_2$ do not appear here, because this image is located outside the interrogation window (see figure 3.5).

It can be seen from equation (3.6) that the displacement correlation is a function of the random variables $(\boldsymbol{X}_i)_{i=1\cdots N}$. Consequently it is a random variable itself and for different realizations at the same overall conditions we will obtain different qualities of the displacement estimation depending on the state of the tracer ensemble. In order to derive rules for a general optimization of the displacement estimation we will determine the expected value of the displacement correlation in section 3.6.

3.5 Correlation of a doubly exposed recording

The correlation function of a doubly (or multiply) exposed recording can be derived by analogy to the correlation for single exposed recordings. Instead of cross-correlating I with I' we will consider the correlation of the intensity field $I^+ = I + I'$ with itself. Assuming identical light sheets and windowing characteristics the intensity field of both exposures I^+ can be written as:

$$
\begin{aligned}
I^{+}(\boldsymbol{x}, \boldsymbol{\Gamma}) &= I(\boldsymbol{x}, \boldsymbol{\Gamma}) + I'(\boldsymbol{x}, \boldsymbol{\Gamma}) \\
&= \sum_{i=1}^{N} \left(V_0(\boldsymbol{X_i})\, \tau(\boldsymbol{x} - \boldsymbol{x_i}) + V_0(\boldsymbol{X_i} + \boldsymbol{D})\, \tau(\boldsymbol{x} - \boldsymbol{x_i} - \boldsymbol{d}) \right) .
\end{aligned}
$$

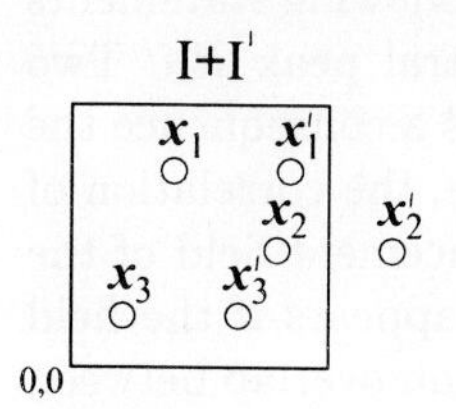

Fig. 3.8. The sum of the intensity fields I and I' (see figure 3.5) as obtained by a recording of the tracer ensemble at t and t' on the same frame

It can be shown that the auto-correlation of I^{+} consists of four terms:

$$
R_{\mathrm{I}^+}(\boldsymbol{s}, \boldsymbol{\Gamma}, \boldsymbol{D}) = R_{\mathrm{I}}(\boldsymbol{s}, \boldsymbol{\Gamma}) + R_{\mathrm{I}'}(\boldsymbol{s}, \boldsymbol{\Gamma}) + R_{\mathrm{II}}(\boldsymbol{s}, \boldsymbol{\Gamma}, \boldsymbol{D}) + R_{\mathrm{II}}(-\boldsymbol{s}, \boldsymbol{\Gamma}, \boldsymbol{D}) .
$$

It is therefore appropriate to decompose the estimator into the following terms:

$$
\begin{aligned}
R_{\mathrm{I}^+}(\boldsymbol{s}, \boldsymbol{\Gamma}, \boldsymbol{D}) &= R_{\mathrm{C}}(\boldsymbol{s}, \boldsymbol{\Gamma}, \boldsymbol{D}) + R_{\mathrm{F}}(\boldsymbol{s}, \boldsymbol{\Gamma}, \boldsymbol{D}) + R_{\mathrm{P}}(\boldsymbol{s}, \boldsymbol{\Gamma}) \\
&+ R_{\mathrm{D}^+}(\boldsymbol{s}, \boldsymbol{\Gamma}, \boldsymbol{D}) + R_{\mathrm{D}^-}(\boldsymbol{s}, \boldsymbol{\Gamma}, \boldsymbol{D}) \qquad (3.7)
\end{aligned}
$$

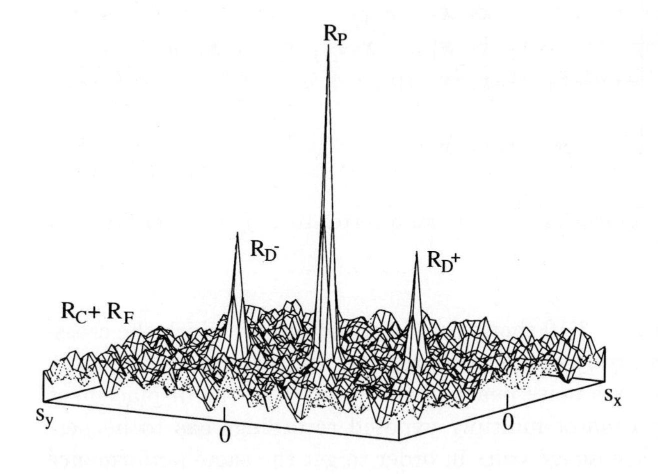

Fig. 3.9. Components of the auto-correlation function

where $R_{\mathrm{C}}(\boldsymbol{s}, \boldsymbol{\Gamma}, \boldsymbol{D})$ is the convolution of the mean intensity of I^{+} and $R_{\mathrm{F}}(\boldsymbol{s}, \boldsymbol{\Gamma}, \boldsymbol{D})$ is the fluctuating noise component. $R_{\mathrm{P}}(\boldsymbol{s}, \boldsymbol{\Gamma})$ is the self-correlation peak located in the center of the correlation plane. It results from the components that correspond to the correlation of each particle image with itself. $R_{D^+}(\boldsymbol{s}, \boldsymbol{\Gamma}, \boldsymbol{D})$ and $R_{D^-}(\boldsymbol{s}, \boldsymbol{\Gamma}, \boldsymbol{D})$ represent the components of the correlation function which correspond to the correlation of images of particles obtained from the first exposure with that of images of identical particles obtained from the second exposure and vice versa.

When comparing the correlation of a doubly exposed recording with the correlation of a pair of two singly exposed recordings, the following statements can be made: R_{I+} is symmetric with respect to its central peak R_P. Two identical displacement peaks R_{D+} and R_{D-} appear and as a consequence the sign of the displacement cannot be determined. Therefore, the correlation of a doubly exposed recording is not conclusive if the displacement field of the whole recording is not unidirectional. Another problem appears if the field contains displacements close to zero, which would lead to an overlap between the displacement peaks with the central peak. However, these problems have to be solved during recording. Precautions have to be made that the images of identical particles due to the different exposures do not overlap and that the sign of their displacement is determined. If the flow field under investigation contains areas of reverse flow or of relative slow velocities image shifting has to be used (see section 4.3). It can be seen from figure 3.10 that the correlation of doubly exposed recordings contains more than twice the number of randomly distributed noise peaks.

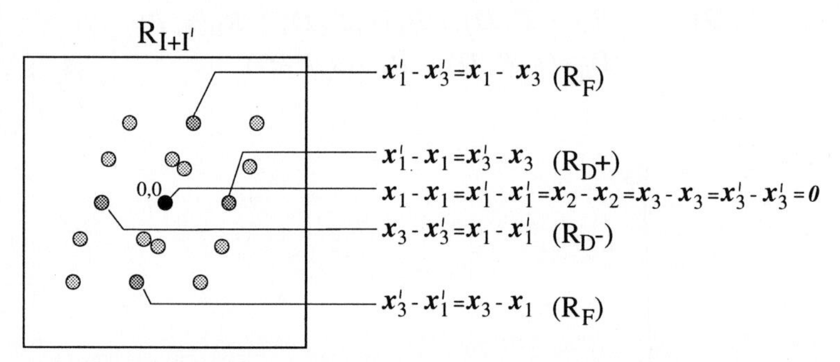

Fig. 3.10. Schematic representation of the auto-correlation of the intensity field $I + I'$ given in figure 3.8

The example given in figure 3.10 shows that in situations for which the cross-correlation of single exposure yields good results, the correlation of doubly exposed recordings contains noise peaks of the strength of the displacement peak. Hence, the evaluation of multiply exposed recordings has to be performed with more particle image pairs in order to get the same performance as that of single exposure evaluation. This can be done by different methods: the seeding density, the number of exposures, or the light sheet thickness can be increased. Besides other problems related to these methods their application is restricted due to the limited number of particle images that can be stored on the sensor without a significant overlap. Therefore, in most cases the size of the interrogation areas has to be increased compared to the evaluation of single exposures resulting in a lower spatial resolution of the measurement at the same sensor size.

3.6 Expected value of displacement correlation

In order to derive rules for a general optimization of the displacement estimation we will determine the expected value of the displacement correlation $E\{R_D\}$ for all realizations of $\boldsymbol{\Gamma}$. More concretely: we want to calculate the mean correlation function of all possible "patterns" that can be realized with N particles. From equation (3.6), it follows that

$$\begin{aligned} E\{R_D\} &= E\left\{ R_\tau(\boldsymbol{s}-\boldsymbol{d}) \sum_{i=1}^{N} V_0(\boldsymbol{X}_i) V_0(\boldsymbol{X}_i + \boldsymbol{D}) \right\} \\ &= R_\tau(\boldsymbol{s}-\boldsymbol{d})\, E\left\{ \sum_{i=1}^{N} V_0(\boldsymbol{X}_i) V_0(\boldsymbol{X}_i + \boldsymbol{D}) \right\} \end{aligned}$$

Defining $f_1(\boldsymbol{X}) = V_0(\boldsymbol{X})\, V_0(\boldsymbol{X} + \boldsymbol{D})$ yields:

$$E\{R_D\} = R_\tau(\boldsymbol{s}-\boldsymbol{d})\, E\left\{ \sum_{i=1}^{N} f_1(\boldsymbol{X}_i) \right\} . \tag{3.8}$$

We prove in appendix B.4 that:

$$E\left\{ \sum_{i=1}^{N} f_1(\boldsymbol{X}_i) \right\} = \frac{N}{V_F} \int_{V_F} f_1(\boldsymbol{X})\, \mathrm{d}\boldsymbol{X}$$

where $\int_{V_F} f_1(\boldsymbol{X})\, \mathrm{d}\boldsymbol{X}$ is the volume integeral

$$\int\!\!\int\!\!\int f_1(X,Y,Z)\, \mathrm{d}\boldsymbol{X} \mathrm{d}\boldsymbol{Y} \mathrm{d}\boldsymbol{Z} .$$

Thus:

$$E\{R_D\} = \frac{N}{V_F} R_\tau(\boldsymbol{s}-\boldsymbol{d}) \int_{V_F} f_1(\boldsymbol{X})\, \mathrm{d}\boldsymbol{X} . \tag{3.9}$$

Since we defined N to be the number of all particles of the ensemble, V_F has to be interpreted as the whole volume of fluid that has been seeded with particles. According to the above definition of $f_1(\boldsymbol{X})$ we can say in a more practical sense that the integration has to be performed over the volume which contained all particles that were inside the interrogation volumes during the first or second exposure. We can rewrite the integral over $f_I(\boldsymbol{X})$ as:

$$\begin{aligned}\int_{V_F} f_I(\boldsymbol{X})\,\mathrm{d}\boldsymbol{X} &= \int I_0(Z)\,I_0(Z+D_Z)\,\mathrm{d}Z \\ &\times \iint W_0(X,Y)\,W_0(X+D_X,Y+D_Y)\,\mathrm{d}X\,\mathrm{d}Y \\ &= \int_{V_F} V_0^2(\boldsymbol{X})\,\mathrm{d}\boldsymbol{X}\cdot F_O(D_Z)\,F_I(D_X,D_Y)\end{aligned}$$

with

$$F_I(D_X,D_Y) = \frac{\displaystyle\iint W_0(X,Y)\,W_0(X+D_X,Y+D_Y)\,\mathrm{d}X\,\mathrm{d}Y}{\displaystyle\iint W_0^2(X,Y)\,\mathrm{d}X\,\mathrm{d}Y} \tag{3.10}$$

and

$$F_O(D_Z) = \frac{\displaystyle\int I_0(Z)\,I_0(Z+D_Z)\,\mathrm{d}Z}{\displaystyle\int I_0^2(Z)\,\mathrm{d}Z}\,. \tag{3.11}$$

KEANE & ADRIAN [64, 65, 66] have defined F_I as a factor expressing the in-plane loss-of-pairs, and F_O as a factor expressing the out-of-plane loss-of-pairs. When no in-plane or out-of-plane loss-of-pairs are present the latter two are unity, respectively. Finally equation (3.9) yields:

$$E\{R_D(\boldsymbol{s},\boldsymbol{D})\} = C_R\,R_\tau(\boldsymbol{s}-\boldsymbol{d})\,F_O(D_Z)\,F_I(D_X,D_Y) \tag{3.12}$$

where the constant C_R is defined as:

$$C_R = \frac{N}{V_F}\int_{V_F} V_0^2(\boldsymbol{X})\,\mathrm{d}\boldsymbol{X}\,.$$

3.7 Optimization of correlation

The first parameter that has to be optimized during a PIV measurement is the pulse separation time between the successive light pulses. Besides technical limitations some general effects have to be considered. According to the principle of PIV the measured velocity is determined by the ratio of two components of the measured particle displacement between successive light pulses D_X and D_Y respectively, and the pulse separation time Δt. Since the particle displacement – which is considered to be a function of Δt in the following – is determined by the particle image displacement with $D_X(\Delta t) = d_x(\Delta t)/M$ and $D_Y(\Delta t) = d_y(\Delta t)/M$ respectively, and the measured image displacements contain certain residual errors $\varepsilon_{\text{resid}}$ we can define the following equation for the magnitude of the locally measured velocity:

$$|\boldsymbol{U}| = \frac{|\boldsymbol{d}(\Delta t)|}{M\Delta t} + \frac{\varepsilon_{\text{resid}}}{M\Delta t} . \tag{3.13}$$

Since the particle image displacement for a given recording configuration reduces linearly with the pulse separation time, the first summand of the above equation stays constant for vanishing pulse separations:

$$\lim_{\Delta t \to 0} \frac{|\boldsymbol{d}(\Delta t)|}{M\Delta t} = |\boldsymbol{U}| .$$

In contrast to that, the residual error contained in the measured image displacement will not be reduced below a certain limit by a reduction of the pulse separation, because the uncertainty in determining the particle image positions will be unaffected. Therefore, the second summand of equation (3.13) – which states that the measurement error is weighted with $1/\Delta t$ – increases rapidly with decreasing pulse separation:

$$\lim_{\Delta t \to 0} \frac{\varepsilon_{\text{resid}}}{M\Delta t} = \infty .$$

From these considerations it can be seen that the accuracy of PIV measurements can be increased by increasing the separation time between the exposures at least within certain limits. However, for high values of Δt the measurement noise increases. This becomes clear when looking at the expectation of the displacement correlation given in equation (3.12). It can be seen that the average signal strength is weighted with the loss of pairs due to the particle displacement $\boldsymbol{D}(\Delta t)$. For a very large separation time the particle displacement, which increases linearly with Δt, will exceed the extent of the interrogation volume. Then, no particle will be illuminated twice and no image correlation would be obtained. What can be done to improve the situation? First of all the pulse separation time can be reduced. This directly reduces the particle displacement and the loss of pairs.

In figure 3.11 we have tried to illustrate the two aspects of the choice of Δt on the quality of the PIV data: the dotted line, curve g, represents the effect of the weighting of the residual error with Δt, the solid line, curve f, represents the influence of the loss of pairs. The optimum Δt could therefore be found by determining the maximum of a quality function Q_{PIV}, for example the product of curves f and g which is represented by the dashed line.

However, the shape of curve f has been chosen arbitrarily, since a general value for the quality of a measurement is difficult to define. When using digital equipment, which allows immediate feedback during the measurement, the optimum can be found interactively by slowly increasing the pulse separation until the number of obvious outliers within the vector map increases. However, the number of valid data yield is only one parameter of the obtained quality, but not an exact measure of it.

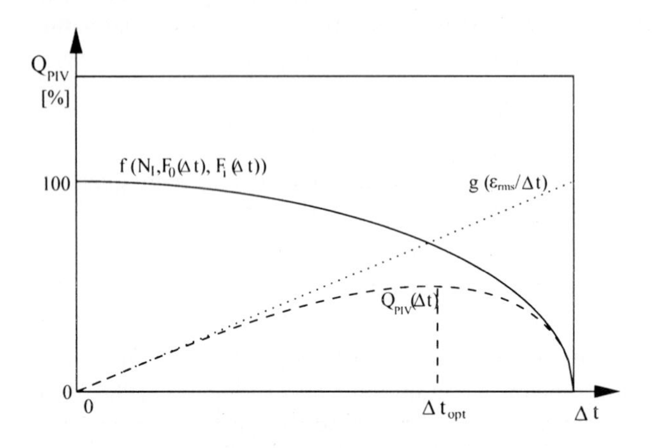

Fig. 3.11. Schematic representation for the optimization of the pulse delay time

Another parameter, which can be used for optimization, if it is made available from the evaluation software, is the normalized strength of the displacement correlation, the so-called cross-correlation coefficient given by:

$$c_{\mathrm{II}} = \frac{C_{\mathrm{II}}}{\sigma_{\mathrm{I}}\,\sigma_{\mathrm{I'}}} = \frac{R_{\mathrm{II}} - \mu_{\mathrm{I}}\,\mu_{\mathrm{I'}}}{\sigma_{\mathrm{I}}\,\sigma_{\mathrm{I'}}}\ .$$

When using photographic recording the choice of all recording parameters merely depends on the experiences of the experimentalist, because the valid data yield, the cross-correlation coefficient, or, in the case of optical evaluation, the visibility of the Young's fringes, can be assessed only after several hours.

Another way to reduce the loss of pairs is to change the size of the interrogation volumes and/or to displace them slightly with respect to each other in order to compensate for the mean particle displacement. The extension of the interrogation volume in the out-of-plane direction is given by the light sheet thickness. This parameter can be increased only if enough laser power is available. If one of the two possible out-of-plane directions is predominant, the light sheet can be displaced between the successive illuminations towards the mean flow. When using double oscillator systems this can be done by a slight "misalignment" of the beam combining optics. In the case of CW lasers a displacement requires additional equipment (see e.g. section 8.4). The extension and location of the interrogation volumes in the in-plane directions is given by the size of the interrogation areas during evaluation and the magnification during recording. In the case of cross-correlation analysis the location of the interrogation windows with respect to each other can be changed. This is one main advantage of cross-correlation and the reason why it is frequently applied also for the evaluation of single frame recordings instead of auto-correlation.

The effects of the interrogation volume locations during the first and second exposure $\boldsymbol{X_0} = (X_0, Y_0, Z_0)$ and $\boldsymbol{X'_0} = (X'_0, Y'_0, Z'_0)$ respectively, can best be described by presenting equation (3.12) in a more generalized form:

$$\begin{aligned} E\{R_D(\boldsymbol{s}, \boldsymbol{D}, \boldsymbol{X'_0} - \boldsymbol{X_0})\} &= C_{\mathrm{R}}\, R_\tau[\boldsymbol{s} - \boldsymbol{d} - (\boldsymbol{x'_0} - \boldsymbol{x_0})] \\ &\times F_{\mathrm{O}}[D_Z - (Z'_0 - Z_0)] \\ &\times F_{\mathrm{I}}[D_X - (X'_0 - X_0), D_Y - (Y'_0 - Y_0)] \,. \end{aligned}$$

From this equation the effect of an interrogation window offset $\boldsymbol{x'_0} - \boldsymbol{x_0}$ can clearly be seen: the peak location has changed by the amount of the offset and the influence of the in-plane loss of correlation on to the peak strength has changed. The significance of the loss of correlation also depends on the absolute extension of the interrogation volume. This is implied in the equations for F_{O} and F_{I} as given in the previous section (equation (3.10) and equation (3.11)), but shall be illustrated by the following equation which has been derived for a top-hat light sheet profile (equation (3.3)) and rectangular interrogation windows (equation (3.4)):

$$\begin{aligned} E\{R_D(\boldsymbol{s}, \boldsymbol{D}, \boldsymbol{X'_0} - \boldsymbol{X_0})\} &= C_{\mathrm{R}}\, R_\tau[\boldsymbol{s} - \boldsymbol{d} - (\boldsymbol{x'_0} - \boldsymbol{x_0})] \\ &\cdot \left(1 - \frac{D_X - (X'_0 - X_0)}{\Delta X_0}\right) \\ &\cdot \left(1 - \frac{D_Y - (Y'_0 - Y_0)}{\Delta Y_0}\right) \\ &\cdot \left(1 - \frac{D_Z - (Z'_0 - Z_0)}{\Delta Z_0}\right) . \end{aligned}$$

Generally speaking, a stronger peak results in a better peak detection probability and in reduced influence of noise components on the determination of the peak location. In many cases prior knowledge of the main displacement due to mean flow or image shifting can be used in order to improve the result of the evaluation. In other cases more sophisticated algorithms are required in order to take advantage of this effect. The software can either work in a multiple path scheme or with different sizes of the interrogation windows to be correlated. In both cases the resolution and accuracy of the measurement can be considerably increased at the cost of more computing time. The different aspects of choosing the right interrogation window sizes and locations for the evaluation by cross-correlation are described in section 5.4 in detail. According to our experience it is advisable to apply a quick-look evaluation during recording for the optimization of the experimental parameters, but also to store the original recordings in order to be able to optimize the evaluation after the experiment.

4. PIV recording techniques

In this chapter different approaches to PIV recording are introduced. It is important to realize that the various recording methods are not necessarily defined by the recording medium. The same approach may for instance be applied using either photography or digital recording. The PIV recording modes can be separated into two main branches: (1) methods which capture the illuminated flow on to a single frame and (2) methods which provide a single illuminated image for each illumination pulse. These branches are referred to as *single frame/multi-exposure* PIV and *multi-frame/single exposure* PIV, respectively [29].

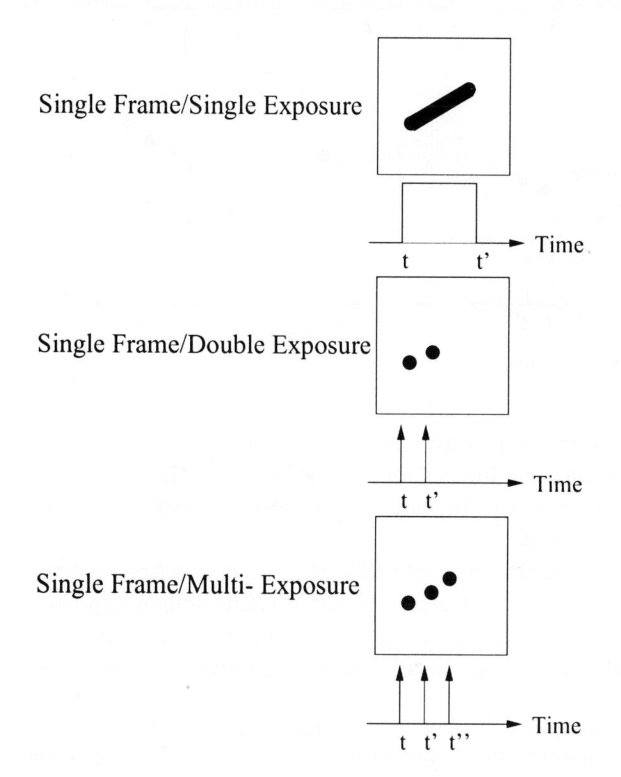

Fig. 4.1. Single frame techniques

The principal distinction between the two branches is that the former method, without additional effort, does not retain information on their temporal order of the illumination pulse giving rise to a directional ambiguity in the recovered displacement vector. This gave rise to a wide variety of schemes to account for the directional ambiguity, such as displacement biasing, the so-called image shifting (i.e. using a rotating mirror or birefringent crystal), pulse tagging or color coding[1] [36, 49, 51, 57, 68, 69].

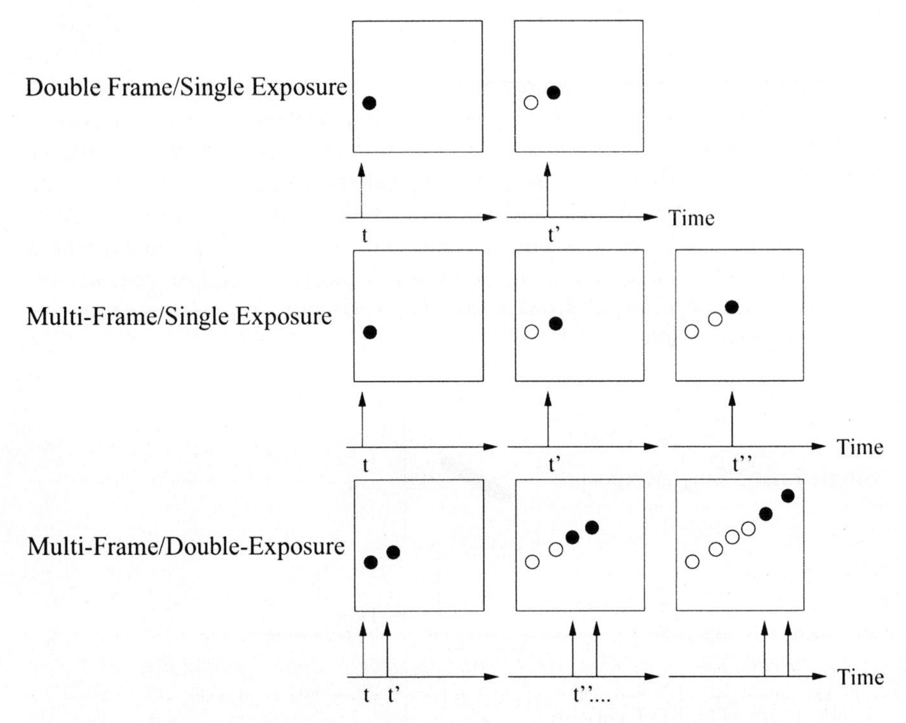

Fig. 4.2. Multiple frame techniques

In contrast, multi-frame/single exposure PIV recording inherently preserves the temporal order of the particle images and hence is the method of choice if the technological requirements can be met. Also in terms of evaluation this approach is much easier to handle.

Historically single frame/multi-exposure PIV recording was first utilized in conjunction with photography. Although multiple frame/single exposure PIV recording is possible using high speed motion cameras, other problems such as interframe registration arise. Continual development over the past

[1] Strictly speaking color coding is a form of *multi-frame/single exposure* PIV: the color recording can be separated into different color channels containing single exposed particle images.

decade in the area of electronic imaging has made multi-frame/single exposure PIV recording possible at flow velocities extending into the transonic domain. This chapter on PIV recording is laid out as follows: after an introduction to the cameras most frequently used, the advantages and associated problems of single frame recording and image shifting will be discussed. The second part will introduce multiple frame PIV recording mainly in the context of digital imaging.

A description of all possible recording techniques that can be used for PIV cannot be given here. A variety of different sophisticated ideas related to this problem has been reported in the literature over the past few years, most of them with special advantages for individual applications. It is clear that a decision on which method is best cannot be made without taking the individual needs of each application into account. Therefore, the recording techniques described here are not complete and not necessarily the best, but the most common.

In summary it can be said that the design of an experimental set-up for PIV is based on a decision of which of the following goals have priority:

- high spatial and/or temporal resolution of the flow field under investigation,
- the required resolution of velocity fluctuations,
- the time interval between the individual PIV measurements, and
- which components are already available in the laboratory or can be obtained at adequate costs.

Depending on the choice of priority an appropriate system for recording can be configured. However, it must be kept in mind that not every requirement can be fulfilled which is mainly due to technical limitations such as the available laser power, pulse repetition rates, camera frame rate, etc. The selection of the recording system also influences the method for directional ambiguity removal and, hence, the evaluation technique to be used. At the present stage of the development of the PIV technique photographic recording and mechanical image shifting seems to be well suited for applications in research where the occupation time of the facilities is not critical but the obtained quality of the recordings has first priority. Video recording offers so many advantages by allowing the user immediate feedback and quality optimization during the course of the experiment that it is well suited for most applications especially if time for the experiment is expensive.

4.1 Photo cameras for PIV recording

The current level of technology allows photographic recording to achieve a higher spatial resolution than can be obtained with current methods of digital video recording, especially if using large formats. Though CCD sensors with increasing resolutions are continuously under development and brought on to the market, it will be some time until video recording techniques can match the spatial resolution of present day film material. Thus, for the time being, the photographic technique is the method of choice for PIV applications requiring high velocity and spatial resolution. One major disadvantage of the photographic technique is that it is difficult to record the images of the tracer particles on to different frames, especially in the case of high speed investigations where pulse separations on the order of a few microseconds are required. That indicates that the problem of directional ambiguity removal has to be solved for photographic PIV in a reliable and flexible manner using a technique such as image shifting which will be described below.

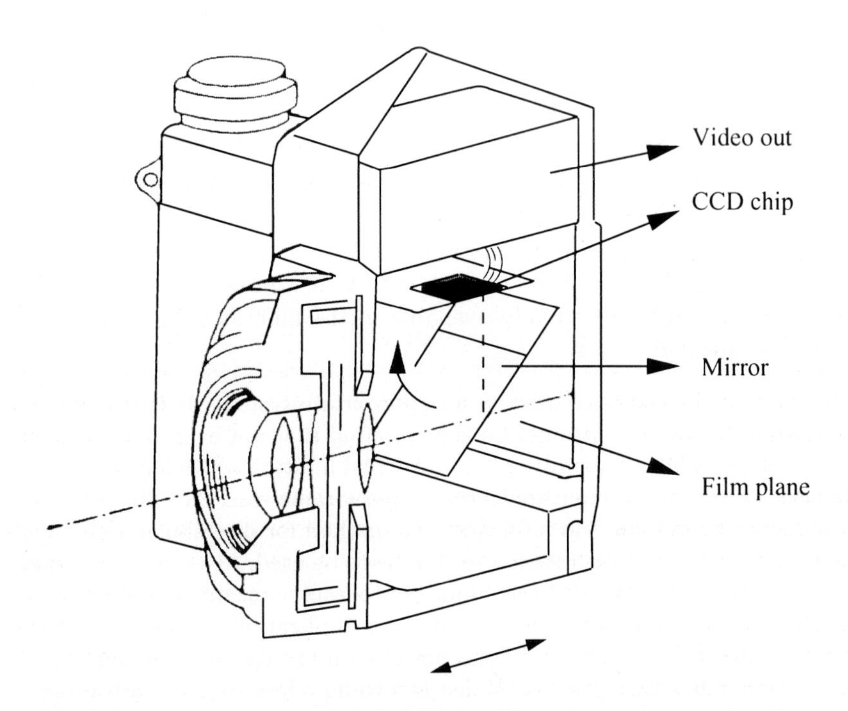

Fig. 4.3. Photo camera with CCD sensor for fast focusing

The first PIV experiments have shown that a high quality and reliable focusing device is necessary, in order to save time in the alignment of the system. For this purpose a photographic camera should be equipped with a device for fast focusing. A small area in the film plane can for example be observed by means of a CCD camera looking through an orifice in the back wall of the camera in order to control focusing [59]. This focusing device worked well and helped to reduce the time necessary for alignment considerably. A different solution which possesses some technical advantages has been suggested in the literature [93, 132].

Therefore, we improved our set-up and mounted a CCD sensor in the viewfinder of our 35 mm photo camera as shown in figure 4.3. The position of the CCD sensor is carefully aligned in such a way that the distance between the lens and the CCD sensor via the mirror is exactly the same as that from the lens to the film plane. The distance between the light sheet and the film plane can be changed by moving the complete camera system by means of a traversing table, thereby observing for minimum diameter of the particle images on a TV monitor.

4.2 Types of CCD sensors and their application to PIV

In section 2.6 the CCD as an imaging sensor was described. Here various types of CCD's are introduced in the context of application to PIV recording. Figure 4.4 schematically describes the layout of a CCD sensor. The individual pixels are typically grouped into a rectangular array to form a light sensitive area (linear, circular or hexagonal formats also exist). It should be pointed out that the pixels in this array cannot be randomly addressed the way memory can be addressed in a computer. Rather, the array has to be read out sequentially in a two-step process: after exposing the sensor the accumulated charge (i.e. electrons) is shifted vertically, one row at a time, into a masked-off analog shift register on the lower edge of the sensor's active area. Each row in the analog shift register is then clocked, pixel by pixel, through a charge-to-voltage converter and thereby provides one voltage for each pixel. The stream of pixel voltages along with a variety of synchronization pulses compose the actual (analog) video signal. Depending on the employed image transmission format the read-out of the sensor can either be sequential (also known as *progressive scan*) or interlaced, in which first all odd rows are read out before the even rows are accessed. The latter is the common format for standard video equipment (see also section 2.8). Since the progressive scan approach preserves the image integrity it is more useful for PIV recording as well as for other imaging applications such as machine vision.

In the following four sections we will concentrate on the operation of the various types of CCD sensors and how these may be utilized in PIV recording. Sections 2.7 and 4.4.1 should be consulted in regard to the utilization of standard (consumer) video equipment for PIV recording.

4.2.1 Full-frame CCD

The full-frame CCD sensor represents the CCD in its classical form (figure 4.4): a photosensitive area of pixels that is first exposed to light and then read out sequentially (progressive scan) on a row-by-row basis without separating the image into two separate interlaced fields such as in standard video. This sensor has been in use in scientific imaging such as astronomy, spectroscopy and remote sensing ever since its introduction in the 1960's. It is characterized by large fill factors which can even reach 100% for special back-thinned, back-illuminated sensors. With adequate cooling and slow read-out speeds, imaging at very low noise levels with high dynamic range (up to 16 bits) is possible. The most striking advantage is that these sensors are available as very large arrays with pixel counts exceeding tens of millions (7000×5000 pixel).

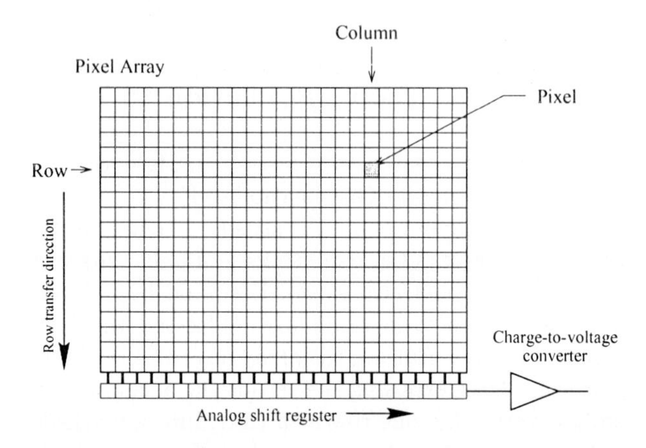

Fig. 4.4. Typical CCD sensor geometry

The use of these sensors does however have some major drawbacks. To achieve the low read-out noise and high dynamic range, the pixel read-out rate has to be kept low. Even at standard video characteristics the data rate is limited to 10 – 20 MHz which results in a decreasing frame rate as the number of pixels increases. Frame rates of less than 1 Hz are not uncommon for larger sensors. For this reason multiple read-out ports are sometimes used, which brings about the problem of calibrating the respective charge-to-voltage converters with respect to one another. Another drawback is that the sensor stays active during read-out. Unless a shutter is placed in front of it, light falling on to the sensor will also be captured resulting in a vertical smear in the final image.

Because of its high spatial resolution the full-frame sensor can be used as a direct replacement of photographic film. These sensors are frequently incorporated into 35 mm single-lens reflex (SLR) camera bodies. As for their use in PIV, single images containing multiple exposed particle images ($n_{\mathrm{exp}} \geq 2$) can be recorded analogous to the photographic method. The same ambiguity removal schemes as in photographic PIV recording (rotating mirror, birefringent crystal) can be employed. If the flow under investigation is sufficiently slow in comparison to the frame rate of a camera based on this sensor, then single exposed PIV recordings can be obtained. In this case the ambiguity removal schemes are not needed. The timing chart given in figure 4.8 (a) and (b) summarizes how the particle illumination pulses have to be placed to produce single exposed or multiple exposed PIV images.

4.2.2 Frame transfer CCD

The pixel architecture of the frame transfer CCD sensor (figure 4.5) is essentially equivalent to that of the full-frame CCD sensor with the difference that the lower half of its rows are masked off and cannot be exposed by incoming light. Once exposed, the rows of accumulated charge are rapidly shifted down into the masked-off area at rates as fast as $\Delta t_{\text{row-shift}} = 1\,\mu\text{s}$ per row. The entire image can thereby be shielded from further exposure within $\Delta t_{\text{transfer}} = 0.5 - 1\,\text{ms}$ depending on the vertical clocking speed and vertical image size. However, the sensor does stay active during the vertical transfer time such that vertical smear is possible. Charge stored within the masked area prior to the shift is lost however. Once the shift has been completed, the sequential read-out is equivalent to that of a full-frame CCD.

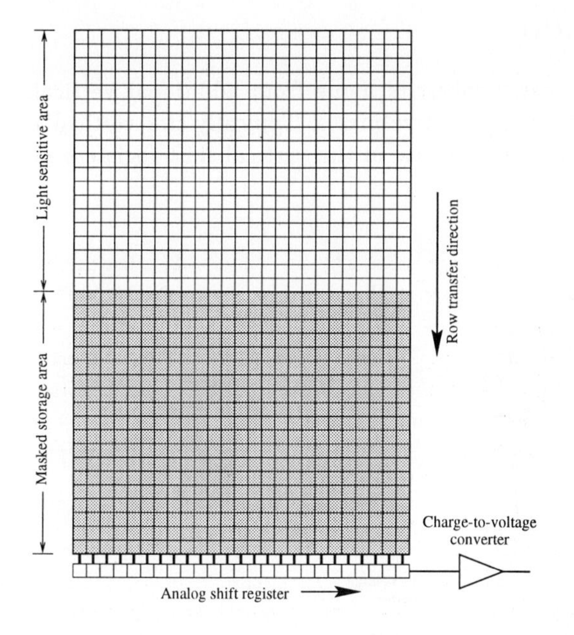

Fig. 4.5. Frame transfer CCD sensor layout

The frame transfer CCD sensor offers two application possibilities in PIV recording. The fast transfer of the accumulated charge into the storage area allows two single exposed PIV images to be captured at a time delay, Δt, slightly longer than the transfer time, e.g. $\Delta t \geq \Delta t_{\text{transfer}}$. To achieve this the illumination pulses are placed such that the first pulse occurs immediately before the frame transfer event (i.e. on frame n), while the second pulse occurs immediately thereafter (i.e. on frame $n+1$, see figure 4.8 c). This placement of the illumination pulses with respect to the CCD sensor's periodic exposure cycles is sometimes referred to as *frame straddling.* At standard video resolution and a field of view of 20 cm the measurement of flow velocities up to the order of 5 m/s is possible. In this case the PIV frame rate is half the camera frame rate (e.g. 15 Hz for NTSC video, 12.5 Hz for PAL video).

The frame transfer CCD sensor can alternatively be used to impose an image shift in order to remove the displacement bias associated with double exposure single frame PIV recording. This is achieved by placing the first illumination pulse just prior to the start or at the beginning of the vertical transfer period (figure 4.8 d). The second light pulse is placed such that it occurs while the collected charge of the first exposure is transferred into the masked area. For example, at a transfer rate of $\Delta t_{\text{row-shift}} = 1\,\mu\text{s}$ per row, a pulse delay of $\Delta t = 10\,\mu\text{s}$ would produce a maximum of 10 pixel image shifts. In this mode of operation the PIV frame rate is equal to the camera's frame rate. In this context it should be mentioned that standard CCD sensors are also capable of performing this type of displacement biasing [100].

4.2.3 Interline transfer CCD

The interline transfer CCD sensor takes its name from additional vertical transfer registers located between the active pixels. Typically two vertically adjacent pixels share a common storage site in the vertical shift registers as shown on the right of figure 4.6.

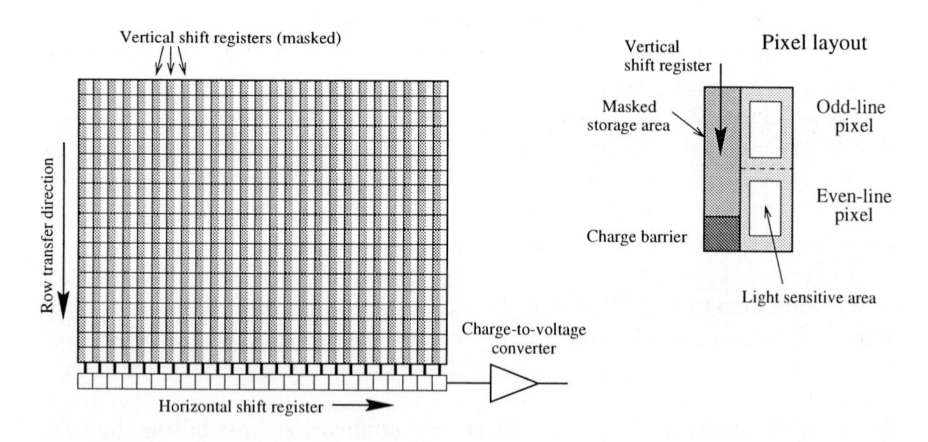

Fig. 4.6. Interline transfer CCD layout

Charge accumulated in the active area of the pixel can be rapidly transferred into the storage area ($\Delta t_{\text{transfer}} < 1\,\mu\text{s}$). This fast charge dumping feature also opens the possibility for full electronic shuttering on the sensor level.

This type of sensor is among the most common in consumer video products and hence readily available at standard video resolutions (see also table 2.6) although high resolutions are also in use. Contrary to the previously described framing CCD sensors, the interline transfer CCD sensor tends to be more sensitive in the blue-green region of the light spectrum (see figure 2.37).

The major drawback of these sensors is their reduced fill-factor due to the additional storage sites next to each light-sensitive area. Additional microlenses on the face of the sensor improve their light gathering capability. The alternative back-thinning approach is not possible with this sensor due to the additional storage sites which have to stay shielded against light exposure.

Since the sensor only provides half as many charge storage sites as there are active pixels, an image can only be stored at half the vertical resolution. This storage mode is an artifact of standard video transmission which separates a full image frame into distinct *fields* containing only even or odd lines (see section 2.8). Thus, the sensor can offer only half the vertical resolution in the shuttered mode of operation. For example if the odd lines of captured image data are read out from the sensor, the even lines are accumulating charge and vice versa. As a result, the odd and even lines are active during different periods of time, resulting in the capture of image data that is staggered by the period of one field for adjoining video lines – captured images of moving objects seem to flicker back and forth.

Cameras based on the interline transfer CCD have two possible applications in electronic PIV recording. The electronic shutter can be used to shutter the light of a continuous wave laser such as an argon-ion laser (figure 4.8 e). This electronic shutter is implemented by means of a clamping voltage on each pixel which inhibits the photon-to-charge conversion of the CCD for most of the framing interval; leaving a short period for photon collection just prior to the charge transfer event. Since the temporal position of the light-sensitive period is fixed relative to the camera's field rate, the effective pulse delay, Δt, will be equivalent to the field rate (e.g. $\Delta t = 20\,\text{ms}$ for CCIR, $\Delta t = 16.7\,\text{ms}$ for NTSC). This limits the application to low speed phenomena which can however be resolved in time.

In the second mode of operation the electronic shutter has to be completely deactivated making the sensor active at all times except for the brief charge transfer event. The illumination is provided by a pulsed laser (figure 4.8 f). In this case the frame straddling method is applied in which the first of the two illumination pulses is placed before the transfer event and the second right afterward (see also section 4.2.2). Thereby two single exposed images with half the vertical resolution (i.e. fields) can be recorded with a very short effective pulse delay which may be as short as the 1–2 μs duration of the charge transfer [60, 71, 89, 98]. Since two fields comprise a frame the effective PIV image frame rate is equivalent to the camera frame rate (e.g. 25 Hz for CCIR, 30 Hz for NTSC), given that the pulse laser can provide pulse pairs at this frequency. In this context it should be observed that PIV recording based on interlaced images is only reliable when the particle image diameter is large enough such that particle images will not be disappear in the second exposure, or vice versa.

4.2.4 Full-frame interline transfer CCD

This sensor is a derivative of the interline transfer CCD described before with the difference that each active pixel has its own storage site (figure 4.7, right side). Introduced in the first half of the 1990's cameras based on these progressive scan sensors rapidly gained popularity in the field of machine vision as they removed all the artifacts associated with interlaced video imaging. The electronic shutter can be applied to the entire image rather than to one of its fields as for standard interline transfer CCD's. Here also microlenses above each pixel help raise the effective fill factor from 20% to up to 60%.

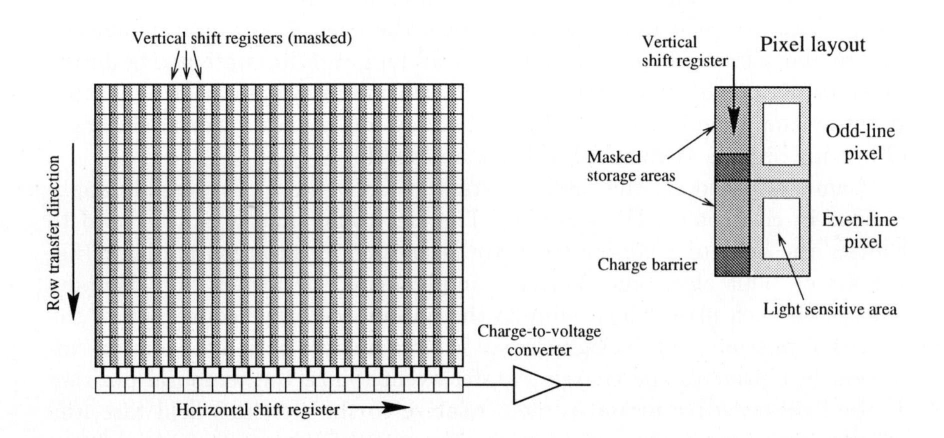

Fig. 4.7. Progressive scan, interline transfer CCD layout

The fast transfer of the entire exposed image into the adjoining storage sites within a few microseconds in conjunction with higher resolution formats, in a departure from the standard video resolutions, has extended the application of single exposure double frame PIV images into the transonic flow velocity domain. Here the maximum PIV image frame rate is half the camera's frame rate with pulse delays as low as $\Delta t = 1\,\mu s$ [150, 156]. As these cameras also frequently have asynchronous reset possibilities, their range of application is the most flexible of the CCD systems described in this chapter. A timing diagram for PIV recording based on this sensor is given in figure 4.8 (g).

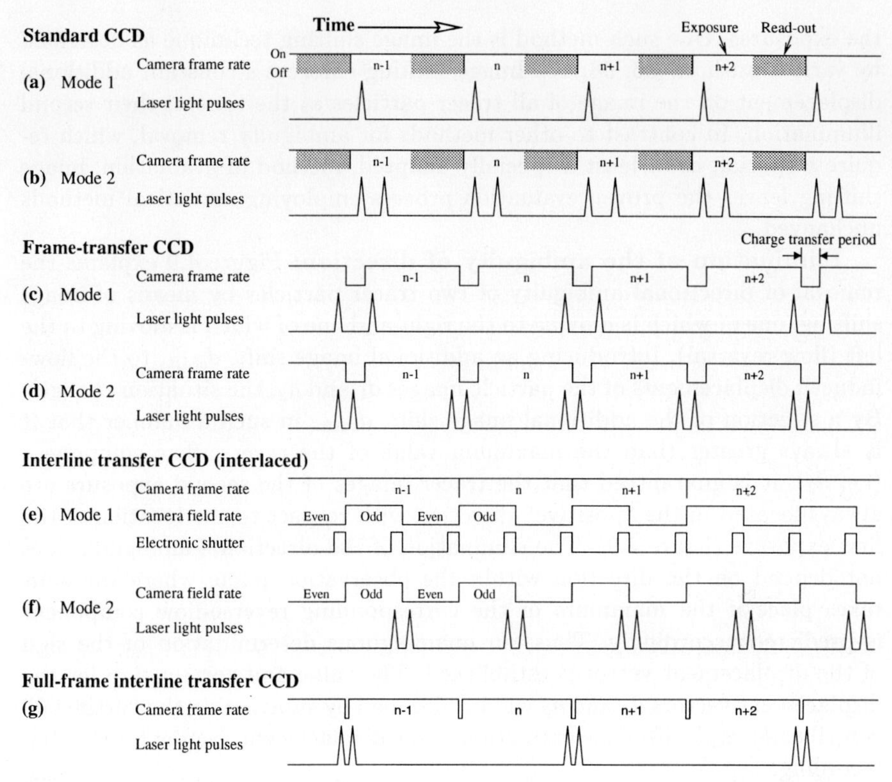

Fig. 4.8. Timing diagrams for PIV recording based on various types of CCD sensors

4.3 Single frame/multi-exposure recording

In conventional PIV two or more exposures of the particles of the same kind are stored on a single frame. Therefore, the sign of the direction of the particle motion within each interrogation spot cannot be determined uniquely, since there is no way to decide which image is due to the first and which is due to the second illumination pulse. Although, for many applications the sign of the velocity vector can be derived from a priori knowledge of the flow, other cases involving flow reversals, such as in separated flows, require a technique by which the sign of the displacement can be determined correctly.

4.3.1 General aspects of image shifting

The great interest in PIV measurements in many different fields of research requires a flexible technique for ambiguity removal that can be applied to a variety of experimental situations. Especially for aerodynamic investigations it is very important to be able to apply this technique to high speed flows, that is with short time intervals on the order of a few microseconds between the exposures. One such method is the image shifting technique as described by various authors [36, 68, 47]. Image shifting enforces a constant additional displacement on the image of all tracer particles at the time of their second illumination. In contrast to other methods for ambiguity removal, which require a special, or at least a specially adapted, method of evaluation, image shifting leaves the proven evaluation process employing statistical methods unchanged.

Elimination of the ambiguity of direction: Figure 4.9 explains the removal of directional ambiguity of two tracer particles by means of image shifting, one of which is moving to the right and one of which is moving to the left (flow reversal). Introducing an additional image shift, d_{shift}, to the flow-induced displacements of the particle images d_1 and d_2, the situation changes: By a selection of the additional image shift, d_{shift}, in such a manner that it is always greater than the maximum value of the reverse-flow component (i.e. d_2), it is guaranteed that the tracer images of the second exposure are always located in the "positive" direction with respect to the location of the first exposure (figure 4.9). The elimination of the directional ambiguity does not depend on the direction within the observation plane where the shift takes place if the maximum of the corresponding reverse-flow component is predicted accordingly. Thus, an unambiguous determination of the sign of the displacement vector is established. The value and correct sign for the displacement vectors d_1 and d_2 will be obtained by subtracting the "artificial" contribution d_{shift} after the extraction of the displacement vectors for the PIV recording.

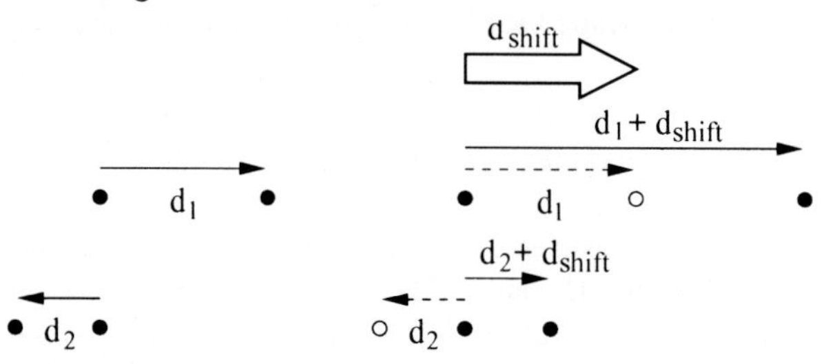

Fig. 4.9. Elimination of the ambiguity of direction of the displacement vector as observed in the recording plane

4.3.2 Optimization of PIV recording for auto-correlation analysis by image shifting

As already mentioned in chapter 3 the application of the cross-correlation technique for two subareas of a single frame/multi-exposure recording instead of performing an auto-correlation on a single subarea, increases the flexibility of the PIV system. This evaluation approach cannot remove the directional ambiguity of the velocity vectors or handle situations where the image displacements are of the order of the particle image diameter. However, the pulse separation time can be adapted in a wider range, because the size of the two interrogation windows and their displacement can be adapted during evaluation.

If the evaluation system is not flexible enough to allow cross-correlation of slightly displaced interrogation areas the evaluation process of doubly exposed PIV recordings has to be analyzed by auto-correlation. This problem often occurs if optical evaluation is used in order to obtain an improved signal-to-noise ratio. Besides resolving the directional ambiguity image shifting is required in order to optimize the recording for later auto-correlation evaluation. This will be illustrated in the following at three typical experimental situations.

In most optical evaluation systems the diameter of the interrogation spot is typically 700 μm. For this condition, experience has shown that best results are obtained if the displacements of the tracer particles as measured on the recording media (photographic film) is $150\,\mu\text{m} \leq d_{\text{opt}} \leq 250\,\mu\text{m}$. This agrees with the results from the simulation of the PIV recording and evaluation process by KEANE & ADRIAN [64].

Measurement of highly three-dimensional flows: Since PIV records the light which is scattered by particles illuminated by a double pulse of light within a thin sheet, the method limits its application to flows with a limited three-dimensionality. This problem is more severe for measurements which require thin light sheets for higher intensity, e.g. in gaseous media where very small particles are required in order to follow the flow faithfully. With decreasing particle size the scattered light decreases as well which necessitates optical arrangements with a small effective depth of focus at recording. Experience has shown that for aerodynamic measurements the thickness of the light sheet and the depth of focus cannot be increased considerably. In order to limit the out-of-plane loss of correlation, the out-of-plane displacement of the particles should not exceed 30% of the light sheet thickness [64]:

$$\frac{W \Delta t}{\Delta Z_0} \leq 0.3 \,.$$

The only practical way to reduce the number of particles leaving the light sheet plane in most practical situations is to decrease the time between the light pulses considerably (figure 4.10).

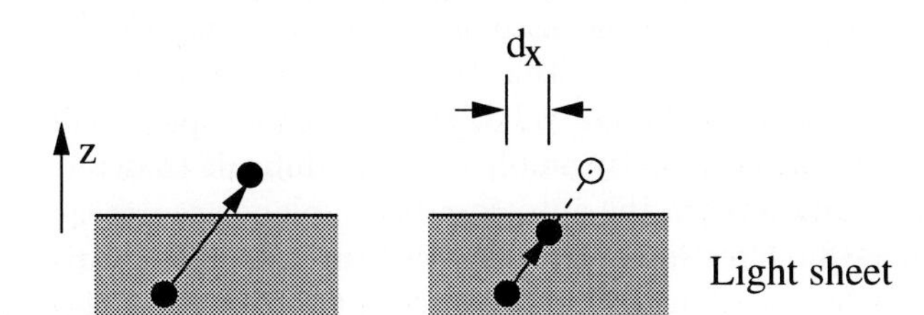

Fig. 4.10. Effect of reducing the time between two pulses as observed in the light sheet (figure is not to scale)

In the same manner the displacement of the images of the tracer particles on the PIV recording will be reduced from d_1 to d_2. However, the optimal distance between two particle images, d_{opt}, that is required for the evaluation by means of the auto-correlation technique can be re-established by applying image shifting. Figure 4.11 describes this situation by showing the virtual positions of a tracer particle at the moment of the first and second exposure in the light sheet. Since the number of matched particle images is increased considerably by this method, the number of valid data can be increased also. Therefore, a much greater third component of the flow velocity is tolerable at the same signal-to-noise ratio in the auto-correlation plane.

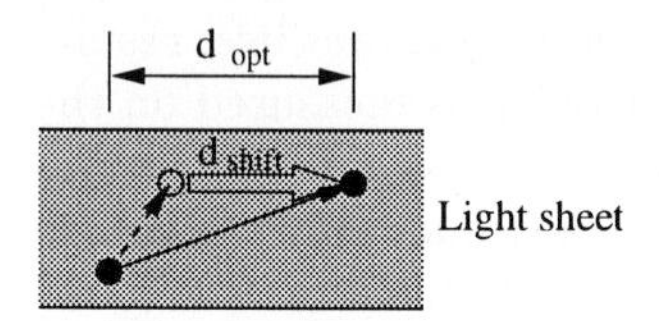

Fig. 4.11. Re-establishing the optimal distance between pulses (projected back into the light sheet)

In section 8.1.8 we show as an example of the application of image shifting the photographic PIV measurement of a highly three-dimensional detached flow above a pitching airfoil. However, it has to be kept in mind that the resolution of the velocity measurement is decreased by reducing the time between the light pulses. This has to be balanced against the improved signal-to-noise ratio in the auto-correlation plane. The maximum attainable accuracy of the PIV method results from the chosen compromise between reduced resolution of the velocity measurement and the improved signal-to-noise ratio due to a short pulse separation (see section 3.7). Additionally, it has to be pointed out that allowing for a significant velocity component normal to the light sheet plane leads to a considerable error due to this component. This error arises from the fact that the camera lens reproduces the tracer particles by perspective projection and not by parallel projection (see section 2.4.3).

Reduction of the variance within an interrogation spot: Another situation where the likelihood of a peak detection in the auto-correlation plane can be increased by image shifting is if high velocity gradients are present in the flow field. During evaluation the displacement vector is determined by averaging over all displacements between image pairs within one interrogation spot. Therefore, it has to be assumed that the flow velocity does not change magnitude and direction within the interrogation area (see chapter 3). If this assumption is no longer valid, the correlation peak will broaden due to the variance of the velocity and it becomes difficult to determine the location of the signal peak in the auto-correlation plane accurately. This means that for successful detection of the displacement vector, the variance of the particle image displacement within the interrogation spot has also to be limited. An estimate for the maximum variance that can be accepted is given by [64]:

$$\frac{M\Delta U\Delta t}{d_\tau} \leq 1 \tag{4.1}$$

where d_τ is the particle image diameter and ΔU the fluctuation of the flow velocity within the interrogation volume. The condition given in equation (4.1) can be fulfilled if the pulse separation time is reduced. However, in order to avoid systematic overlap of the particle images and to adapt the image displacement to the requirement of auto-correlation analysis image shifting has to be employed during recording. It has been shown that a good detection probability of the auto-correlation peaks can be obtained by employing image shifting during the recording of transonic flow fields (see section 8.1.5).

Improving the velocity resolution: In flows with only a small velocity component normal to the main flow direction and with only small fluctuations in the main flow direction, an increase in resolution can be achieved by choosing longer time delays between the two illumination pulses and image shifting "against" the main flow direction (pedestal removal) in order to re-establish optimum particle image displacement for the application of the auto-correlation technique. This situation is demonstrated in figure 4.12. An example of such an application is given in section 8.1.1. There it can be seen that image shifting enables us to resolve very weak turbulent structures.

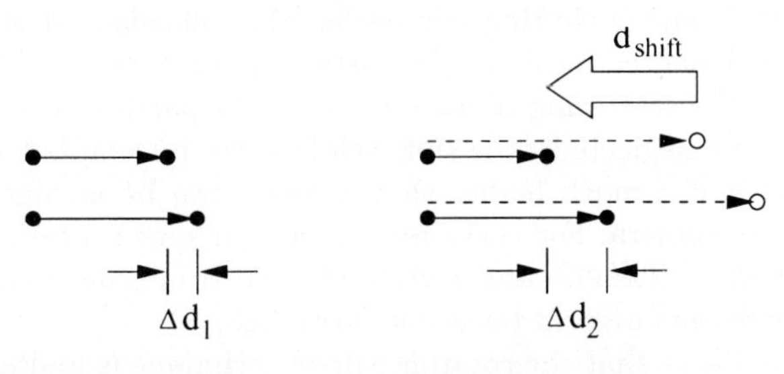

Fig. 4.12. Effect of increasing the resolution by increasing the pulse delay and shifting "against" the main flow direction as observed in the recording plane

4.3.3 Realizations of image shifting

At present the most widely used experimental technique for image shifting involves the use of a rotating mirror over which the observation area within the flow is imaged on to the recording area in the camera. The magnitude of the additional displacement of the images of the tracer particles depends on the angular speed of the mirror, the distance between the light sheet plane and the mirror, the magnification of the imaging system and the time delay between the two illumination pulses [68]. Various experimental set-ups employing rotating mirrors were realized by different authors and show good experimental results, such as the investigation of dynamic flow separation (wake with flow reversals) above profiles [72]. The application of rotating mirror systems to turbulent and convective flows indicated that the attainable resolution in the PIV technique is strongly influenced by the image shifting method [68, 54].

In order to achieve very high shift velocities electro-optical methods employing differently polarized light for illumination have been proposed and applied by LANDRETH & ADRIAN [69], LOURENÇO [74] and MOLEZZI & DUTTON [142]. The constant shift of the particle images is obtained by means of birefringent crystals of appropriate thickness. REUSS describes the problems associated with this method, such as "depolarization effects" [147].

Another scheme by which the directional ambiguity problem may be resolved has been presented by WORMELL & SOPCHAK and involves a CCD camera in which the charge associated with the first illumination is electronically moved by a known distance within the sensor during the time period between the first and the second laser pulse [100] (see section 4.2.2). This arrangement allows a minimum pulse separation of $\approx 40\,\mu s$.

Drawing from our own experiences, we see most practical advantages in the application of a rotating mirror system for image shifting. A detailed description of such a system, which has been successfully applied to a variety of experiments in wind tunnels, is given in the following.

A rotating mirror based image shifting system has the following advantages: it can be easily implemented into already existing apparatus; it makes no additional demands on the scattering characteristics of the particles (light depolarization effects can be neglected); the shift velocity can be adapted to the problem very easily; and a much higher shift velocity can be attained than by moving the entire camera. For these reasons a high speed rotating mirror system as described by RAFFEL was developed at DLR in order to be able to carry out measurements even in transonic flows [143].

It should be mentioned here that the rotating mirror technique is limited by a maximum framing rate which is given by the product of the angular speed and number of mirror surfaces. In most aerodynamic applications the maximum framing rate is higher than the repetition rate of a pulse laser system. Nevertheless, the framing rate may become a limiting factor, for example, when flow fields are observed at low flow (and shift) velocities with high resolution in time. Furthermore, it is not possible to synchronize non-periodic flow events with a mirror that is rotating at a constant speed.

4.3.4 Layout of a rotating mirror system

Figure 4.13 shows the high speed rotating mirror system for image shifting in use at DLR Göttingen. It allows for shift velocities exceeding 500 m/s without any noticeable reduction of the optical quality of the images.

The system shown in figure 4.13 consists of the following components: a shaft mounted in precision bearings, a mirror mount attached to one end of the shaft and an optical encoder connected to the other end. The mirror assembly is driven by a stepper motor with stable revolution frequencies ranging from 1 to 100 Hz. A toothed belt guarantees slip free transmission while a revolving mass attached to the rotating shaft compensates for the velocity fluctuations of the motor drive. The optical encoder is a commercially available angle encoder that is coupled to the shaft using a twist free shaft clutch and supplies the signals for the laser triggering as well as angular frequency monitoring. This precisely machined set-up ensures that the 90° angle required between objective lens and observation plane is exactly reproducible even in a noisy industrial environment.

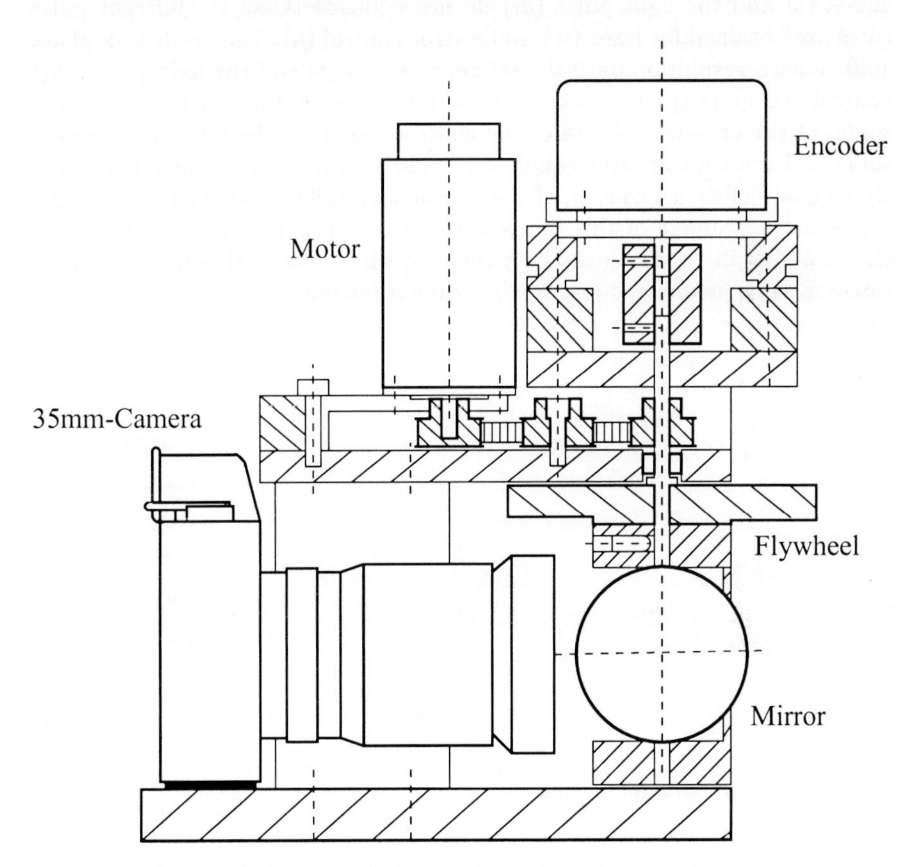

Fig. 4.13. Schematic diagram of the rotating mirror system

The adaption of the signal frequency from the angle encoder to the repetition rate of the pulse laser is performed by digital dividers, see figure 4.14. The angular position of the mirror at the time of image capture is kept constant by means of a digital controller and is phase-locked to the trigger for the laser pulses. The controller itself was designed in such a manner that it is able to handle severe electronic noise due to the stepper motors or the electric drive systems present in many wind tunnel environments. The procedure of controlling the observation angle is as follows. Signal (a) is the increment signal from the encoder. The resolution of 1000 pulses per revolution as delivered by the encoder was sufficient. Signal (b) is the reference signal (recording position) from the encoder which usually is one inverted pulse per revolution. Signals (a) and (b) are combined by a logical AND gate in the angle controller. The resulting signal (c) and a signal obtained from the light pulse of the laser (d) are combined by a logical OR gate. This signal, (e), is used to control the laser and the camera for recording. If the reference signal (b) and the light pulse (d_{I}) do not coincide (Case I), different pulse rates are obtained for laser (e_{I}) and mirror control (a). This leads to a phase shift in each revolution until the reference signal (b) and the light pulse (d_{II}) coincide temporally. In this case (Case II) the pulse rates of the increment signal of the encoder (a) and of the control signal for the laser (e_{II}) are the same and the control error equals zero. The digital dividers included in the signal chain allow a variation of the mirror revolutions and the framing rate. The main advantage of this image shifting set-up is the easy handling and the flexibility in adjusting the shift velocity, which can be chosen from a wide range in very small steps just by "pressing a button".

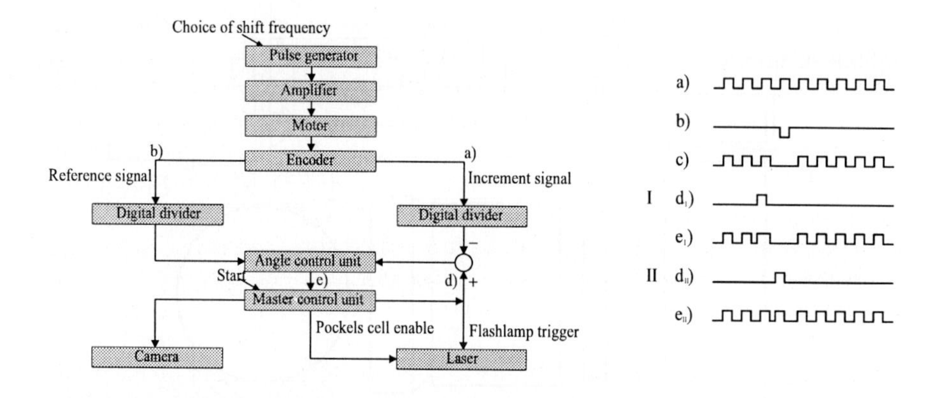

Fig. 4.14. Flow and impulse chart of the angle control

This section described how to build a rotating mirror system which is precisely phase locked with the laser light pulses. However, there exist other error sources in rotating mirror systems. In particular the way in which the virtual image of the tracer particles is moved by the rotation of the mirror has to be taken into account. It will be shown in the following that the movement of the particle images due to the rotating mirror are not uniform over the recording, but vary locally. By means of the equations arising from the system's optical geometry derived in the following, algorithms can be implemented, which allow a full compensation of these errors.

4.3.5 Calculation of the mirror image shift

In the literature, the motion of the virtual image due to the rotation of the mirror is derived from geometric relations in two dimensions with assumptions that are not valid in general [53]. Only the displacement of the depth-of-field center from the light sheet was determined in order to estimate the effect of this parameter on the defocusing of the images of the particles [144]. In most cases the shift of the particle images d_{shift} is assumed to be constant in the entire observation area:

$$d_{\mathrm{shift}} = 2\omega_{\mathrm{m}} Z_{\mathrm{m}} M \Delta t \tag{4.2}$$

where ω_{m} is the angular velocity of the rotating mirror, Z_{m} the distance between the light sheet and mirror-axis (Z_{m} = const.), M the magnification (image size/object size) and $\Delta t = t' - t$ the time delay between the light pulses.

A more detailed model of the imaging geometry associated with a rotating mirror is shown in figure 4.15. The matrices – given in appendix A – are written in four-dimensional form, for reasons stated in section 2.4.3. The model can again be described completely in terms of linear algebra. The explanation starts at point $\boldsymbol{X}_i$ located in the light sheet. The mirror image of point $\boldsymbol{X}_i$ moves within the virtual light sheet plane from $\boldsymbol{X}_{\mathrm{v}}$ to $\boldsymbol{X'}_{\mathrm{v}}$ due to the rotation of the mirror. This movement can be determined by considering the reflection of $\boldsymbol{X}_{\mathrm{v}}$ with respect to the plane $M(t)$ (which yields $\boldsymbol{X}_{\mathrm{H}} = \mathsf{R}_{\mathrm{refl}}(\theta = \pi/4)_{\mathrm{H}} \circ \boldsymbol{X}_{\mathrm{v,H}}$) and the reflection of $\boldsymbol{X}$ with respect to the plane $M(t')$ (which yields $\boldsymbol{X'}_{\mathrm{v,H}} = \mathsf{R}_{\mathrm{refl}}(\theta = \pi/4 + \omega_{\mathrm{m}} t)_{\mathrm{H}} \circ \boldsymbol{X}_{\mathrm{H}}$). The resulting movement of such a virtual image point is a rotation around the mirror-axis with twice the angular velocity ω. This result is obtained by concatenating the two reflection matrices into a single matrix of rotation around the y-axis $\mathsf{Rot}_{y,\mathrm{H}} = \mathsf{R}_{\mathrm{refl}}(\theta = \pi/4 + \omega_{\mathrm{m}} t)_{\mathrm{H}} \circ \mathsf{R}_{\mathrm{refl}}(\theta = \pi/4)_{\mathrm{H}}$. This statement can be proven by using the addition formulas for trigonometric functions and the transformation matrix $\mathsf{R}_{\mathrm{refl}}(\theta)_{\mathrm{H}}$ that describes a reflection in a plane that is tilted against the yz-plane by an angle of θ.

The fact that the distance between the mirror axis and a point of the virtual light sheet is not constant but a function of the x-coordinate leads to an error of more than 1% of the mean image shift at typical experimental conditions. A further, additional error results from the direction of movement of the virtual image. Since the z component of the virtual particle image shift increases towards the edges of the virtual observation area (see figure 4.15), there is also an influence on the x and y components due to the perspective projection of the virtual particle image on to the recording plane. In order to fully describe the effects of a shift component perpendicular to the virtual light sheet on the position of the image points in the coordinate system x, y, z (see figure 4.15), the imaging through the lens must also be considered. In this case we have to define two different homogeneous coordinate systems, one with respect to the mirror coordinates and one with respect to the camera coordinates. As explained above the movement of a virtual image point due to the rotation of the mirror is described within the mirror coordinate system $x^\star, y^\star, z^\star$ (see figure 4.15) as follows: $\boldsymbol{X'}_{\mathrm{v,H}} = \mathsf{Rot}_{y,\mathrm{H}} \circ \boldsymbol{X}_{\mathrm{v,H}}$.

As $\mathsf{Rot}_{y,\mathrm{H}}$ describes the rotation in the mirror coordinate system and not in the camera coordinate system, a transformation T_{H} from camera coordinates to mirror coordinates is required. The inverse transformation to convert from mirror coordinates back to camera coordinates is $\mathsf{T}_{\mathrm{H}}^{-1}$. The transformation matrices are given in detail in the appendix. The equation of motion for a virtual image point in camera coordinates can be described as follows:

$$\boldsymbol{X'}_{\mathrm{v,H}} = \mathsf{T}_{\mathrm{H}}^{-1} \circ \mathsf{Rot}_{y,\mathrm{H}} \circ \mathsf{T}_{\mathrm{H}} \circ \boldsymbol{X}_{\mathrm{v,H}} \ .$$

The following operators describe the required perspective transformations: P_{H} for imaging a point in the virtual light sheet plane on to the recording plane, $\mathsf{P}_{\mathrm{H}}^{-1}$ for the transformation of an image point from the recording plane on to the virtual light sheet plane (assuming a mirror position $M(t)$ with a rotation by $\pi/4$ with respect to the optical axis as presented in figure 4.15). Finally, we obtain the following relation between the recording plane positions of a point P in the light sheet when imaged via a rotating mirror:

$$\boldsymbol{x'}_{\mathrm{H}} = \mathsf{P}_{\mathrm{H}} \circ \mathsf{T}_{\mathrm{H}}^{-1} \circ \mathsf{Rot}_{y,\mathrm{H}} \circ \mathsf{T}_{\mathrm{H}} \circ \mathsf{P}_{\mathrm{H}}^{-1} \circ \boldsymbol{x}_{\mathrm{H}} \ .$$

After converting into the (camera-)world coordinate system both components of the image displacement $\boldsymbol{d} = \boldsymbol{x'} - \boldsymbol{x}$ due to mirror rotation can be obtained:

$$d_x(x,y) = \frac{x\cos(2\omega_{\mathrm{m}}\Delta t) - MX_{\mathrm{m}}[1-\cos(2\omega_{\mathrm{m}}\Delta t)] - MZ_{\mathrm{m}}\sin(2\omega_{\mathrm{m}}\Delta t)}{\frac{x}{z_0}\sin(2\omega_{\mathrm{m}}\Delta t) + \frac{X_{\mathrm{m}}}{Z_0}\sin(2\omega_{\mathrm{m}}\Delta t) - \frac{Z_{\mathrm{m}}}{Z_0}[1-\cos(2\omega_{\mathrm{m}}\Delta t)] + 1} - x$$

$$d_y(x,y) = \frac{y}{\frac{x}{z_0}\sin(2\omega_{\mathrm{m}}\Delta t) + \frac{X_{\mathrm{m}}}{Z_0}\sin(2\omega_{\mathrm{m}}\Delta t) - \frac{Z_{\mathrm{m}}}{Z_0}[1-\cos(2\omega_{\mathrm{m}}\Delta t)] + 1} - y \ .$$

Equation (4.2), which is given in the literature for the shift of the tracer images due to a rotating mirror, can be derived from the exact solution by the following assumptions:

- Δt approaches zero
  $\Rightarrow \quad \sin(2\omega_{\mathrm{m}}\Delta t) \cong 2\omega_{\mathrm{m}}\Delta t; \ \cos(2\omega_{\mathrm{m}}\Delta t) \cong 1;$
- the axis of rotation of the mirror lies on the optical axis
  $\Rightarrow \quad X_{\mathrm{m}} = 0;$
- the particle images are located close to the center of the image
  $\Rightarrow \quad x \cong 0; y \cong 0.$

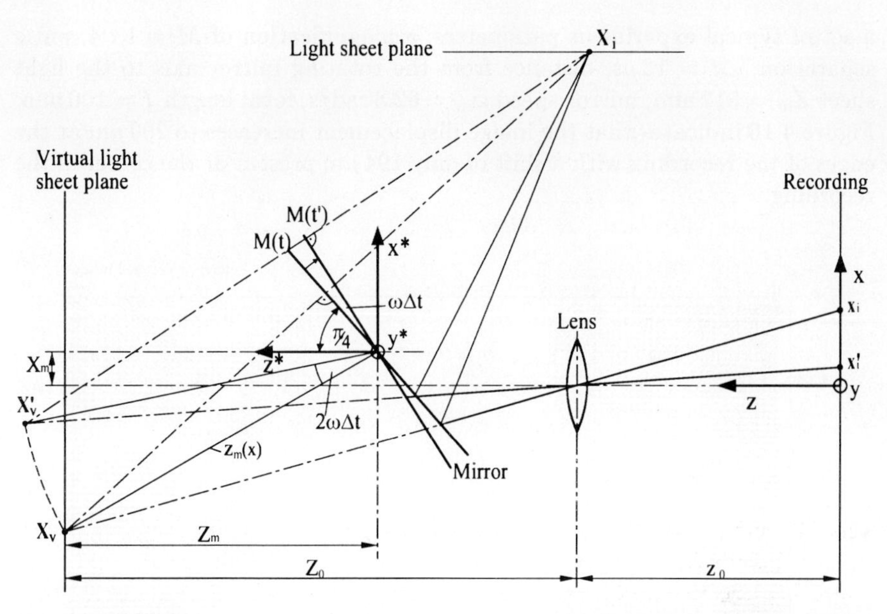

Fig. 4.15. Movement of the virtual particle image and its projection on to the recording plane

The first assumption leads to a negligible error due to the fact that the shift angle $2\omega_\mathrm{m}\Delta t$ is usually less than 0.1°. The relations $\sin(2\omega_\mathrm{m}\Delta t) = 2\omega_\mathrm{m}\Delta t$ and $\cos(2\omega_\mathrm{m}\Delta t) = 1$ lead to the following formulas for the calculation of the particle image shift, when the formulas $1/f = 1/z_0 - 1/Z_0$ and $M = z_0/Z_0$ (see section 2.4.1) are used:

$$d_x(x,y) = \frac{x - M \cdot Z_\mathrm{m}\, 2\omega_\mathrm{m}\Delta t}{(x + X_\mathrm{m}\, M)\, 2\omega_\mathrm{m}\Delta t\, f^{-1}\,(1+M)^{-1} + 1} - x \tag{4.3}$$

$$d_y(x,y) = \frac{y}{(x + X_\mathrm{m}\, M)\, 2\omega_\mathrm{m}\Delta t\, f^{-1}\,(1+M)^{-1} + 1} - y\,. \tag{4.4}$$

These formulas are recommended for the practical use of image shifting by means of rotating mirror systems.

In many cases the distance X_m from the mirror axis to the optical axis of the lens can be adjusted to zero. However, asymmetric configurations with X_m greater than zero allow the use of a smaller mirror.

The assumption that the particle images are located close to the center of the recording ($x \cong 0$; $y \cong 0$) leads to a systematic shift error in the measured displacement data. This error $\epsilon = (d_x - d_\mathrm{shift}, d_y)$, which can be calculated when using equations (4.3) and (4.4), must be accounted for in the evaluation of PIV images. As an example, the difference between the local tracer image shift resulting from the mirror rotation, (d_x, d_y), and the tracer image shift in the center of the observation field, d_shift, was calculated given a set of typical experiment parameters: a magnification of $M = 1:4$, pulse separation, $\Delta t = 12\,\mu s$, distance from the rotating mirror axis to the light sheet $Z_m = 512\,mm$, mirror speed $\omega_m = 62.8\,rad/s$, focal length $f = 100\,mm$. Figure 4.16 indicates that the image displacement increases to 200 μm at the edges of the recording with a shift of only 194 μm present at the center of the recording.

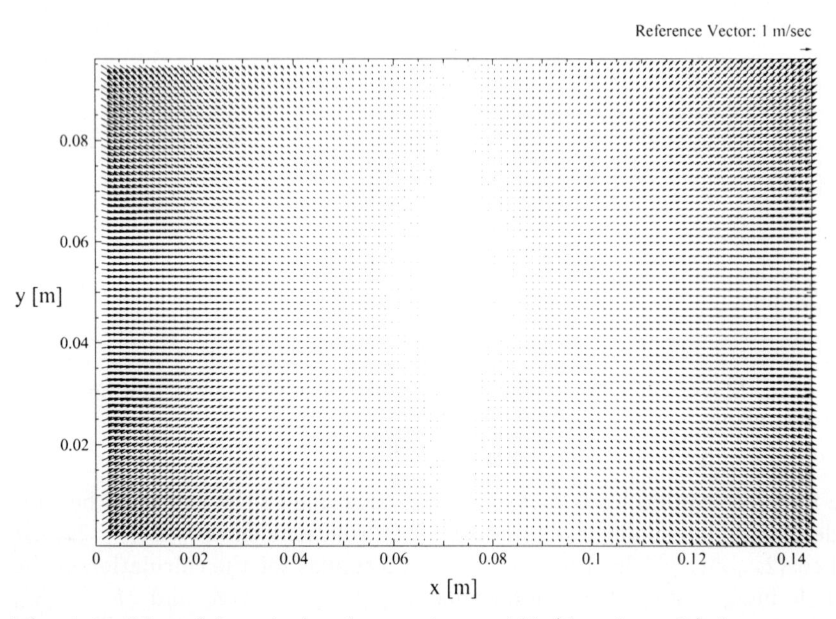

Fig. 4.16. Map of the calculated tracer image shift $(d_x - d_{\text{shift}}, d_y)$ due to perspective transformation

4.3.6 Experimental determination of the mirror image shift

The error in the particle image shift on the PIV recording due to the use of a rotating mirror system for image shifting has also been determined experimentally. For this purpose PIV measurements were performed in air at rest. Tracer particles were injected into the air in the area of the observation field. The experimental set-up is defined by the same parameters as given in the example calculation in the previous section. Particle motion resulting from convection and/or gravity effects was minor and therefore is negligible compared to the shift velocity of 64 m/s. Figure 4.17 displays the difference between the local tracer image shift and the tracer image shift in the center of the observation field. The scaling of the vector field is identical to figure 4.16.

The deviations between the experimental values and the theoretical values as calculated with equations (4.3) and (4.4) lie within the measurement resolution of our PIV evaluation system. Therefore we can safely assume that the rotating mirror-camera system was modeled correctly. The conventional practice – as described in the literature – of assuming that the value for the tracer image shift is constant over the entire image leads to systematic errors of 2–3% in the actual values of tracer image shift even for small observation angles. As will be described below, a virtual shift greater than the mean flow velocity is chosen frequently. This results in errors of up to 10% of the mean flow velocity. Equations (4.3) and (4.4) indicate that the shift of the particle images depends on the magnification, M, the focal length of the lens, f, as well as the position of the mirror axis. These parameters stay constant during an experiment. Therefore, in order to be able to correct the measurement results appropriately, the locally varying tracer image shift has to be determined only once for a given recording configuration by means of the equations given above.

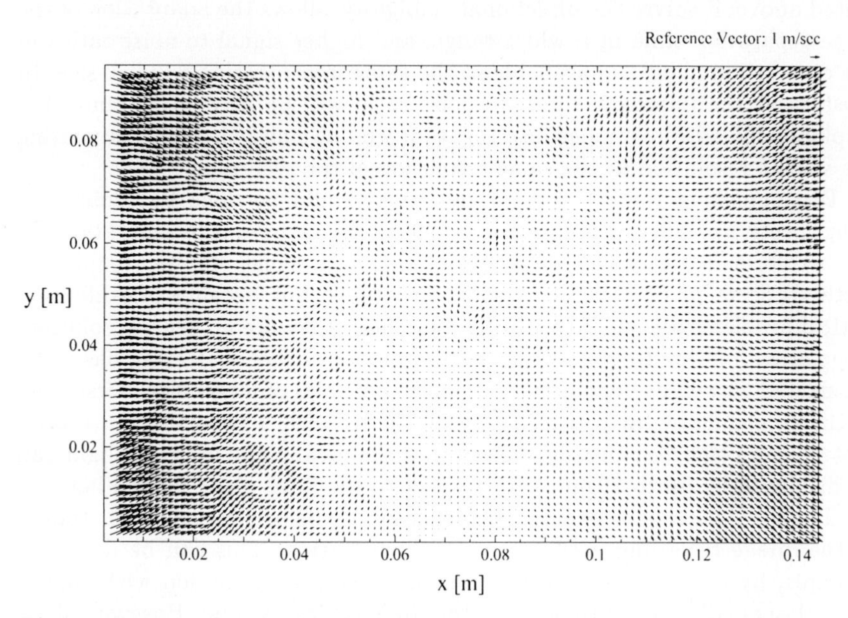

Fig. 4.17. Map of the experimentally determined tracer image shift ($d_x - d_{\text{shift}}$, d_y) of particles at rest on the recording. Same parameters were used for calculation and experiment

An alternative practical approach has been implemented recently and avoids the need for an exact measurement of the parameters in equations (4.3) and (4.4). Usually after completion of the experiments, the fluid is left to come to rest while the shifting mirror continues to operate at constant angular speed. A PIV recording of the quiescent fluid ($U_{\text{residual}} \ll U_{\text{shift}}$) only records the effects of the image shift. By performing a second order least squares fit to the obtained virtual flow field the distortions brought about by the mirror can be very well estimated and can subsequently be used to correct the actual PIV data of the original experiment. Such a direct calibration with subsequent fit to the expected function can also be used to analyze and compensate for the nonuniform image shift of other devices (e.g. nonuniform image shift by a birefringent crystal as reported by Reuss [147]).

4.4 Multi-frame PIV recording

In the following a short summary of different techniques for multi-frame PIV recording is given. A more detailed description of the most frequently used video-based implementations of double frame/single exposure recording has been given in a subsequent section. The main advantages of separating the light of the subsequent recordings on to different frames have already been stated above: it solves the directional ambiguity, allows the adaptation of the pulse separation time in a wider range, and higher signal-to-noise ratios in the correlation plane are available at the same interrogation window size. In most cases the improved signal-to-noise ratio will be used to compute the displacement within smaller interrogation windows and therefore increasing the spatial resolution at the same resolution of the recording.

Distinguishing the different illuminations could be done by coding the light sheets by different polarizations [109, 119], or by image separation using a color video camera and a color coded light sheet [49, 48, 103]. Both methods seem to be feasible in some situations, but because of additional optical problems associated with these methods they are no general solution. Depolarization due to glass windows and model surface scattering has to be considered as well as depolarization due to particles of larger size. First tests of the recording of small tracer particles in air with a two-color pulse laser system and color film already showed that the focal plane for green light can be displaced by up to 20 mm with respect to the focal plane for red light.

The separation of different exposures can also be performed by the timing of the image recording with respect to illumination. This can be done, for example, by means of high speed film cameras in combination with copper vapor lasers [137] or multiple oscillator Nd:YAG lasers [149]. However, these experimental set-ups are very difficult to handle and are suitable only for special applications as for example for flow investigations in piston engines. More general solutions based on the proper timing of video cameras will be described in the following.

4.4.1 Video-based implementation of double frame/single exposure PIV

The low cost and frequent availability of standard video equipment and associated PC-based frame grabbers makes their use for PIV especially attractive if spatial resolution is not of primary concern. In the following, three schemes capable of providing image pairs of single exposed particle image recordings are briefly described.

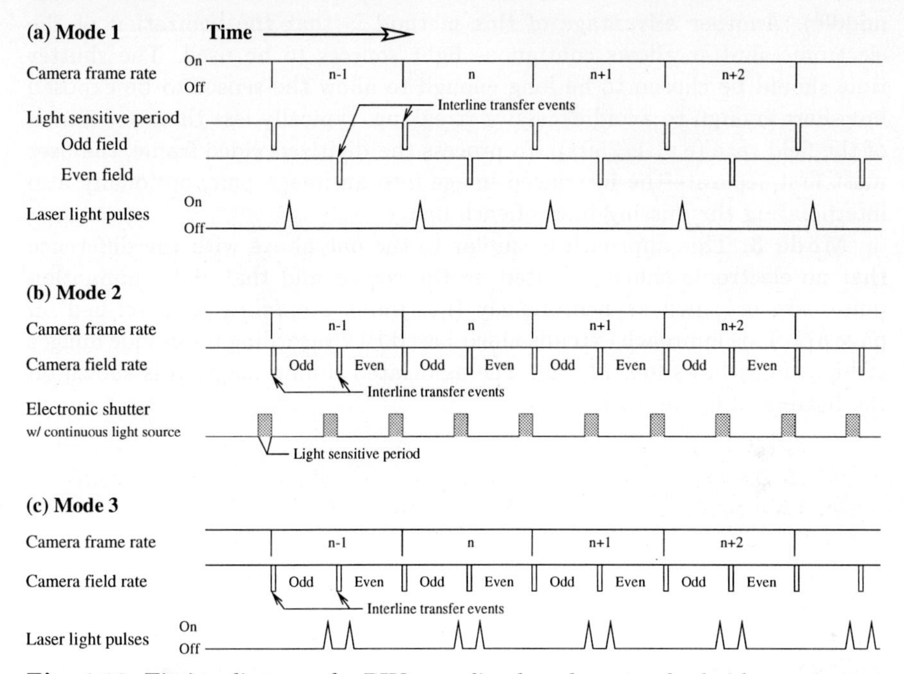

Fig. 4.18. Timing diagrams for PIV recording based on standard video equipment

Mode 1: The first approach can only be applied to rather slow flows ($U_{\mathbf{max}} < 5\,\mathrm{cm/s}$), which makes it useful only in water applications and is essentially equivalent to the original implementation of DPIV described by WILLERT & GHARIB [97]. As shown in the timing diagram in the top of figure 4.18, the laser light is strobed exactly in phase with the frame rate of the camera. If a vertical synchronization pulse is not directly available from the camera it may be obtained using a synchronization stripper. The duration of the light pulse from a continuous wave laser should not exceed one fourth the frame period ($\approx 8\,\mathrm{ms}$) to avoid excessive streaking of the particle images in the recording.

Mode 2: The second method of PIV recording based on video imaging utilizes the electronic shutter frequently available in today's video cameras. However, as mentioned before, these shutters generally only work on a field-by-field basis such that the recorded PIV images will have only half the vertical resolution. The interlacing nature of video can thus provide the user with video frames each containing a PIV image pair. The odd video field (i.e. all odd lines) comprise the first PIV recording, while the even field (i.e. all even lines) correspond to the second PIV recording. The time delay between the recordings and hence the light pulse delay, Δt, is equal to the field interval of 1/50th or 1/60th of a second, which doubles the temporal resolution as well as the maximum recordable fluid velocity compared to **Mode 1** (figure 4.18, middle). Another advantage of this method is that the utilization of the electronic shutter allows continuous light sources to be used. The shutter time should be chosen to be long enough to allow the sensor to be exposed but short enough to avoid excessive streaking, typically less than one fourth of the field rate (e.g. 1/250 s). To process the digitized video frame, the user must first separate the interlaced image into an image pair, optionally also interpolating the missing lines of each image.

Mode 3: This approach is similar to the one above with the difference that no electronic shutter is used on the sensor and that the illumination pulses are provided asynchronously (i.e. *frame straddling* as described on page 87). This approach extends video-based PIV recording to provide images of high-speed flows [60, 71, 89]. The associated timing diagram is shown on the bottom of figure 4.18.

5. Image evaluation methods for PIV

This chapter treats the fundamental techniques for statistical PIV evaluation. In spite of the fact that most realizations of PIV evaluation systems are quite similar – in nearly every case they are based on digitally performed Fourier algorithms – we will also consider optical techniques because they are still important for the classification and understanding of existing set-ups.

In order to extract the displacement information from a PIV recording some sort of interrogation scheme is required. Initially, this interrogation was performed manually on selected images with relatively sparse seeding which allowed the tracking of individual particles [158, 160]. With computers and image processing becoming more commonplace in the laboratory environment it became possible to automate the interrogation process of the particle track images [162, 50, 169, 163]. However, the application of tracking methods, i.e to follow the images of an individual tracer particle from exposure to exposure, is only practicable in the low image density case, see figure 1.5(a). The low image density case often appears in strongly 3D high speed flows (e.g. turbomachinery) where it is not possible to provide enough tracer particles or in two phase flows, where the transport of the particles themselves is investigated. Additionally, the Lagrangian motion of a fluid element can be determined by applying tracking methods [161, 127, 167].

In principle, however, a high data density is required on the PIV vector maps, especially for the comparison of experimental data with the results of numerical calculations. This demand requires a medium concentration of the images of the tracer particles in the PIV recording. (In particular, in air flows it is not possible to achieve a high image density, because beyond a certain level the number of detectable images cannot be increased by further increasing the loads with tracers in the flow [39].) Medium image concentration is characterized by the fact that matching pairs of particle images – due to subsequent illuminations – cannot be detected by visual inspection of the PIV recording, see figure 1.5(b). Hence statistical approaches, which will be described in the next sections, had to be developed. After a statistical evaluation has been performed first, tracking algorithms can be additionally applied in order to achieve subwindow spatial resolution of the measurement: *super resolution* PIV [110]. However, since the extraction of the displacement information from individual particle images requires spatially well resolved recordings of the particle images, those techniques are more appropriate to increase the spatial resolution of photographic PIV recordings than that of digital recordings.

Tracking algorithms have continuously been improved during the past decade. Methods like the application of neural networks [52, 43] seem to be very promising. Thus, for some applications particle tracking might be an interesting alternative to statistical PIV evaluation methods on which we focus in this book. Readers interested in obtaining a survey of tracking methods are referred to the survey paper by GRANT [30], to the lecture notes on *Three-dimensional velocity and vorticity measuring and image analysis techniques*, edited by TH. DRACOS [2], and to the section on low image density PIV in the SPIE milestone series on PIV [4].

5.1 Correlation and Fourier transform

5.1.1 Correlation

The main objective of the statistical evaluation of PIV recordings at medium image density is to determine the displacement between two patterns of particle images, which are stored as a 2D distribution of gray levels. Looking around in other areas of metrology shows that it is common practice in signal analysis to determine, for example, the shift in time between two (nearly) identical time signals by means of correlation techniques. Details about the mathematical principles of the correlation technique, the basic relations for correlated and uncorrelated signals and the application of correlation techniques in the investigation of time signals can be found in many textbooks [8, 22]. The theory of correlation can be extended straightforwardly from the one dimensional (1D time signal) to the two-dimensional (2D gray value distribution) case [9]. In chapter 3 the use of auto- and cross-correlation techniques for statistical PIV evaluation has already been explained. Similarly as for time signals also for a 2D spatial signal $I(x, y)$ its power spectrum $|\hat{I}(r_x, r_y)|^2$ can be determined with r_x, r_y being spatial frequencies in orthogonal directions. The basic theorems for correlation and Fourier transform known from the theory of time signals are also valid for the 2D case (with appropriate modifications) [9].

For the calculation of the auto-correlation function two possibilities exist: either direct numerical calculation or indirectly (numerically or optically), using the Wiener–Kinchin theorem [8, 9]. This theorem states that the Fourier transform of the auto-correlation function R_{I} and the power spectrum $|\hat{I}(r_x, r_y)|^2$ of an intensity field $I(x, y)$ are Fourier transforms of each other.

Figure 5.1 illustrates that the auto-correlation function can either be determined directly in the spatial domain (upper half of the figure) or indirectly by Fourier transform FT (left hand side), multiplication, i.e. the calculation

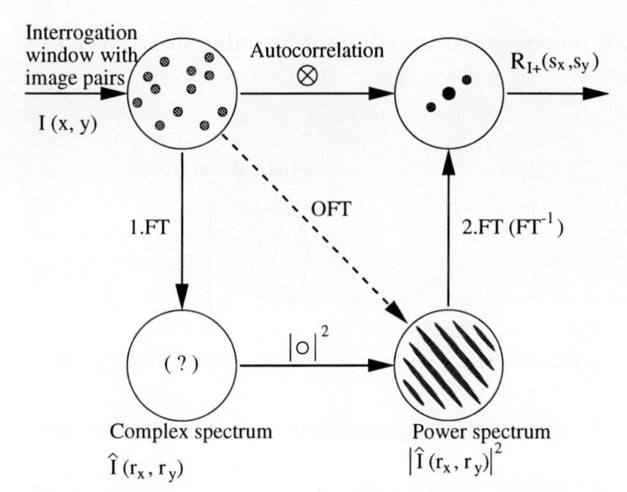

Fig. 5.1. Sketch of relation between 2D correlation function and spatial spectrum by means of the Wiener–Khinchin theorem. FT – Fourier transform, FT^{-1} – inverse Fourier transform, OFT – optical Fourier transform

of the squared modulus, in the frequency plane (lower half of the figure), and by inverse Fourier transform FT^{-1} (right hand side).

5.1.2 Optical Fourier Transform

As already mentioned in section 2.4 the far field diffraction pattern of an aperture transmissivity distribution is represented by its Fourier transform [12, 24, 19]. A lens can be used to transfer the image from the far field close to the aperture. For a mathematical derivation of this result some assumptions have to be made, which are described by the so-called Fraunhofer approximation. These assumptions (large distance between object and image plane, phase factors) can be fulfilled in practical optical set-ups for Fourier transforms.

Figure 5.2 shows two different set-ups for such optical Fourier processors. In the arrangement on the left hand side the object, which would consist of a transparency to be Fourier transformed (e.g. the photographic PIV recording), is placed in front of the so-called Fourier lens. In the second set-up (right hand side) the object is placed behind the lens. As derived in the book of GOODMAN [12] both arrangements differ only by the phase factors of the complex spectrum and a scale factor. Light sensors (photographic plate as well as CCD sensor) are only sensitive to the light intensity. The intensity corresponds to the squared modulus of the complex distribution of the electromagnetic field: phase differences in the light wave cannot be detected. Therefore, both of the arrangements shown in figure 5.2 can be used for PIV evaluation. The result of the optical Fourier transform (OFT, dashed line in figure 5.1) is directly the power spectrum of the gray value distribution of the transparency.

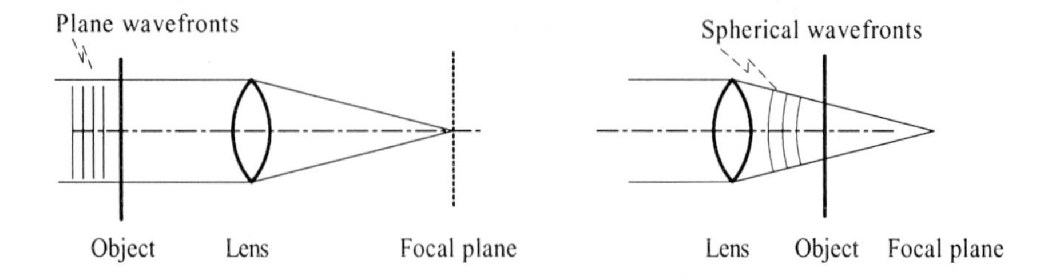

Fig. 5.2. Optical Fourier processor, different positions of object and Fourier lens

In the following this will be illustrated for the case of a pair of two particle images. White (transparent) images of a tracer particle on a black (opaque) background will form a double aperture on the photographic PIV recording. With good lens systems the diameter of an image of a tracer particle on the recording is of the order of 20 to 30 μm. The spacing between the two images of a tracer particle should be approximately 150–250 μm, in order to obtain optimum conditions for optical evaluation (compare section 4.3). Figure 5.3 shows a cross-sectional cut through the diffraction pattern of a double aperture (parameters are similar to those of the PIV experiment). The figure at the left side shows several peaks of the light intensity distribution under an envelope. The envelope represents the diffraction pattern of a single aperture with the same diameter (i.e. the Airy pattern, see section 2.4). The intensity distribution will extend in the 2D presentation in the vertical direction, thus forming a fringe pattern, i.e. the so-called Young's fringes. The fringes are oriented normal to the direction of the displacement of the apertures (tracer images). The displacement between the fringes is inversely proportional to the displacement of the apertures (tracer images). If the distance between the apertures (tracer images) is decreased, the distance between the fringes will increase inversely. This is illustrated in the center of figure 5.3, where the distance between the two apertures is only half that of the example on the left side. It can be seen that the distance between the fringes is increased by a factor of two. The same inverse relation, which is due to the scaling theorem of the Fourier transform, is valid for the envelope of the diffraction pattern: if the diameter of the aperture (particle images) decreases, the extension of the Airy pattern will increase inversely (see figure 5.3, right side). As a consequence, more fringes can be detected in those fringe patterns which are generated by smaller apertures (particle images). This is one possibility to explain why small and well focused particle images will increase the quality and detection probability in the evaluation of PIV recordings. Due to another property of the Fourier transform, i.e. the shift theorem, the characteristic shape of the intensity pattern does not change if the position of the particle image pairs is changed inside the interrogation spot. Increasing the number of particle image pairs also does not change the Young's fringe pattern significantly. Of course this is not true for the case of just two image pairs: two fringe systems of equal intensity will overlap, allowing no unambiguous evaluation.

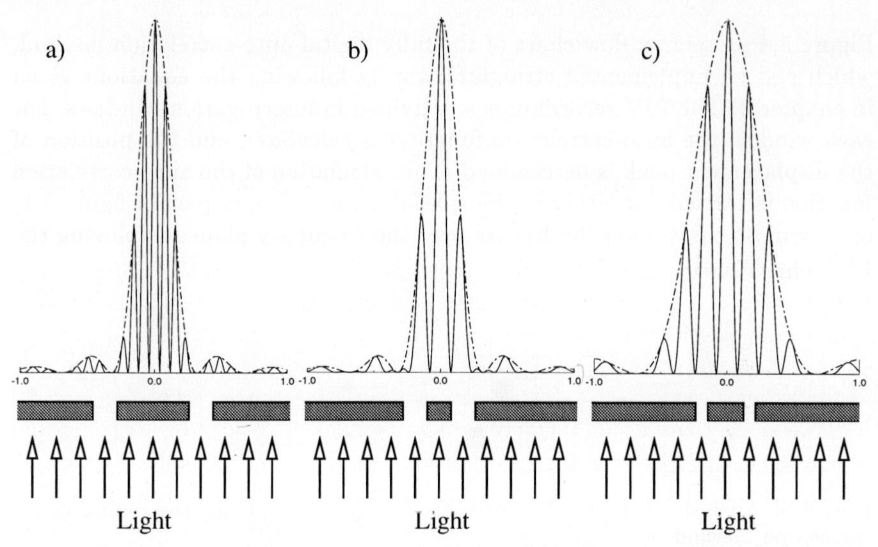

Fig. 5.3. Fraunhofer diffraction pattern of three different double apertures, from left to right, first the separation between the apertures has been decreased, then – on the right hand side – the diameter of the apertures has been decreased

5.1.3 Digital Fourier Transform

The digital Fourier transform is the basic tool of modern signal and image processing. A number of textbooks describe the details [8, 27, 9, ?]. The breakthrough of the digital Fourier transform is due to the development of fast digital computers and to the development of efficient algorithms for its calculation (Fast Fourier Transformation, FFT) [8, 9, 10, 27]. Those aspects of the digital Fourier transform relevant for the understanding of digital PIV evaluation will be described in section 5.4.

5.2 Summary of PIV evaluation methods

In the following the different methods for the evaluation of PIV recordings by means of correlation and Fourier techniques will be summarized.

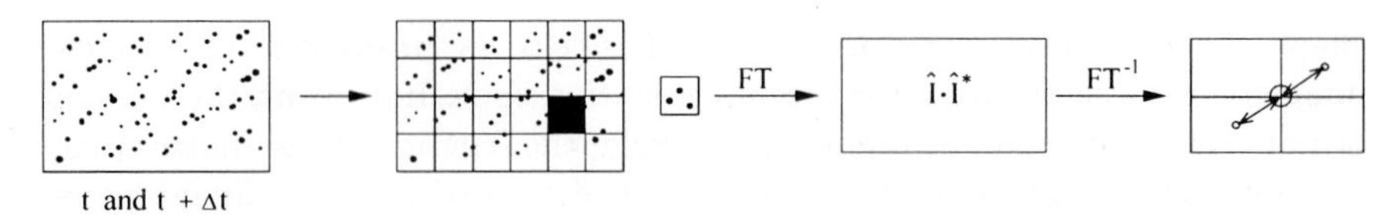

Fig. 5.4. Analysis of single frame/double exposure recordings: the fully digital auto-correlation method

Figure 5.4 presents a flow chart of the fully digital auto-correlation method, which can be implemented straightforwardly following the equations given in chapter 3. The PIV recording is subdivided in interrogation windows. For each window the auto-correlation function is calculated and the position of the displacement peak is determined. The calculation of the auto-correlation function is carried out either in the spatial domain (upper part of figure 5.1) or – in most cases – via the bypass over the frequency plane, employing the FFT algorithms.

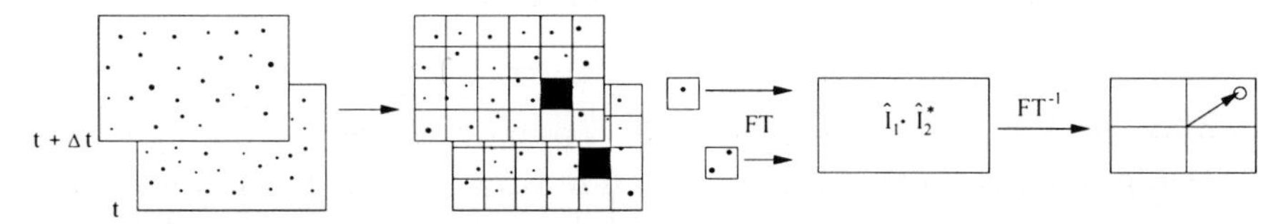

Fig. 5.5. Analysis of double frame/single exposure recordings: the digital cross-correlation method

If the PIV recording system allows the employment of the double frame/single exposure recording technique (see figure 4.2) the evaluation of the PIV recordings is performed by cross-correlation (figure 5.5). In this case, the cross-correlation between two interrogation windows out of the two recordings is calculated. As will be explained later in section 5.4, it is advantageous to set off both windows according to the mean displacement of the tracer particles between the two illuminations. This reduces the in-plane loss of correlation (see section 5.4) and therefore increases the correlation peak strength. The calculation of the cross-correlation function is done numerically with a computer, by means of the FFT algorithms in most cases.

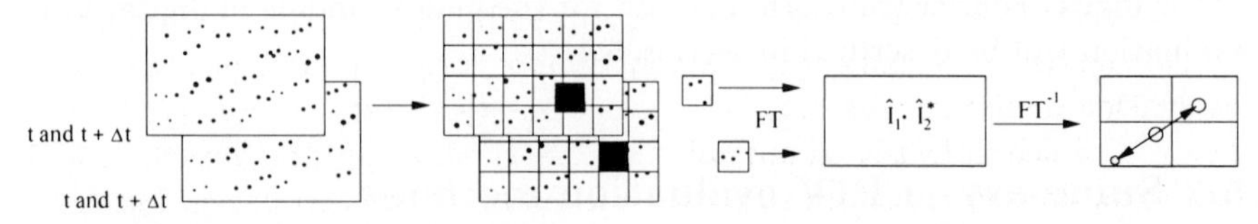

Fig. 5.6. Single frame/double exposure cross-correlation method flow chart

However, in most applications using high resolution imaging it is difficult or impossible to store the two exposures on two different frames. In the case of photographic PIV the recording should be digitized in order to utilize cross-correlation techniques for evaluation (figure 5.6). Then, the interrogation windows can be chosen of different size and/or slightly displaced with respect to each other in order to compensate for the in-plane loss of correlation due to the mean displacement of particle images. Depending on the different parameters auto-correlation peaks will also appear in the correlation plane beside the cross-correlation peak.

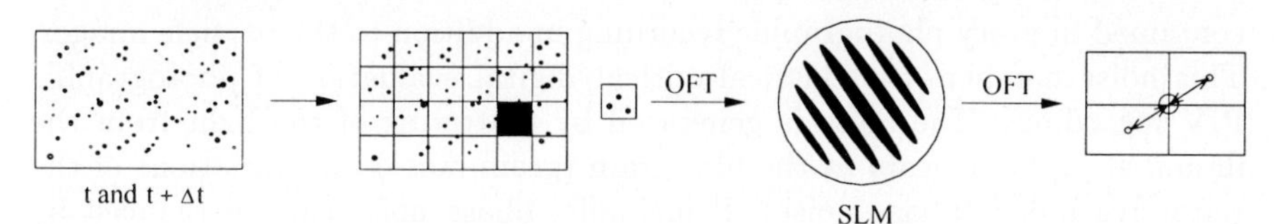

Fig. 5.7. Analysis of single frame/double exposure recordings: the fully optical method

The counterpart of the fully digital evaluation by means of auto-correlation is a system employing optical Fourier transform (OFT) for evaluation. In order to obtain the auto-correlation function a set-up with two optical Fourier processors has to be implemented, following the bypass through the frequency plane as outlined in figure 5.1. A spatial light modulator is required to store the output of the first Fourier processor and to serve as input of the second Fourier processor. This is shown in figure 5.7. Up to now no optical set-ups giving the 2D cross-correlation function for PIV evaluation have been described in literature.

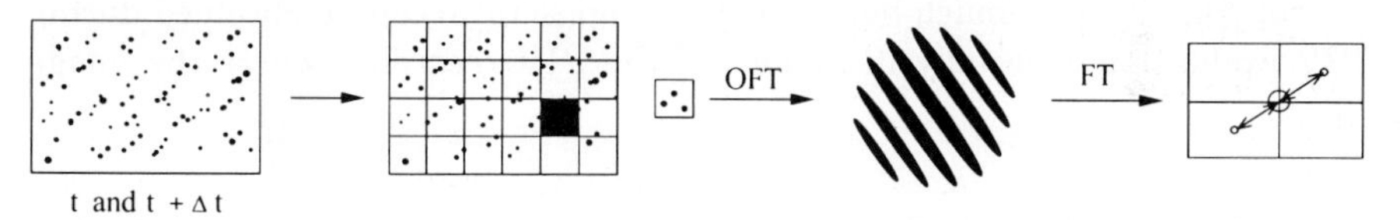

Fig. 5.8. Analysis of single frame/double exposure recordings: the hybrid (optical/digital) method utilizing the Young's fringes technique

Computer memory and computation speed being limited in the beginning of the eighties, PIV work was strongly promoted by the existence of optical evaluation methods. The most widely used method was the Young's fringes method, which in fact is an optical-digital method, employing optical as well as digital Fourier transforms for the calculation of the correlation function. The flow chart of this evaluation method is shown in figure 5.8.

In the next section the fully optical method of PIV evaluation will be treated in order to give an introduction to the problems of this "old-fashioned" technique which still offers some advantages compared to digital techniques. The most commonly used and very flexible digital evaluation methods will be discussed in the sections thereafter in more detail.

5.3 Optical PIV evaluation

In order to achieve high quality in optical PIV evaluation some preprocessing of the recordings is required.

Due to the granular composition of the emulsion, photographic noise is contained in every photographic recording in addition to the particle images. This noise hampers the classical optical/digital evaluation of photographic PIV recordings. The noise is generated by scattering of the light from the illuminating laser beam at the film grain (grain noise) and variations of the refractive index (phase noise). Principally, phase noise can be reduced by immersing the negative in an index matching liquid as reported by PICKERING & HALLIWELL [81]. However, a much better improvement of the Young's fringes visibility and, thus, of the probability of detecting valid velocity data during PIV evaluation can be achieved by a two step photographic process [82]. When interrogating the original PIV negative (images of tracer particles: dark, background: bright) the noise in the Fourier plane where the Young's fringes are formed reaches a considerable level, because the areas on the negative which have the highest transmittance (background) maintain the gross fog (figure 2.33 on page 52). By preparing a contact copy from the negative a positive transparency can be obtained (images of tracer particles: bright, background: dark) which reduces the bias transmittance. This process prevents noise being transferred to the contact copy, which will be employed for evaluation, by taking advantage of the nonlinear behavior of the film used for copying. Thus, a much better signal-to-noise ratio can be obtained during PIV evaluation especially in regions of the PIV recording where the image density is low.

5.3.1 Young's fringes method

An experimental set-up for the implementation of the Young's fringes technique is shown in figure 5.9. In this set-up only the first Fourier transform (compare figure 5.8) is performed optically.

In order to determine the local auto-correlation the input to the first optical Fourier processor is achieved by simply illuminating a small area (i.e. the interrogation spot) of the photographic negative of the PIV exposure with a He-Ne laser light beam. After optical Fourier transform by means of an arrangement already shown on the right side of figure 5.2 the Young's fringe pattern is obtained in the Fourier plane. The light intensity distribution in the Fourier plane is recorded by means of a video camera. Its image is

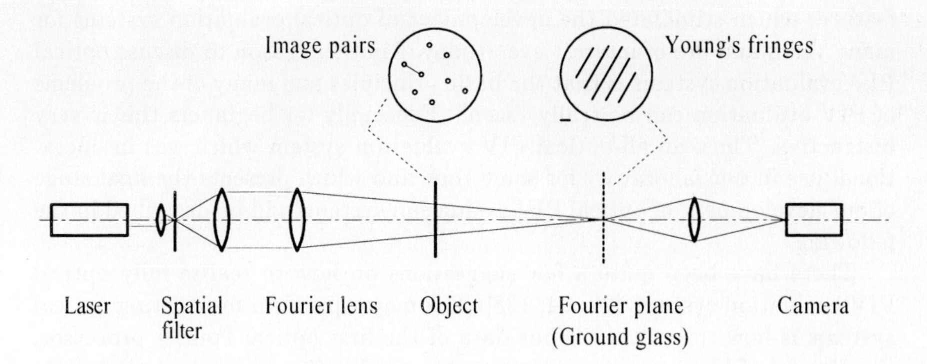

Fig. 5.9. Set-up for PIV evaluation employing the Young's fringes technique

digitized and stored in a computer. As explained the spacing of the fringes is inversely proportional to the displacement of matched image pairs. The direction of the fringes is perpendicular to the direction of the displacement. Thus, if evaluating the distance between the fringes and their direction, the magnitude and velocity of the tracer particles in the flow can be determined. The second, better and more widely used method to evaluate the Young's fringes pattern is to perform a second Fourier transform by means of the FFT algorithm and the peak is found numerically in the computer. The major advantage of this procedure was the increased speed (in the eighties) and the higher accuracy of the optical-digital method as compared to digital-digital methods. Set-ups like the one described here have been widely used in the first ten years of the development of PIV. Details about the different optical evaluation techniques for PIV and the Young's fringes method can be found in [31].

The problem in the development of fully optical evaluation systems was to store the output of the first optical processor (i.e. the Young's fringes) in such a way that it can be used as input to a second optical Fourier processor. Only after the development of easy to use and cheap spatial light modulators (SLM's) was it possible to set up operational fully optical PIV evaluation systems.

5.3.2 Fully optical PIV evaluation

In spite of the nearly exclusive utilization of digital evaluation methods for PIV today, it is still of interest to compare the performance of optical and digital evaluation systems with respect to their accuracy. An optical-optical PIV evaluation system works to the greatest part fully analog. It is not influenced by problems associated with digitization. The calculation of the 2D Fourier transformation is done at the speed of light. The method can easily be adapted to large format photographic films. These are very attractive features which stimulated the development of optical evaluation systems for many years and are of interest even today. Another reason to discuss optical PIV evaluation systems is that the basic principles and many of the problems of PIV evaluation can be really "seen". Especially for beginners this is very instructive. Thus, an all-optical PIV evaluation system which was in operational use in our laboratory for some time and which presents the final stage of our development of optical PIV evaluation systems will be described in the following.

There have been quite a few suggestions on how to realize fully optical PIV evaluation systems [79, 94, 138]. The major problem for existing optical systems is how to get the output data of the first optical Fourier processor, the Young's fringes pattern, into the second Fourier processor. In principle the output of the first processor can be recorded with a video camera and subsequently be fed to a liquid crystal spatial light modulator (LC-SLM) [94, 138]. Optical set-ups with *electronically addressable* liquid crystal light modulators are easy to handle and allow frame rates (data input) of about 100 Hz, but due to the finite size of their pixels (currently about $60 \times 76\,\mu m^2$) their spatial resolution is limited. Furthermore the input data to these devices, the fringes, have to be digitized before they are fed into this kind of light modulator. Another possibility is to write the fringe pattern optically to an appropriate light modulator. The *optically addressable* $Bi_{12}SiO_{20}$ (BSO) spatial light modulators have a higher spatial resolution but at the same time they require some effort for their operation (high voltage, flashlamps for erasing the data, etc.) [79].

The optically working evaluation system presented in this section combines the advantages of the above mentioned techniques by simultaneously avoiding their disadvantages. The system utilizes an optically addressed liquid crystal spatial light modulator (type 300 p/01 of Jenoptik Technologie) capable of frame rates in the order of the video frequency and with a high spatial resolution of 40 lp/mm (in the frequency domain). The device is extremely easy to handle (operating voltage ca. 5 V a.c., frequency of 100–2 000 Hz) so that the entire optical set-up itself is also quite simple and compact.

Set-up for all-optical evaluation. On the left hand side of figure 5.10 the light source, i.e. a He-Ne laser, is located with a subsequent spatial filter which removes the imperfections of the original laser beam. This spatial filter helps to provide a low signal-to-noise ratio in the frequency plane as well as in the auto-correlation plane. Behind the spatial filter the beam is split. The rays passing the beam splitter (the write beam) are focused by the subsequent Fourier lens on to the light modulator. The PIV recording which is to be analyzed is placed between the Fourier lens and Fourier plane, i.e. in the converging part of the beam [12]. (This arrangement allows an adjustable interrogation spot size on the PIV recording.) Thus, the Young's fringes pattern (i.e. the two-dimensional spatial spectrum of the light intensity distribution inside the interrogation spot) is imaged onto the light sensitive layer of the light modulator.

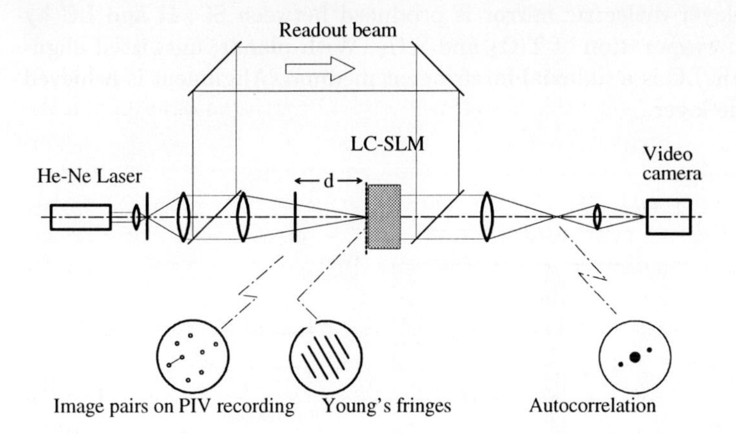

Fig. 5.10. All-optical auto-correlator for the evaluation of photographic PIV recordings

The readout beam is reflected on to the back plane of the SLM by means of two mirrors and a second beam splitter. This beam impinges on to the right hand side of the SLM and is phase modulated by the device according to the Young's fringes pattern. (The working principle of the optically addressable spatial light modulator will be explained below.) Due to a mirror inside the SLM the readout beam is reflected back again. The part of the now modulated readout beam passing the beam splitter is then also Fourier transformed by another Fourier lens thus yielding the two-dimensional spectrum of the Young's fringes pattern, i.e. the auto-correlation of the gray value distribution inside the interrogation spot. The auto-correlation function is recorded by a video camera.

Optically addressable LC spatial light modulator. The optically addressed LC-SLM (see also [170, 171, 172]) is a sandwich system consisting of a photoconductor, a liquid crystal layer, a dielectric mirror, alignment layers, and transparent electrodes on glass substrates (figure 5.11). The alternating voltage applied to the transparent electrodes is divided according to the impedance of the different layers. As light impinges on to the photoconductive material, the conductivity of the latter increases so that the correspondingly increasing voltage drop across the liquid crystal results in a variation of the orientation of the molecules. This SLM may therefore be used to transform two-dimensional intensity distributions into suitable distributions of refractive index on the output side, if the selection and control of suitable photoconductors do not produce any noticeable lateral charge balancing.

In the SLMs Plasma-CVD-deposited Si:H on indium-tin-oxide (ITO)-coated glass substrates are used as a photoconductive layer. To reflect the light, a multilayer dielectric mirror is produced between Si : H and LC by electron-beam evaporation of TiO_2 and SiO_2. With planar, untwisted alignment a nematic LC is a uniaxial birefringent medium. Alignment is achieved by a polyimide layer.

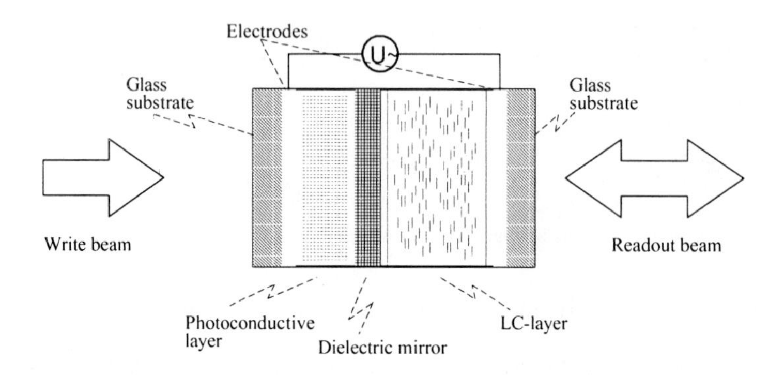

Fig. 5.11. Principle of the optically addressable liquid crystal light modulator

Due to the birefringent properties of the molecules light passing the liquid crystal is phase modulated according to the pattern of the more or less tilted molecules. The write beam with the input pattern is incident from the left on to the device whereas the readout beam falls on to the device from the right hand side. The readout beam passes the liquid crystal and is reflected by the dielectric mirror so that it passes the crystal twice. Leaving the device the readout beam is thus phase modulated corresponding to the input pattern.

Digital part of the evaluation system. After the auto-correlation function has been recorded and digitized the coordinates of the peaks in the auto-correlation plane representing the particle image displacement have still to be determined. This step is done digitally. For peak detection the center of gravity method as discussed in section 5.4 below was employed. With this optical set-up the center of the correlation plane is not known a priori, and must be calibrated (e.g. by taking into account both signal peaks) which is different from the digital correlation.

A detailed quantitative comparison of the performance of optical-digital, fully digital and fully optical evaluation of PIV recordings is given in [99]. The results can be summarized as follows: in principle the lowest measurement uncertainty (i.e. highest precision) can be expected from the fully optical interrogation system as described above. However, the same investigation showed that for applications of PIV in complex flow fields it is advantageous to employ the cross-correlation technique with displaced interrogation windows also in the case of single frame/double exposure (photographic) recordings. Such a feature has not yet been implemented for optical PIV evaluation systems. As far as speed of evaluation is concerned the fully digital method will outperform the optical methods since the latter typically make use of an XY translation stage to move the transparency and a frame grabber to record each of the many thousands of correlation planes. In terms of turn around time (from recording to evaluation) photographical and hence optical methods have another disadvantage: in addition to the time required to develop the film even more time is required to prepare contact copies from the original PIV negatives in order to remove noise (see section 5.3).

Generally speaking, the main advantage of the fully optical method is that it allows the user to detect small velocity fluctuations and very weakly pronounced spatial structures in the flow. An example for an all-optical evaluation of a PIV recording is given in section 8.1.1, figure 8.3.

5.4 Digital PIV evaluation

The complexity of these optical assemblies not only made a thorough understanding of Fourier optics necessary but also required the use of electromechanical parts (e.g. translation stages), digital imaging as well as computer interfaces. With computer speed and storage capacity increasing almost inversely proportional to its cost, it was merely a matter of time before the interrogation could be performed fully digitally. In the case of a photographic recording a desktop slide scanner for its digitization and a personal computer for the subsequent analysis can completely replace the optical interrogation assembly. Recent advances in electronic imaging further allow a replacement of the rather cumbersome photographic recording process. In the following we describe the necessary steps in the fully digital analysis of PIV recordings using statistical methods. The focus initially is on the analysis of single exposed image pairs, that is, single exposure/double frame PIV, by means of cross-correlation. The analysis of multiple exposure/single frame PIV recordings can be viewed as a special case of the former.

5.4.1 Digital spatial correlation in PIV evaluation

Before introducing the cross-correlation method in the evaluation of a PIV image, the task at hand should be defined from the point of view of linear signal or image processing. First of all let us assume we are given a pair of images containing particle images as recorded from a light sheet in a traditional PIV recording geometry. The particles are illuminated stroboscopically so that they do not produce streaks in the images. The second image is recorded a short time later during which the particles will have moved according to the underlying flow (ignoring effects such as particle lag, three-dimensionality, etc. for the moment). Given this pair of images, the most we can hope for is to measure the straight-line displacement of the particle images since the curvature information between the recording instances is lost. (Acceleration information also cannot be obtained from a single image pair.) Further, the seeding density is too homogenous that it is difficult to match up discrete particles. In some cases the spatial translation of groups of particles can be observed. The image pair can yield a field of linear displacement vectors where each vector is formed by analyzing the movement of localized groups of particles. In practice this is accomplished by extracting small samples or interrogation windows and analyzing them statistically (figure 5.12).

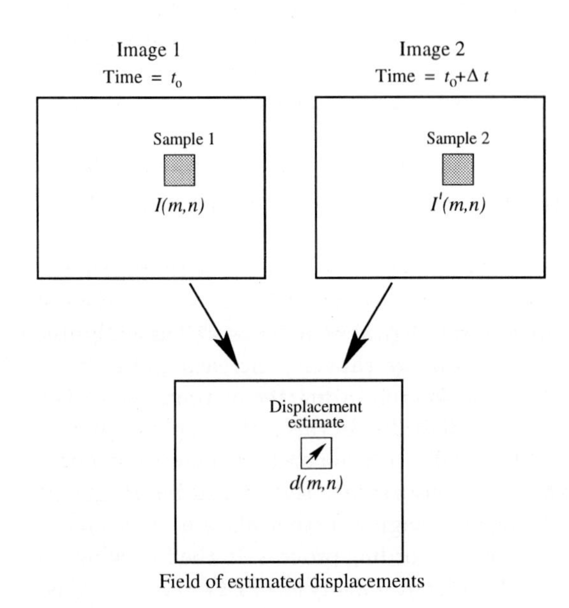

Fig. 5.12. Conceptual arrangement of frame-to-frame image sampling associated with double frame/single exposure particle image velocimetry

From a signal (image) processing point of view, the first image of the pair may be considered the input to a system whose output produces the second image of the pair (figure 5.13). The system's transfer function, $\boldsymbol{H}$, converts the input image I to the output image I' and is comprised of the displacement function $\boldsymbol{d}$ and an additive noise process, $\boldsymbol{N}$. The function of interest is a shift by the vector $\boldsymbol{d}$ as it is responsible for displacing the particle images from one image to the next. This function can be described, for instance, by a convolution with $\delta(\boldsymbol{x} - \boldsymbol{d})$. The additive noise process, $\boldsymbol{N}$, in figure 5.13 models effects due to recording noise and three-dimensional flow among other things. If both $\boldsymbol{d}$ and $\boldsymbol{N}$ are known, it should then be possible to use them as transfer functions for the input image I to produce the output image I'. With both images I and I' known the aim is to estimate the displacement

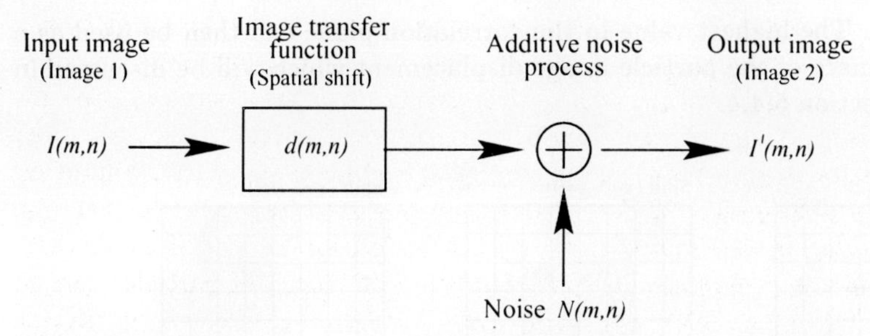

Fig. 5.13. Idealized linear digital signal processing model describing the functional relationship between two successively recorded particle image frames.

field $\boldsymbol{d}$ while excluding the effects of the noise process $\boldsymbol{N}$. The fact that the signals (i.e. images) are not continuous, that is, the dark background cannot provide any displacement information, makes it necessary to estimate the displacement function $\boldsymbol{d}$ using a statistical approach based on localized interrogation windows (or samples).

One possible scheme to recover the local displacement function would be to deconvolve the image pair. In principle this can be accomplished by dividing the respective Fourier transforms by each other. This method works when the noise in the signals is insignificant, however the noise associated with realistic recording conditions quickly degrades the data yield. Also the signal peak is generally too sharp to allow a reliable subpixel estimation of the displacement.

Rather than estimating the displacement function $\boldsymbol{d}$ analytically the method of choice is to locally find the best match between the images in a statistical sense. This is accomplished through the use of the discrete cross-correlation function, whose integral formulation was already described in section 3:

$$R_{\mathrm{II}}(x,y) = \sum_{i=-K}^{K} \sum_{j=-L}^{L} I(i,j) I'(i+x, j+y) \,. \tag{5.1}$$

The variables I and I' are the samples (e.g. intensity values) as extracted from the images where I' is larger than the template I. Essentially the template I is linearly 'shifted' around in the sample I' without extending over edges of I'. For each choice of sample shift (x, y), the sum of the products of all overlapping pixel intensities produces one cross-correlation value $R_{\mathrm{II}}(x, y)$. By applying this operation for a range of shifts $(-M \leq x \leq +M, -N \leq y \leq +N)$ a correlation plane the size of $(2M+1) \times (2N+1)$ is formed. This is shown graphically in figure 5.14. For shift values at which the samples' particle images align with each other, the sum of the products of pixel intensities will be larger than elsewhere, resulting in a high cross-correlation value R_{II} at this position (see also figure 5.15). Essentially the cross-correlation function statistically measures the degree of match between the two samples for a given shift. The highest value in the correlation plane can then be used as a direct estimate of the particle image displacement which will be discussed in detail in section 5.4.4.

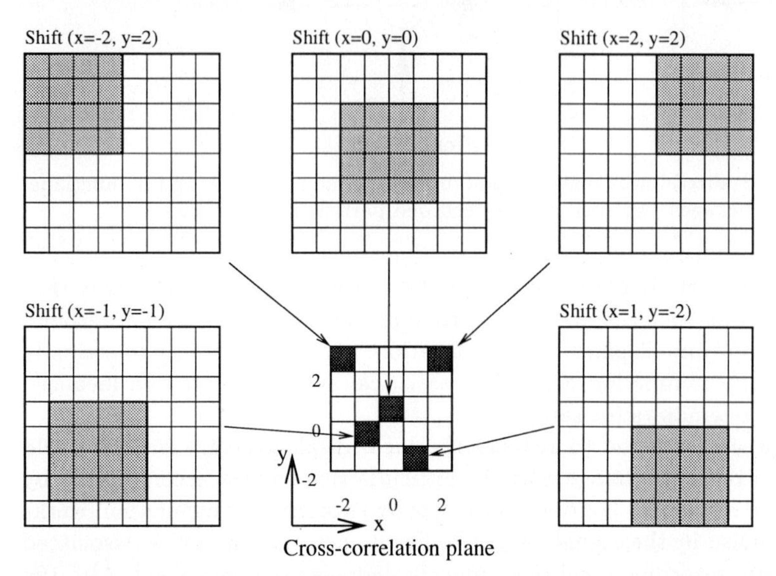

Fig. 5.14. Example of the formation of the correlation plane by direct cross-correlation: here a 4×4 pixel template is correlated with a larger 8×8 pixel sample to produce a 5×5 pixel correlation plane.

Upon examination of this direct implementation of the cross-correlation function two things are obvious: 1) the number of multiplications per correlation value increases in proportion to the interrogation window (or sample) area, and 2) the cross-correlation method inherently recovers linear shifts only. No rotations or deformations can be recovered by this first order method. Therefore, the cross-correlation between two particle image samples will only yield the displacement vector to first order, that is, the average linear shift of the particles within the interrogation window. This means that the interrogation window size should be chosen sufficiently small such that the second order effects (i.e. displacement gradients) can be neglected. In a later section this will be discussed in more detail.

The first observation concerning the quadratic increase in multiplications with sample size imposes a quite substantial computational effort. In a typical PIV interrogation the sampling windows cover on the order of several thousand pixels while the dynamic range in the displacement may be as large as ± 10 to ± 20 pixels which would require up to one million multiplications and summations to form only one correlation plane. Clearly, taking into account that several thousand displacement vectors can be obtained from a single

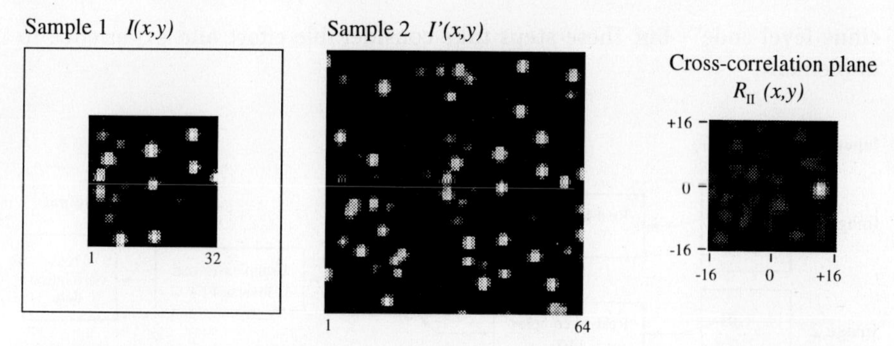

Fig. 5.15. The cross-correlation function R_{II} (right) as computed from real data by correlating a smaller template I (32×32 pixel) with a larger sample I' (64×64 pixel). The mean shift of the particle images is approximately 12 pixels to the right.

PIV recording, a more efficient means of computing the correlation function is required.

Frequency domain based correlation. The alternative to calculating the cross-correlation directly using equation (5.1) is to take advantage of the correlation theorem which states that the cross-correlation of two functions is equivalent to a complex conjugate multiplication of their Fourier transforms:

$$R_{II} \Longleftrightarrow \hat{I} \cdot \hat{I'}^{*} \tag{5.2}$$

where $\hat{I}$ and $\hat{I}'$ are the Fourier transforms of the functions I and I', respectively. In practice the Fourier transform is efficiently implemented for discrete data using the fast Fourier transform or FFT which reduces the computation from $O[N^2]$ operations to $O[N \log_2 N]$ operations [10, 21, ?]. The tedious two-dimensional correlation process of equation (5.1) can be reduced to computing two two-dimensional FFT's on equal sized samples of the image followed by a complex-conjugate multiplication of the resulting Fourier coefficients. These are then inversely Fourier transformed to produce the actual cross-correlation plane which has the same spatial dimensions, $N \times N$, as the two input samples. Compared to $O[N^4]$ for the direct computation of the two-dimensional correlation the process is reduced to $O[N^2 \log_2 N]$ operations. The computational efficiency of this implementation can be increased even further by observing the symmetry properties between real valued functions and their Fourier transform, namely, that the real part of the transform is symmetric: $\mathrm{Re}(\hat{I}_i) = \mathrm{Re}(\hat{I}_{-i})$, while the imaginary part is antisymmetric: $\mathrm{Im}(\hat{I}_i) = -\mathrm{Im}(\hat{I}_{-i})$. In practice two real-to-complex, two-dimensional FFT's and one complex-to-real inverse, two-dimensional FFT are needed, each of which require approximately half the computation time of standard FFT's (figure 5.16). A further increase in computation speed can of course be achieved by optimizing the FFT routines such as using lookup tables for the required data reordering and weighting coefficients or fine tuning the machine level code – but these steps take considerable effort and experience to implement.

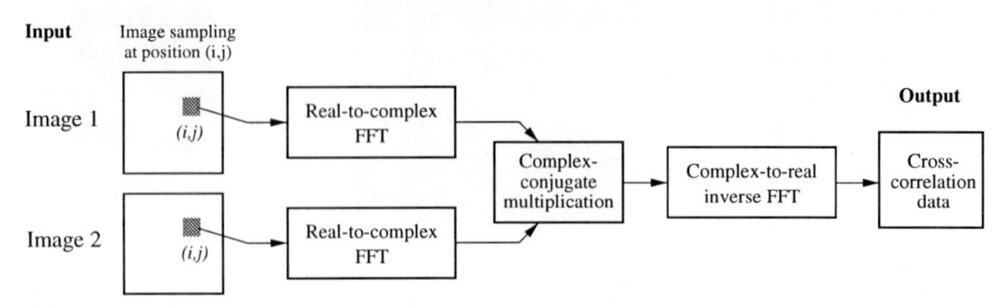

Fig. 5.16. Implementation of cross-correlation using fast Fourier transforms.

The use of two-dimensional FFT's for the computation of the cross-correlation plane has a number of properties whose effects have to be dealt with.

Fixed sample sizes: The FFT's computational efficiency is mainly derived by recursively implementing a symmetry property between the even and odd coefficients of the discrete Fourier transform (*the Danielson–Lanczos lemma* [10, 21]). The most common FFT implementation requires the input data to have a base-2 dimension (i.e. 32×32 pixel or 64×64 pixel samples). While other implementations (radix-4, radix-3, split radix FFT) are possible, their greater complexity does not always justify the additional effort. For reasons explained below it generally is not possible to simply pad a sample with zeroes to make it a base-2 sized sample.

Periodicity of data: By definition, the Fourier transform is an integral (or sum) over a domain extending from negative infinity to positive infinity. In practice however, the integrals (or sums) are computed over finite domains which is justified by assuming the data to be periodic, that is, the signal (i.e. image sample) continually repeats itself in all directions. While for spectral estimation there exist a variety of methods to deal with the associated artifacts, such as windowing, their use in the computation of the cross-correlation will introduce systematic errors or will even hide the correlation signal in noise.

One of these methods, zero padding, which entails extending the sample size to four times the original size by filling in zeroes, will perform poorly because the data (i.e. image sample) generally consists of a nonzero (noisy) background on which the signal (i.e. particle images) is overlaid. The edge discontinuity brought about in the zero padding process contaminates the spectra of the data with high frequency noise which in turn deteriorates the cross-correlation signal. The slightly more advanced technique of FFT data windowing removes the effects of the edge discontinuity, but leads to a nonuniform weighting of the data in the correlation plane and to a bias of the recovered displacement vector. The treatment of this systematic error is described in more detail below.

Aliasing: Since the input data sets to the FFT-based correlation algorithm are assumed to be periodic, the correlation data itself is also periodic. If the data of length N contains a signal (i.e. displacements) exceeding half the sample size $N/2$, then the correlation peak will be folded back into the correlation plane to appear on the opposite side. For a displacement $d_{\mathbf{x},\mathbf{true}} > N/2$, the measured value will be $d_{\mathbf{x},\mathbf{meas.}} = d_{\mathbf{x},\mathbf{true}} - N$. In this case the sampling criterion (Nyquist theorem) has been violated causing the measurement to be *aliased.* The proper solution to this problem is to either increase the interrogation window size, or, if possible, reduce the laser pulse delay, Δt.

Bias error: Another side-effect of the periodicity of the correlation data is that the correlation estimates are biased. With increasing shifts less data are actually correlated with each other since the periodically continued data of the correlation template makes no contribution to the actual correlation value. Values on the edge of the correlation plane are computed from only the overlapping half of the data and should be weighted accordingly. Unless the correlation values are weighted accordingly, the displacement estitmate will be biased to a lower value (figure 5.17). The proper weighting function for this purpose will be described in section 5.4.4.

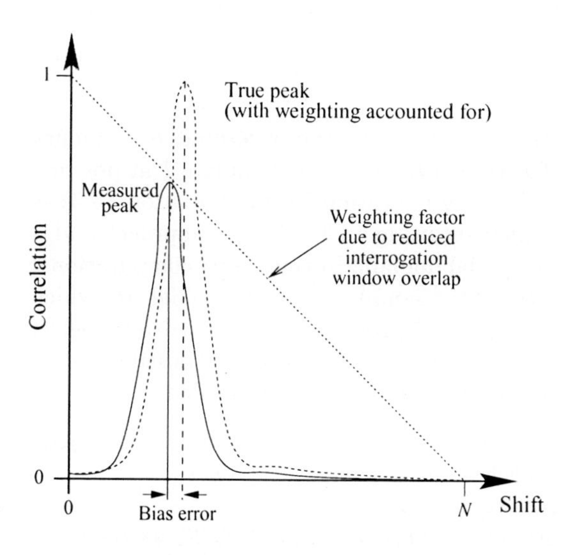

Fig. 5.17. Bias error introduced in the calculation of the cross-correlation using FFT's

If all of the above points are properly handled, an FFT-based interrogation algorithm as shown in figure 5.16 will reliably provide the necessary correlation data from which the displacement data can be retrieved. For the reasons given above, this implementation of the cross-correlation function is sometimes referred to as circular cross-correlation compared to the linear cross-correlation of equation (5.1).

Calculation of the correlation coefficient. For a number of cases it may be useful to quantify the degree of correlation between the two image samples. The standard cross-correlation function equation (5.1) will yield different maximum correlation values for the same degree of matching because the function is not normalized. For instance, samples with many (or brighter) particle images will produce much higher correlation values than interrogation windows with fewer (or weaker) particle images. This makes a comparison of the degree of correlation between the individual interrogation windows impossible. The cross-correlation coefficient function normalizes the cross-correlation function equation (5.1) properly:

$$c_{\mathrm{II}}(x,y) = \frac{\mathrm{C}_{\mathrm{II}}(x,y)}{\sqrt{\sigma_{\mathrm{I}}x,y)}\sqrt{\sigma_{\mathrm{I}}x,y)}} \tag{5.3}$$

where

$$\mathrm{C}_{\mathrm{II}}(x,y) = \sum_{i=0}^{M}\sum_{j=0}^{N}[I(i,j)-\mu_I\,][I'(i+x,j+y)-\mu_{\mathrm{I}'}(x,y)\,] \tag{5.4}$$

$$\sigma_{\mathrm{I}}(x,y) = \sum_{i=0}^{M}\sum_{j=0}^{N}[I(i,j)-\mu_I\,]^2 \tag{5.5}$$

$$\sigma_{\mathrm{I}}(x,y) = \sum_{i=0}^{M}\sum_{j=0}^{N}[I'(i,j)-\mu_{\mathrm{I}'}(x,y)\,]^2\,. \tag{5.6}$$

The value μ_I is the average of the template and is computed only once while $\mu_{\mathrm{I}'}(x,y)$ is the average of I' coincident with the template I at position (x,y). It has to be computed for every position (x,y). Equation (5.3) is considerably more difficult to implement using an FFT-based approach and is usually computed directly in the spatial domain. In spite of its computational complexity, the equation does permit the samples to be of unequal size which can be very useful in matching up small groups of particles. Nevertheless a first order approximation to the proper normalization is possible if the interrogation windows are of equal size and are not zero-padded:

Step 1: Sample the images at the desired locations and compute the mean and standard deviations of each.
Step 2: Subtract the mean from each of the samples.
Step 3: Compute the cross-correlation function using 2D-FFT's as described in figure 5.16.
Step 4: Divide the cross-correlation values by the standard deviations of the original samples. The resulting values will fall in the range $-1 \le c_{\mathrm{II}} \le 1$, that is, they are normalized.
Step 5: Proceed with the correlation peak detection taking into account all artifacts present in FFT-based cross-correlation.

5.4.2 Auto-correlation on doubly exposed PIV images

Although the current trends in technology suggest that the standard PIV recording method will be of the single exposure/multiple frame type, recordings with multiple exposed particle images may still be utilized. This is especially the case when photographic recording is used whose spatial resolution will stay unsurpassed for years to come. In previous sections, optical techniques for extracting the displacement information from the photographs were described. However, desktop slide scanners now also make it possible to digitize the photographic negatives and thus enable a purely digital evaluation. Alternatively, high resolution, single frame CCD sensors can directly provide multiple exposed digital PIV recordings.

Essentially the same components utilized in the digital evaluation of PIV image pairs described before can be used with minor modifications to extract the displacement field from multiple exposed recordings. The major difference between the evaluation modes arises from the fact that all information is contained on a single frame – in the trivial case a single sample is extracted from the image for the displacement estimation (figure 5.18, case I). From this sample the auto-correlation function is computed by the same FFT method described earlier. In effect the auto-correlation can be considered to be a special case of the cross-correlation where both samples are identical. Unlike the cross-correlation function computed from different samples the auto-correlation function will always have a self-correlation peak located at the origin (see also the mathematical description in section 3.5). Located symmetrically around the self-correlation peak two peaks with less than one fourth the intensity describe the mean displacement of the particle images in the interrogation area. The two peaks arise as a direct consequence of the directional ambiguity in the double (or multiple) exposed/single frame recording method.

To extract the displacement information in the auto-correlation function the peak detection algorithm has to ignore the self-correlation peak, R_P, located at the origin, and concentrate on the two displacement peaks, R_{D+} and R_{D-}. If a preferential displacement direction exists, either from the nature of the flow or through the application of displacement biasing methods (e.g. image shifting), then the general search area for the peak detection can be predefined. Alternatively a given number of peak locations can be saved from which the correct displacement information can be extracted using a global histogram operator (section 6.1.1).

The digital evaluation of multiple exposed PIV recordings can be significantly improved by sampling the image at two positions which are offset with respect to each other according to the mean displacement vector. This has the advantage of increasing the number of paired particle images while decreasing the number of unpaired particle images. This minimization of the in-plane loss-of-pairs increases the signal-to-noise ratio, and hence detection of the principle displacement peak R_{D+}. However, the interrogation window

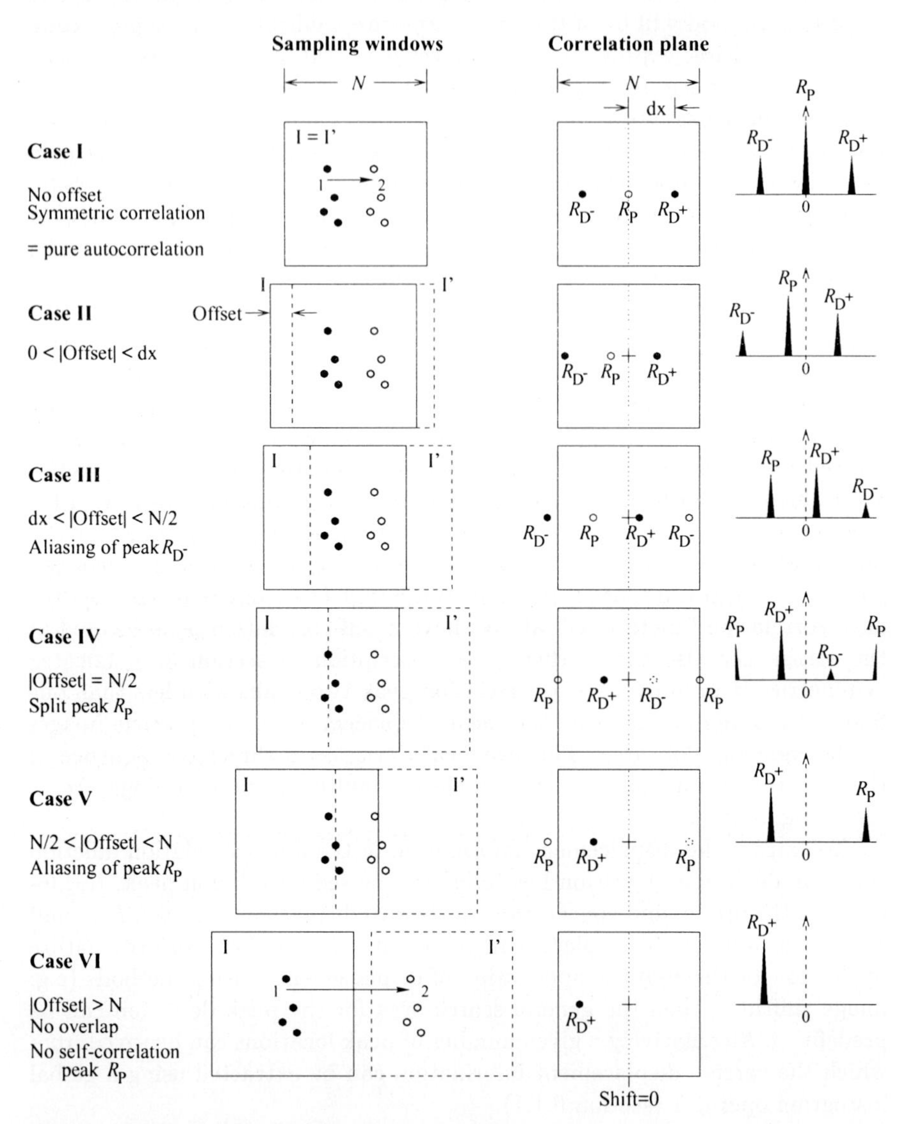

Fig. 5.18. The effect of interrogation window offset on the position of the correlation peaks using FFT based cross-correlation on double exposed images. R_{D+} marks the displacement correlation peak of interest, R_P is the self-correlation peak. In this case a horizontal shift is assumed

offset also shifts the location of the self-correlation peak, R_{P}, away from the origin as illustrated in figure 5.18 (Case II–Case V).

The use of FFT's for the calculation of the correlation plane introduces a few additional aliasing artifacts that have to be dealt with. As the offset of the interrogation window is increased, first the negative correlation peak, R_{D^-}, and then the self-correlation peak, R_{P}, will be aliased, that is, folded back into the correlation plane (figure 5.18 Case III – Case V). In practice, detection of the two strongest correlation peaks by the procedure described in section 5.4.4 is generally sufficient to recover both the positive displacement peak, R_{D^+}, and the self-correlation, R_{P}. The algorithm can be designed to automatically detect the self-correlation peak because it generally falls within a one pixel radius of the interrogation window offset vector.

5.4.3 Advanced digital interrogation techniques

The data yield in the interrogation process can be significantly increased by using a window offset equal to the local integer displacement in a second interrogation pass [96]. By offsetting the interrogation windows the fraction of matched particle images to unmatched particle images is increased, thereby increasing the signal-to-noise ratio of the correlation peak (see section 5.5.3). Also, the measurement noise or uncertainty in the displacement, ϵ, reduces significantly when the particle image displacement is less than half a pixel (i.e. $|\boldsymbol{d}| < 0.5\,\text{pixel}$) where it scales proportional to the displacement [96]. The interrogation window offset can be relatively easily implemented in an existing digital interrogation software for both single exposure/double frame PIV recordings or multiple exposure/single frame PIV recordings described in the previous section. The interrogation procedure could take the following form:

Step 1: Perform a standard digital interrogation with an interrogation window offset close to the mean displacement in the data.

Step 2: Scan the data for outliers using a predefined validation criterion as described in section 6.1. Replace outlier data by interpolating from the valid neighbors.

Step 3: Use the displacement estimates to adjust the interrogation window offset locally to the nearest integer.

Step 4: Repeat the interrogation until the integer offset vectors converge to ± 1 pixel. Typically three passes are required.

The speed of this multiple pass interrogation can be significantly increased by comparing the new integer window offset to the previous value allowing unnecessary correlation calculations to be skipped. The data yield can be further increased by limiting the correlation peak search area in the last interrogation pass.

The multiple pass interrogation algorithm can be further improved by using a hierarchical approach in which the sampling grid is continually refined while the interrogation window size is reduced simultaneously. This procedure has the added capability of successfully applying interrogation window sizes smaller than the particle image displacement. Effectively the dynamic spatial range can be increased by this procedure. This is especially useful in PIV recordings with both a high image density and a high dynamic range in the displacements. In such cases standard evaluation schemes cannot use small interrogation windows without losing the correlation signal due to the larger displacements. However, a hierarchical grid refinement algorithm is more difficult to implement than a standard interrogation algorithm. Such an algorithm may look as follows:

Step 1: Desample the image by consolidating neighboring pixels. For instance, place the sum of 3×3 pixel blocks into one pixel for a desampling factor of 3.

Step 2: Perform a standard interrogation using little or no interrogation window overlap.

Step 3: Scan for outliers and replace by interpolation. As the recovered displacements serve only as estimates for the next higher resolution level, the outlier detection criterion can be more stringent than usual. Data smoothing may also be useful.

Step 4: Project the estimated displacement data on to the next higher resolution level. Use this displacement data to offset the interrogation windows with respect to each other.

Step 5: Increment the resolution level and repeat steps **1** through **4** until the standard image resolution is reached.

Step 6: Finally perform an interrogation at the desired interrogation window size and sampling distance (without outlier removal and smoothing). By limiting the search area for the correlation peak the final data yield may be further increased.

In the final interrogation pass the window offset vectors have generally converged to ± 1 pixel of the measured displacement thereby guaranteeing that the PIV image was optimally evaluated. The choice for the final interrogation window size depends on the particle image density. Below a certain number of matched pairs in the interrogation area (typically $\mathcal{N}_{\mathrm{I}} < 4$) the detection rate will rapidly decrease (see section 5.5.4). Figure 5.19 shows the displacement data of each step of the grid and interrogation refinement.

The spatial resolution in PIV evaluation can be even further increased by eventually tracking the individual particle images, a procedure referred to as *super-resolution PIV* by KEANE et al. [110]. A similar procedure was also implemented by COWEN & MONISMITH for the study of a flat plate turbulent boundary layer [45].

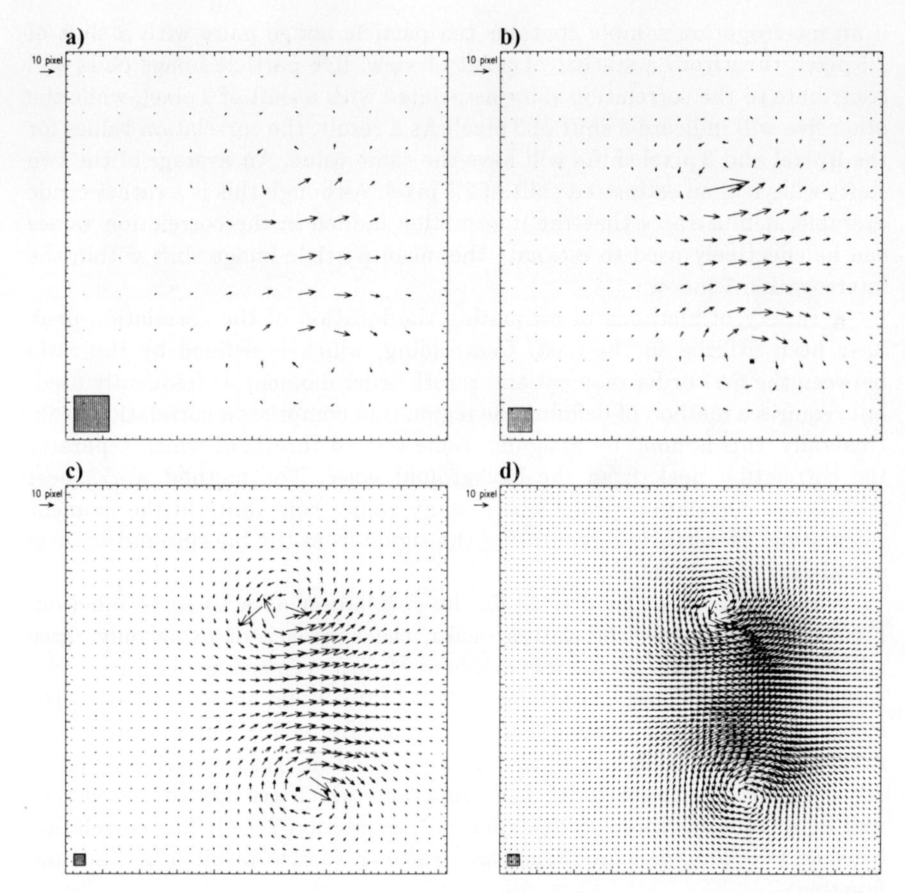

Fig. 5.19. Iteration steps used in a multiple pass, multi-grid interrogation process. The gray squares in the lower left of each data set indicate the size of the utilized interrogation window. In the first pass the original image was desampled by a factor of three

5.4.4 Peak detection and displacement estimation

One of the most crucial, but not necessary easily understood features, of digital PIV evaluation, is that the position of the correlation peak can be measured to subpixel accuracy. Estimation accuracies on the order of 1/10th to 1/20th of a pixel are realistic with 8-bit digital imaging. Simulation data such as those presented in section 5.5 can be used to quantify the achievable accuracy for a given imaging set-up.

Since the input data itself is discretized, the correlation values exist only for integral shifts. The highest correlation value then permits the displacement to be determined with an uncertainty of $\pm 1/2$ pixel. However, with the cross-correlation function being a statistical measure of best match, the correlation values themselves also contain useful information. For example, if an interrogation sample contains ten particle image pairs with a shift of 2.5 pixel, then from a statistical point of view, five particle image pairs will contribute to the correlation value associated with a shift of 2 pixel, while the other five will indicate a shift of 3 pixel. As a result, the correlation values for the 2 pixel and 3 pixel shifts will have the same value. An average of the two shifts will yield an estimated shift of 2.5 pixel. Although this is a rather crude example, it illustrates that the information hidden in the correlation values can be effectively used to estimate the mean particle image shift within the interrogation window.

A variety of methods of estimating the location of the correlation peak have been utilized in the past. Centroiding, which is defined by the ratio between the first order moment and zeroth order moment, is frequently used, but requires a method of defining the region that comprises a correlation peak. Generally, this is done by assigning some sort of threshold which separates the correlation peak from the background noise. The method works best with broad correlation peaks where many values contribute in the moment calculation. Nevertheless separating the signal from the background noise is not always unambiguous.

A more robust method is to fit the correlation data to some function. Especially for narrow correlation peaks, the approach of using only three adjoining values to estimate a component of displacement has become widespread. The most common of these so-called three-point estimators are listed in table 5.1, with the Gaussian peak fit most frequently implemented. The reasonable explanation for this is that the particle images themselves, if properly focused, describe Airy intensity functions which are approximated very well by a Gaussian intensity distribution (see section 2.4.1). The correlation between two Gaussian functions can be shown to also result in a Gaussian function.

The three-point estimators typically work best for rather narrow correlation peaks formed from particle images in the 2–3 pixel diameter range. Simulations such as those shown in figure 5.23 indicate that for larger particle images the achievable measurement uncertainty increases which can be explained by the fact that, while the noise level on each correlation value stays nearly the same, the differences between the three adjoining correlation values are too small to provide reliable shift estimates. In other words, the noise level becomes increasingly significant while the differences between the neighboring correlation values decrease. In this case, a centroiding approach which makes use of more values around the peak will provide more reliable data. If in turn, the particle images are too small ($d_\tau < 1.5$ pixel), the three-point estimators will also perform poorly, mainly because the values adjoining the peak are hidden in noise.

Described here is the use and implementation of the three-point estimators, which were used for almost all the data sets presented as examples in this book. The following procedure can be used to detect a correlation peak and obtain a subpixel accurate displacement estimate of its location:

Step 1: Scan the correlation plane $R = R_{\mathrm{II}}$ for the maximum correlation value $R_{(i,j)}$ and store its integer coordinates (i,j).

Step 2: Extract the adjoining four correlation values: $R_{(i-1,j)}$, $R_{(i+1,j)}$, $R_{(i,j-1)}$ and $R_{(i,j+1)}$.

Step 3: Use three points in each direction to apply the three point estimator, generally a Gaussian curve. The formulas for each function are given in table 5.1.

Table 5.1. Three-point estimators for determining the displacement from the correlation data at the subpixel level

Fitting function	Estimators
Peak centroid $f(x) = \frac{\text{first order moment}}{\text{zero order moment}}$	$x_0 = \frac{(i-1)R_{(i-1,j)} + iR_{(i,j)} + (i+1)R_{(i+1,j)}}{R_{(i-1,j)} + R_{(i,j)} + R_{(i+1,j)}}$ $y_0 = \frac{(j-1)R_{(i,j-1)} + jR_{(i,j)} + (j+1)R_{(i,j+1)}}{R_{(i,j-1)} + R_{(i,j)} + R_{(i,j+1)}}$
Parabolic peak fit $f(x) = Ax^2 + Bx + C$	$x_0 = i + \frac{R_{(i-1,j)} - R_{(i+1,j)}}{2\,R_{(i-1,j)} - 4\,R_{(i,j)} + 2\,R_{(i+1,j)}}$ $y_0 = j + \frac{R_{(i,j-1)} - R_{(i,j+1)}}{2\,R_{(i,j-1)} - 4\,R_{(i,j)} + 2\,R_{(i,j+1)}}$
Gaussian peak fit $f(x) = C \exp\left[\frac{-(x_0-x)^2}{k}\right]$	$x_0 = i + \frac{\ln R_{(i-1,j)} - \ln R_{(i+1,j)}}{2\ln R_{(i-1,j)} - 4\ln R_{(i,j)} + 2\ln R_{(i+1,j)}}$ $y_0 = j + \frac{\ln R_{(i,j-1)} - \ln R_{(i,j+1)}}{2\ln R_{(i,j-1)} - 4\ln R_{(i,j)} + 2\ln R_{(i,j+1)}}$

Multiple peak detection. To detect a given number of peaks, n, within the same correlation plane, a different search algorithm is necessary which sorts out only the highest peaks. In this case it is necessary to extract local maxima based on some sort of neighborhood comparison. This procedure is especially useful for correlation data obtained from single frame/multiple exposed PIV recordings. Also, multiple peak information is useful in cases where the strongest peak is associated with an outlier vector. An easily implemented recipe based on looking at the adjoining five or nine (3×3) correlation values is given here:

Step 1: Allocate a list to store the pixel coordinates and values of the n highest correlation peaks.

Step 2: Scan through the correlation plane and look for values which define a local maximum based on the local neighborhood, that is, the adjoining 4 or 8 correlation values.

Step 3: If a detected maximum can be placed into the list, reshuffle the list accordingly, such that the detected peaks are sorted in the order of intensity. Continue with **Step 2** until the scan through the correlation plane has been completed.

Step 4: Apply the desired three-point peak estimators of table 5.1 for each of the detected n highest correlation peaks, thereby providing n displacement estimates.

Displacement peak estimation in FFT-based correlation data. As already described in section 5.4.1 the assumption of periodicity of both the data samples and resulting correlation plane brings in a variety of artifacts that need to be dealt with properly.

The most important of these is that the correlation plane, due to the method of calculation, does not contain unbiased correlation values, and results in the displacement to be biased to lower magnitudes (i.e. bias error, page 123). This displacement bias can be determined easily by convolving the sampling weighting functions, generally unity for the size of the interrogation windows, with each other. For example, the circular cross-correlation between two equal sized uniformly weighted interrogation windows results in a triangular weighting distribution in the correlation plane. This is illustrated in figure 5.20 for the one-dimensional case.

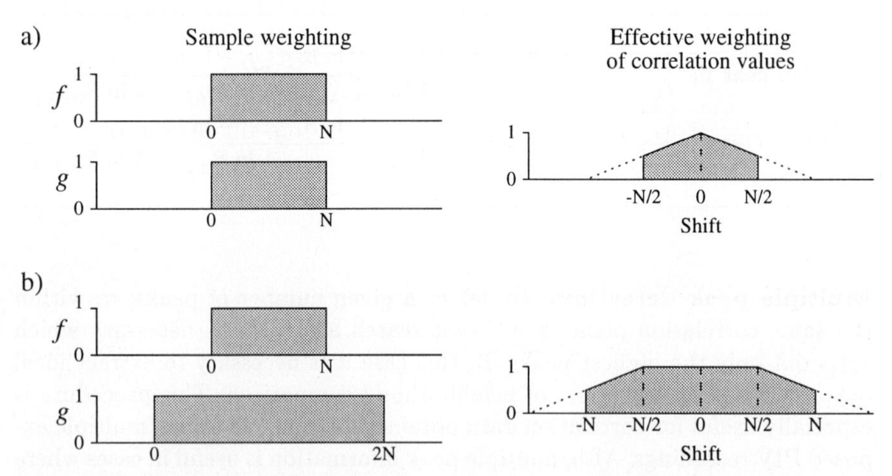

Fig. 5.20. Effective correlation value weighting in FFT-based 'circular' cross-correlation calculation: (a) for interrogation windows of equal size, and (b) for interrogation windows of unequal size (using zero-padding)

The central correlation value will always have unity weight. For a shift value of $N/2$ only half the interrogation windows' data actually contribute to the correlation value such that it carries only a weight of $1/2$. When a three-point estimator is applied to the data, the correlation value closer to the origin is weighted more than the value further out and hence the magnitude of the estimated displacement will be too small. The solution to this problem is very straightforward: before applying the three-point estimator, the correlation values R_{II} have to be adjusted by dividing out the corresponding weighting factors. The weight factors can be obtained by convolving the image sampling function with itself – generally a unity weight, rectangular function – as illustrated in figure 5.20 (a). In the case where the two interrogation windows are of unequal size a convolution between these two sampling functions will yield a weighting function with unity weighting near the center (figure 5.20 b). The extension of the method to nonuniform interrogation windows is of course also possible.

On a related note it should be mentioned that many FFT implementations result in the output data to be shuffled. Oftentimes the DC-component is found at index (0) with increasing frequencies up to index $(N/2-1)$. The next index, $(N/2)$, is actually both the highest positive frequency and highest negative frequency. The following indices represent the negative frequencies in descending order such that index $(N-1)$ is the lowest negative frequency component. By periodicity, the DC component reappears at index (N). In order to achieve a frequency spectrum with the DC component in the middle, the entire data set has to be rotated by $(N/2)$ indices. As illustrated in figure 5.21 two-dimensional FFT-data has to be unfolded in a corresponding manner.

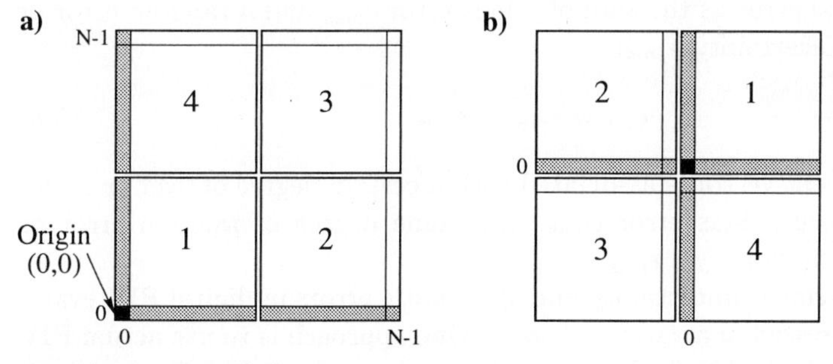

Fig. 5.21. Spatially folded output from a two-dimensional FFT routine (a) requires unfolding to the place the origin back at the center of the correlation plance (b)

For correlation planes calculated by means of a two-dimensional FFT, the zero-shift value (i.e. origin) would initially appear in the lower left corner which would make a similar unfolding of the resulting (periodic) correlation data necessary. Without unfolding, negative displacement peaks will actually appear on the opposite side. However, a careful implementation of the peak finding algorithm allows proper peak detection and shift estimation without having to unscramble to the correlation plane first.

5.5 Measurement noise and accuracy

The overall measurement accuracy in PIV is a combination of a variety of aspects extending from the recording process all the way to the methods of evaluation. This section is devoted to analyzing the contributing factors in the digital evaluation of the PIV record. The absolute measurement error in the estimation of a single displacement vector, ϵ_{tot}, can be decomposed into a group of systematic errors, ϵ_{sys}, and a group of residual errors, $\epsilon_{\mathrm{resid}}$:

$$\epsilon_{\mathrm{tot}} = \epsilon_{\mathrm{sys}} + \epsilon_{\mathrm{resid}} \ . \tag{5.7}$$

The systematic errors comprise all errors which arise due to the inadequacy of the statistical method of cross-correlation in the evaluation of a PIV record, such as its application in gradient regions or the use of an inappropriate subpixel peak estimator. The nature of these errors is that they follow a consistent trend which makes them predictable. By choosing a different analysis method or modifying an existing one to suit the specific PIV recording, the systematic errors can be reduced or even removed.

The second type of errors, the residual errors, remain in the form of a measurement uncertainty even when all systematic errors have been removed. In practice, however, it is not always possible to completely separate the systematic errors, ϵ_{sys}, from the residual errors, $\epsilon_{\mathrm{resid}}$, such that we choose to express the total error as the sum of a bias error ϵ_{bias} and a random error or measurement uncertainty, ϵ_{rms}:

$$\epsilon_{\mathrm{tot}} = \epsilon_{\mathrm{bias}} + \epsilon_{\mathrm{rms}} \ . \tag{5.8}$$

Each displacement vector is associated with a certain degree of over or under estimation, hence a bias error ϵ_{bias}, and some degree of random error or measurement uncertainty $\pm\epsilon_{\mathrm{rms}}$.

The measurement uncertainty and systematic errors in digital PIV evaluation can be assessed in a variety of ways. One approach is to use actual PIV recordings for which the displacement data is known reliably. For instance, PIV recordings obtained of a static (quiescent) flow were used in determining the measurement uncertainty in the cross-correlation evaluation of single exposed particle image pairs [97, 98] as well as double exposed single images [99]. Although this approach is likely to provide the most realistic estimate for the measurement uncertainty, it only permits a limited study of how specific parameters, such as particle image diameter and background noise, influence the measurement precision.

An alternative approach to assessing the measurement precision in PIV evaluation is based in numerical simulation which is an approach taken by a number of researchers [45, 64, 65, 66, 110, 7, 99]. By varying only a single parameter at a time, artificial particle image recordings of known content can be generated, evaluated and compared with the known result. Random positioning of particle images and a high number of simulations (O[1000]) per choice of parameters are crucial in providing reliable measurement precision estimates. The predictions of these Monte Carlo simulations can then be compared with either theory [7, 96] or existing data sets [99]. In the following sections the methodology for Monte Carlo simulation in the assessment of the measurement uncertainty of PIV along with a few important results will be explained.

5.5.1 Artificial particle image generation

The core of the Monte Carlo based measurement error estimation in digital PIV evaluation lies in the generation of adequate particle image recordings. The particle image generator has to fulfil the requirement of providing artificial images with known characteristics: diameter, shape, dynamic range, spatial density and image depth, among others. For most of the simulations presented here, the individual particle images are described by a Gaussian intensity profile:

$$I(x, y) = I_o \exp\left[\frac{-(x - x_o)^2 - (y - y_o)^2}{(1/8)\, d_\tau^2}\right] \tag{5.9}$$

where the center of the particle image is located at (x_o, y_o) with a peak intensity of I_o. For simplicity the magnification factor between object plane and image plane is chosen to be unity, such that $(x, y) \equiv (X, Y)$. The particle image diameter, d_τ, is defined by the e^{-2} intensity value of the Gaussian bell which by definition contains 95% of the scattered light. When the particle image diameter is reduced to zero, the particle images will be represented as delta functions. The factor I_o is a function of the particle's position, Z_o, within the light sheet and the efficiency q with which the particle scatters the incident light. For a light sheet centered at $Z = 0$ with a Gaussian intensity profile, typical of continuous wave argon-ion lasers, I_o could be expressed as:

$$I_o(Z) = q \exp\left[-\frac{Z^2}{(1/8)\, \Delta Z_0^2}\right] \tag{5.10}$$

where ΔZ_0 is the thickness of the light sheet measured at the e^{-2} intensity waist points. Also it is assumed that the particle diameter is much smaller than the light sheet thickness, ΔZ_0. For a top-hat intensity profile the expression for I_o would take the following form:

$$I_o(Z) = q \cdot \begin{cases} 1 & \text{if } |Z| \leq \frac{1}{2}\Delta Z_0 \\ 0 & \text{otherwise}\,. \end{cases} \tag{5.11}$$

To generate a particle image a random number generator specifies the particle's position (X_1, Y_1, Z_1) within a three-dimensional slab containing the light sheet (figure 5.22). The peak intensity $I_o(Z_1)$ is estimated using either equation (5.10), equation (5.11) or any other intensity profile. This value is then substituted into equation (5.9) for the calculation of the light captured by each pixel. Here the integration of equation (5.9) across each pixel can be greatly simplified by computing the product of the error functions (closed form integral of the Gaussian function) along both X and Y. To generate a displacement an artificial flow then moves the particle location to a new position (X_2, Y_2, Z_2) for which a new particle image intensity distribution is calculated. This operation is repeated until a desired particle image density $\mathcal{N}$ is reached. The image is then quantized to the desired image depth (i.e. bits per pixel) and noise may be added to simulate the sensor's background noise.

The next sections illustrate the use of Monte Carlo simulation to test which parameters affect the measurement uncertainty, that is the random error, in digital PIV evaluation. The aim here is not to predict the measurement uncertainty or bias error for a specific set of parameters. Rather, the behavior of these errors with respect to the variation of a given parameter is to be illustrated.

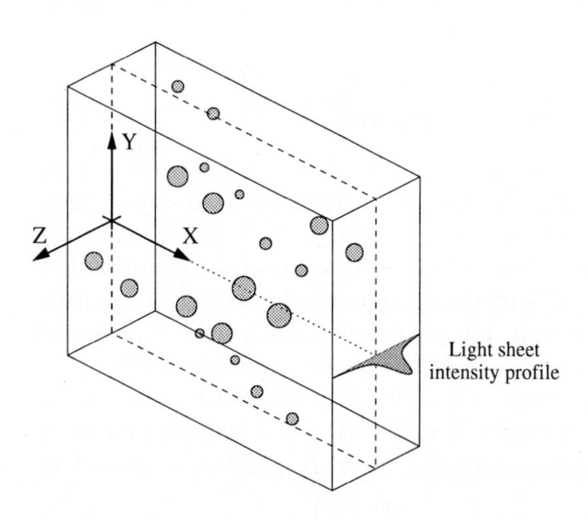

Fig. 5.22. Three-dimensional volume containing a light sheet and particles used in the generation of artificial particle images

5.5.2 Optimization of particle image diameter

Figure 5.23 (a) and (b) predict the existence of an optimum particle image diameter for digital PIV evaluation employing three-point Gaussian peak approximators. For the cross-correlation between two images this diameter is slightly more than 2.0 pixel, while double exposed PIV recordings have an ideal particle image diameter of $d_\tau \approx 1.5$ pixel. Although the same software modules for particle image generation and evaluation (FFT's, peak finder, etc.) were used for both Monte Carlo simulations, there is a discrepancy in the optimum particle image diameter for which no plausible explanation is known.

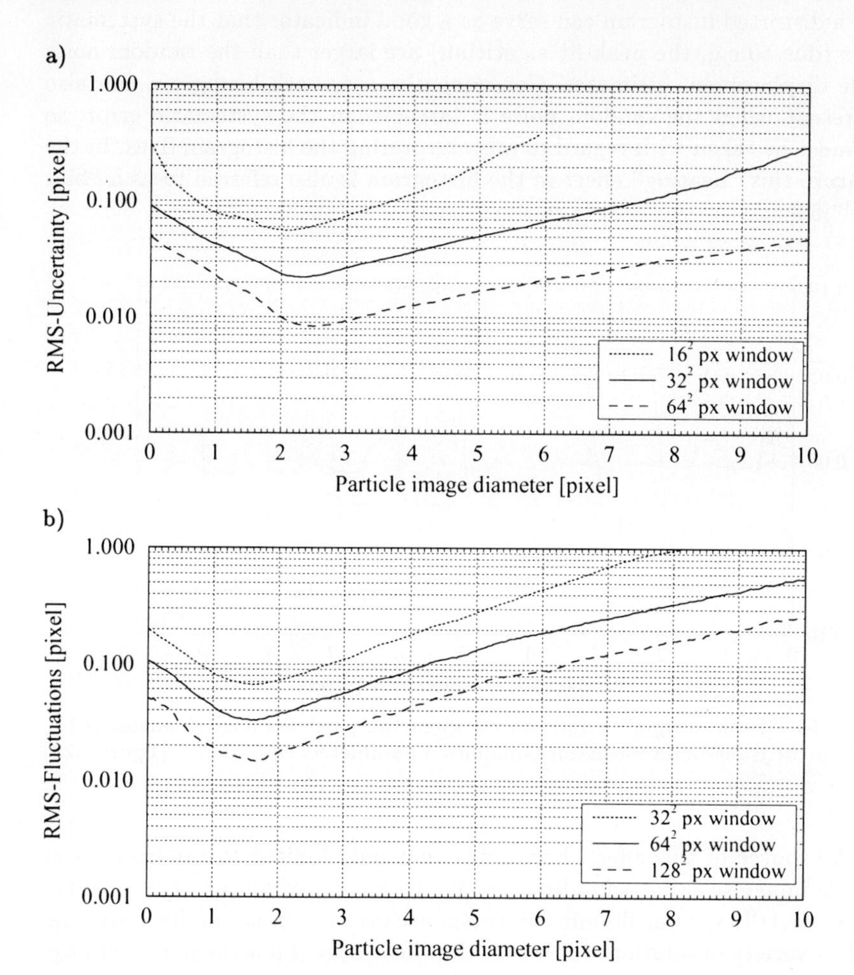

Fig. 5.23. Measurement uncertainty (RMS random error) in digital cross-correlation PIV evaluation with respect to varying particle image diameter: (a) single exposure/double frame PIV imaging, (b) double exposure/single frame PIV imaging. Simulation parameters: $QL = 8$ bits/pixel, no noise, optimum exposure, top-hat light sheet profile, $\mathcal{N} = 1/64\,\text{pixel}^{-1}$

When the particle images become too small another effect arises which can also be observed in the simulation data (figure 5.24): the displacements tend to be biased toward integral values. The effect increases as the particle image diameter is reduced which is a clear indication that the chosen subpixel peak estimator – here a three-point Gaussian peak fit – is unsuited at these particle image diameters. Other three-point fits may perform even worse [7, 96]. In actual displacement data the presence of this "peak-locking" effect can be detected by plotting a displacement histogram such as given in figure 5.25. Such a distorted histogram can serve as a good indicator that the systematic errors (due to e.g. the peak fit algorithm) are larger than the random noise in the displacement estimates. Consequently, a smooth histogram can also be present when the random noise is larger than the systematic error, so care must be taken with regard to misinterpreting the histogram data. In the literature this "beating" effect in the histogram is also referred to as a "bias error" [84].

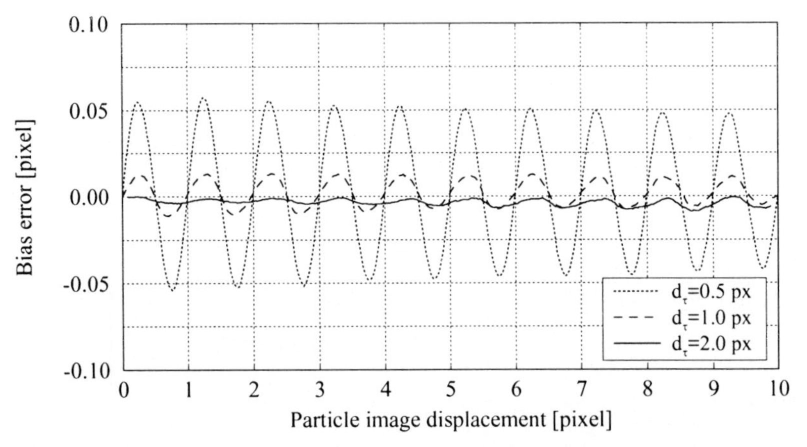

Fig. 5.24. "Peak locking" is introduced when the particle image diameter is too small for the three-point estimator (simulation parameters equivalent to figure 5.23)

The source of this effect, however, is not only limited to the insufficient particle image size, but can also arise due to a reduced fill factor or possibly even a spatially varying illumination response over the extent of the individual pixel. A variety of solutions to remove the effect exist. First the particle image diameter d_τ can be increased during the recording process by increasing the sampling rate or maybe even by defocusing the particle image. The second choice is to choose a different peak estimator which is better suited for smaller particle image diameters. The third alternative is to pre-condition the images using filters which optimize the particle image diameter with respect to the peak estimator.

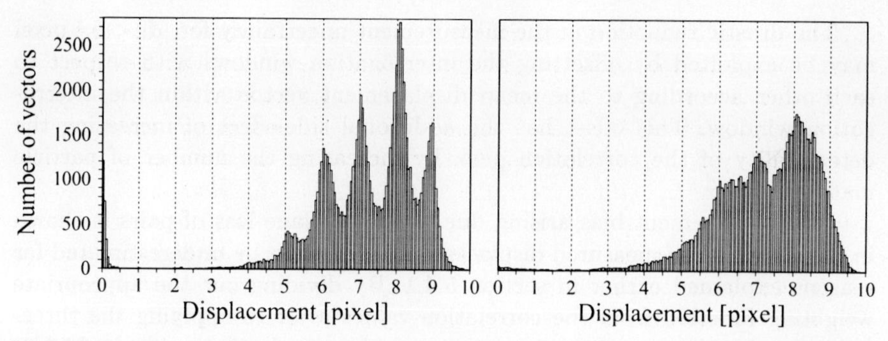

Fig. 5.25. Histograms of actual PIV displacement data obtained from a 10-image sequence of a turbulent boundary layer illustrating the "peak locking" associated with insufficient particle image size (left). Image preconditioning can reduce this effect (right). Histogram bin-width = 0.05 pixel

5.5.3 Optimization of particle image shift

Figure 5.26 shows the simulation results for the measurement uncertainty (RMS random error) as a function of the displacement. For most of the displacements the uncertainty is nearly constant except for displacements less than 0.5 pixel where a linear dependency can be observed (see also figure 5.29). This behavior can also be observed in the experimentally obtained error estimates given in [97]. Theory may be used to explain this behavior [96].

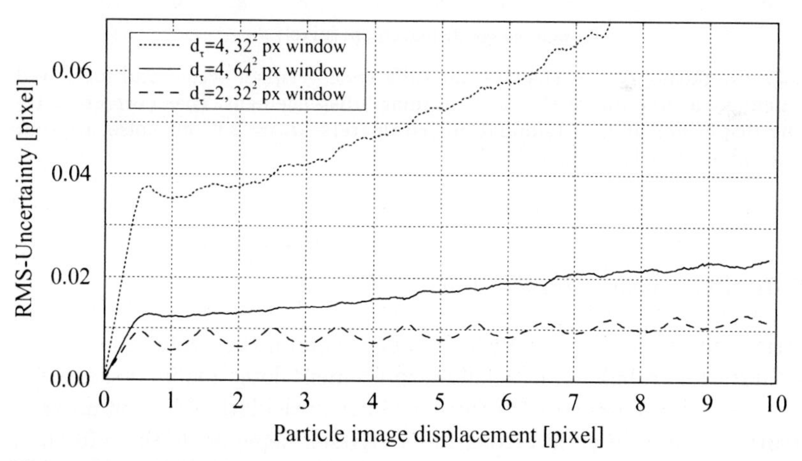

Fig. 5.26. Monte Carlo simulation results for the measurement uncertainty in digital cross-correlation PIV evaluation as a function of particle image displacement

The drastic reduction in the measurement uncertainty for $|\boldsymbol{d}| < 0.5$ pixel may be exploited by offsetting the interrogation windows with respect to each other according to the mean displacement vector within the interrogation window. This offset has the additional side-effect of increasing the detectability of the correlation peak by increasing the number of particle matches [66].

The displacement bias arising due to the in-plane loss of pairs is shown in figure 5.27. The measured displacement will always be underestimated for reasons explained earlier in section 5.4.1. By dividing out the appropriate weighting function from the correlation values prior to applying the three-point fit, this displacement bias can be nearly completely removed which is also shown in figure 5.27.

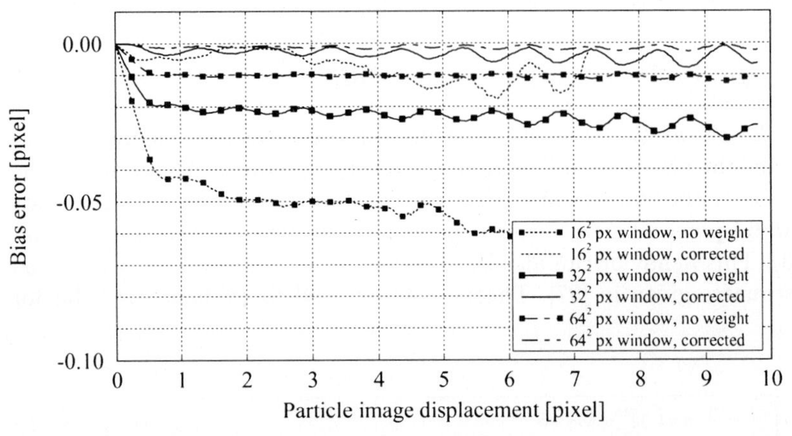

Fig. 5.27. Simulation results showing the difference between actual and measured displacement as a function of the particle image displacement. Bias correction removes the displacement bias (simulation parameters: $d_\tau = 2.0$, no noise, top-hat intensity profile, $\mathcal{N} = 1/64\,\text{pixel}^{-1}$)

5.5.4 Effect of particle image density

The particle image density has two primary effects in the evaluation of PIV images. First, the probability of a valid displacement detection increases when more particle image pairs enter in the correlation calculation. The number of image pairs captured in an interrogation area itself depends on three factors, namely, the overall particle image density, $\mathcal{N}$, the amount of in-plane displacement and the amount of out-of-plane displacement. KEANE & ADRIAN [64, 65, 66] have defined these three quantities as the effective particle image pair density within the interrogation spot, $\mathcal{N}_\mathrm{I}$, a factor expressing the in-plane loss of pairs, F_i and a factor expressing the out-of-plane loss of pairs,

F_o. When no in-plane or out-of-plane loss of pairs are present the latter two are unity, respectively. The product of the three quantities $\mathcal{N}_I F_i F_o$ expresses the mean effective number of particle image pairs in the interrogation spot. Monte Carlo simulations performed by KEANE & ADRIAN showed that for $\mathcal{N}_I F_i F_o > 8$ the valid detection probability exceeds 95 percent in double exposure/single frame PIV, while triple-pulse single frame PIV require only $\mathcal{N}_I F_i F_o > 4$. In contrast, single exposure/double frame PIV requires that $\mathcal{N}_I F_i F_o > 5$, which is consistent with the data shown in figure 5.28. However, depending on the chosen method for validation the probability curves may be shifted up or down. The theoretical Poisson distribution curves describing the presence of at least a given number of particle image pairs, $P[n \geq i]$, are also plotted in figure 5.28 and indicate that the presence of at least three particle image pairs in the interrogation spot matches the simulation data. In practice, the data yield can be easily optimized by ensuring the presence of at least three or four particle image pairs. Further optimization is possible by offsetting the interrogation windows as described in the previous section (section 5.5.3) which minimizes the in-plane loss of pairs, that is, $F_i \to 1$.

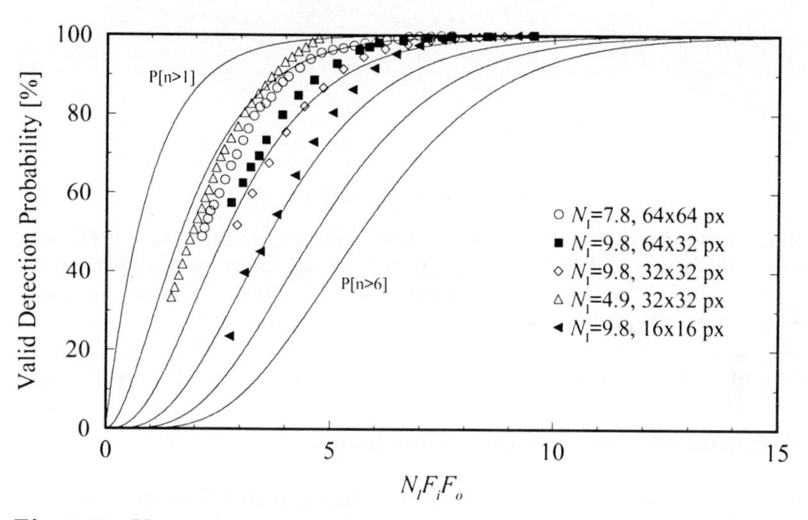

Fig. 5.28. Vector detection probability as a function of the product of image density $\mathcal{N}_I$, in-plane loss of pairs F_i and out-of-plane loss of pairs F_o. The solid line represents the probability for having at least a given number of particle images in the interrogation spot (see also figure 4 in KEANE & ADRIAN [66])

The second effect the particle image density has for the evaluation of PIV images is its direct influence on the measurement uncertainty. In figure 5.29 the measurement uncertainty is plotted as a function of particle image displacement for various particle image densities, $\mathcal{N}_I$. The displacement range was limited to the one pixel range which can be ensured by an interrogation window offset. For displacements less than 1/2 pixel the same linear trend as in figure 5.26 can be observed for all $\mathcal{N}_\mathrm{I}$. For $|\boldsymbol{d}| > 0.5$ pixel the uncertainty remains approximately constant. The principle effect of the particle image density, $\mathcal{N}_\mathrm{I}$, is that it can reduce the measurement uncertainty substantially, which can be explained by the simple fact that more particle image pairs increase the signal strength of the correlation peak.

Together the effects described above indicate that if a flow can be densely seeded then both a high valid detection rate as well as a low measurement uncertainty can be achieved using small interrogation windows, which in turn allows for a high spatial resolution.

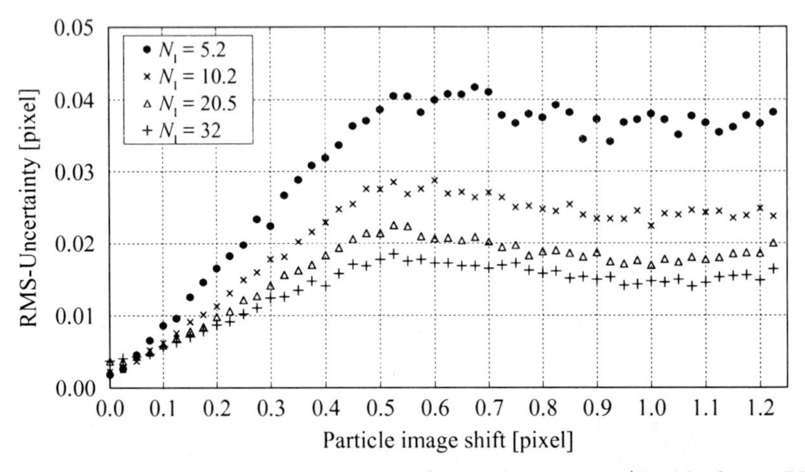

Fig. 5.29. Measurement uncertainty for single exposure/double frame PIV as a function of particle image shift for various particle image densities $\mathcal{N}_\mathrm{I}$. (Simulation parameters: $d_\tau = 2.2$ pixel, $QL = 8$ bits/pixel, 32×32 pixel, no noise, optimum exposure, top-hat light sheet profile.)

5.5.5 Variation of image quantization levels

Monte Carlo simulations for double exposed single frame PIV described in WILLERT [98] already indicated that the image quantization (i.e. bits/pixel) has only little influence on the measurement uncertainty or displacement bias error. This is further confirmed for single exposed double frame PIV as shown in figure 5.30.

Interestingly, a reduction from $QL = 8$ bits/pixel to $QL = 4$ bits/pixel practically has no influence on the RMS error for the given particle image density. This effect can be explained by the fact that the noise introduced by the FFT-based correlation calculation dominates, which also implies that a high number of image quantization levels does not necessarily warrant better PIV measurement accuracy unless the noise effects due the correlation algorithm are removed at the same time (such as through a direct, linear correlation). At lower image quantization levels, $QL < 4$ bits/pixel, the measurement uncertainty does however increase up to ten-fold.

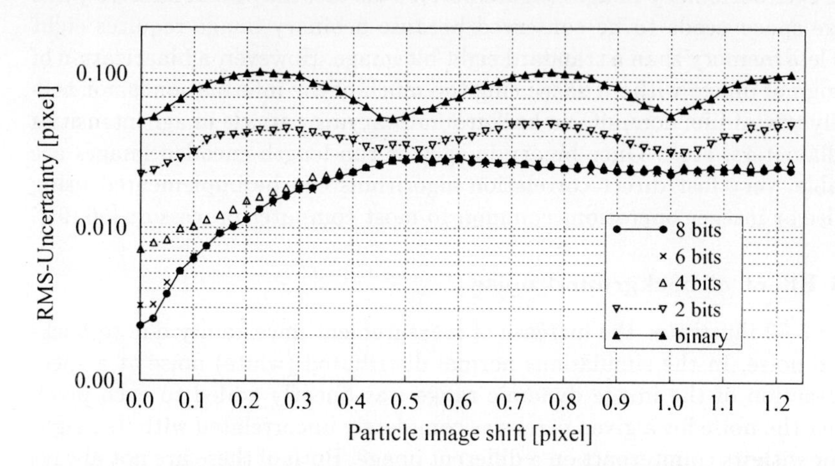

Fig. 5.30. Measurement uncertainty for single exposure/double frame PIV as a function of displacement and image quantization (simulation parameters: d_τ = 2.2 pixel, $\mathcal{N}_I = 10.2$, 32×32 pixel, no noise, optimum exposure, top-hat light sheet profile)

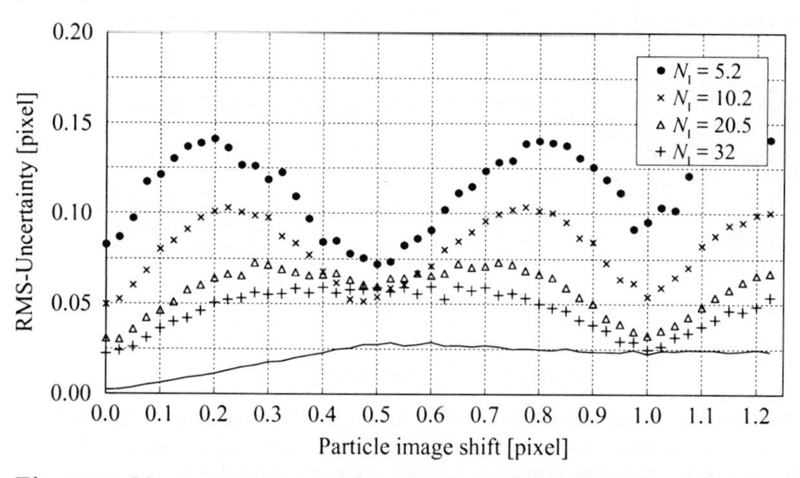

Fig. 5.31. Measurement uncertainty for binary image pairs (i.e. $QL = 1$ bits/pixel) as a function of displacement and particle image density, $\mathcal{N}_I$. The solid line indicates the measurement uncertainty at $QL = 8$ bits/pixel (simulation parameters: same as in figure 5.30 except $\mathcal{N}_I$)

The number of quantization levels is however of importance when single particle images are measured such as in particle tracking velocimetry or super-resolution PIV. In this case the position of a particle image can be more accurately estimated when its pixel intensity values are better resolved. Another noteworthy observation with regard to the image quantization level is that sufficient particle image density ($\mathcal{N}_{\mathrm{I}} > 30$) allows for low-noise measurements even for binary images (figure 5.31). This fact may be of interest when storage space needs to be conserved because a binary image requires eight times less memory than a standard eight bit image. However, a binarization of the original image without sacrificing too much of the information is not necessarily trivial (i.e. nonuniform background, varying particle image intensities and diameters, etc.). Once binary images or run-length-encoded images are available, very fast direct correlation algorithms can be implemented using bit-wise or integer operations common to most computer processors [56, 99].

5.5.6 Effect of background noise

Figure 5.32 illustrates the increase of measurement uncertainty due to background noise. In the simulations normal-distributed (white) noise at a specified fraction of the image dynamic range was linearly added to each pixel. Further the noise for a given pixel was completely uncorrelated with its neighbors or with its counterpart on a different image. Both of these are not always the case for actual image sensors.

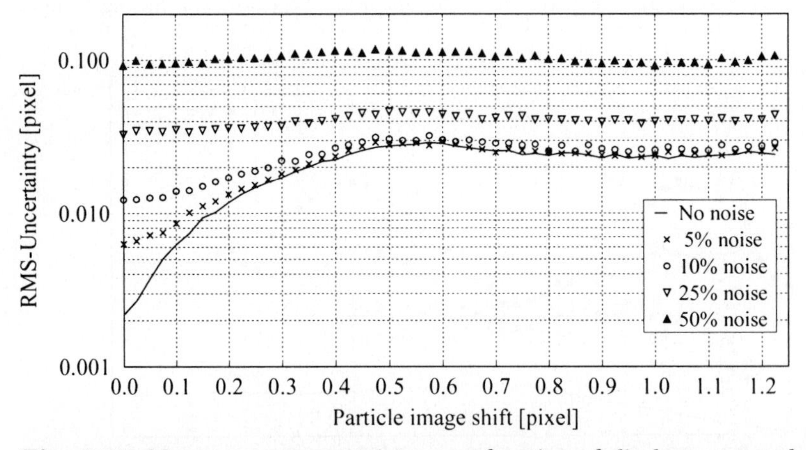

Fig. 5.32. Measurement uncertainty as a function of displacement and various amounts of white background noise (simulation parameters: $d_\tau = 2.2$ pixel, $\mathcal{N}_{\mathrm{I}} = 10.2$, 32×32 pixel, optimum exposure, top-hat light sheet profile)

Although this simulation is not quite realistic, it shows that minor noise has little effect on the measurement uncertainty. Again the noise contributions due to the FFT-based correlations dominate. For the chosen simulation parameters, 10% of noise, which corresponds to roughly $QL = 4$ bits/pixel, cause little deterioration in the measurements. This agrees with the observations made in figure 5.30 for the variation of the image quantization levels.

In that case an image quantization of $QL \geq 4$ bits/pixel shows little effect on the RMS-error.

5.5.7 Effect of displacement gradients

Since PIV is based on a statistical measurement of the displacement using the correlation between two interrogation windows, a displacement gradient across the window is likely to result in biased data since not all of the particle images present in the first interrogation window will also be present in the second interrogation window, even if the mean particle image displacement is accounted for. For interrogation windows without an offset the displacement will be biased to a lower value because particle images with small displacements will be present more frequently than those with higher displacements (e.g. in-plane loss of pairs, [66]). This measurement error does not arise in particle tracking methods because they measure the displacements of individual particle images [45].

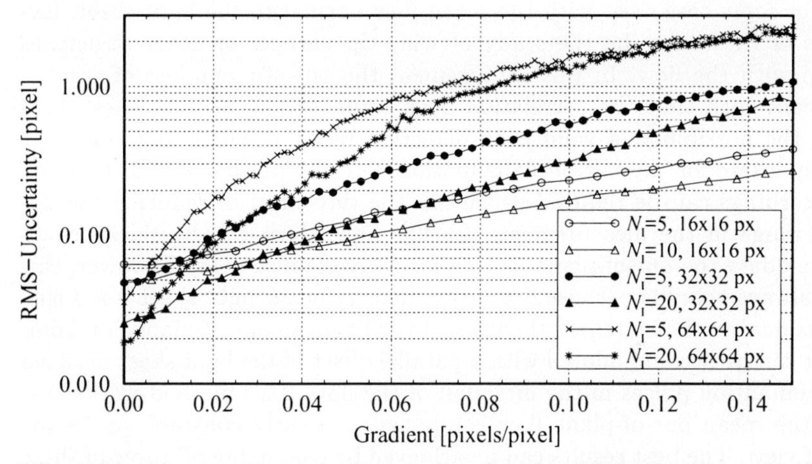

Fig. 5.33. Measurement uncertainty as a function of displacement gradient for various particle image densities and interrogation window sizes (simulation parameters: $d_\tau = 2.0$ pixel, $QL = 8$ bits/pixel, no noise, optimum exposure, top-hat light sheet profile)

In figure 5.33 the measurement uncertainty is plotted as a function of the displacement gradient for two different interrogation window sizes. One interesting observation that can be concluded from this figure is that smaller interrogation windows can tolerate much higher displacement gradients. Even at the same normalized particle image density, $\mathcal{N}$, where the larger window contains four times as many image pairs, this effect cannot be compensated. The reason for this behavior lies in the wider spread of the correlation peak in the larger window: for the same displacement gradient the dynamic range of the displacements scales linearly with the dimension of the interrogation window, resulting in a proportional increase of the correlation peak width. As a consequence, smaller interrogation windows should be favored, provided the particle image density is sufficiently high.

The bias error due to displacement gradients can be accounted for by estimating the mean particle image location in each interrogation window and assigning the displacement estimate to this point. A subsequent bilinear interpolation scheme can then be used to estimate the local displacement vectors at the center of the interrogation windows [154]. In principle the estimation of the mean particle image location requires that only paired particle images which contribute to the correlation are used. As this is difficult to implement, it generally suffices to threshold the data in each sample and calculate the centroid of the remaining pixel intensities (e.g. particle images).

5.5.8 Effect of out-of-plane motion

Frequently the PIV method has to be applied in highly three-dimensional flows, in some case even with the mean flow normal to the light sheet. Examples of these may be the study of wing-tip vortices or other structures aligned with the flow. In this arrangement the out-of-plane loss of pairs is significant such that the correlation peak signal strength diminishes. As a result, the possibility of valid peak detection reduces. Three methods exist to compensate for the out-of-plane motion. (1) The pulse delay Δt between the recordings can be reduced which has the side-effect of reducing the dynamic range in the measurement. (2) The light sheet can be thickened to accommodate the out-of-plane motion for a given pulse delay. However, this is not always possible because the energy density in the light sheet is reduced proportional to the increased thickness. (3) The mean out-of-plane flow component can be accommodated with a parallel offset of the light sheet between the illumination pulses in the direction of the flow. This method works best when the mean out-of-plane flow component is nearly constant across the field of view. The best results can be achieved by combining all three of these approaches. Just as with the in-plane loss of pairs, the general guideline is to keep the out-of-plane loss of pairs small enough to still ensure the presence of a minimum number of particle image pairs within the interrogation window (typically $\mathcal{N}_{\mathrm{I}} \geq 4$, see section 5.5.4).

6. Post-processing of PIV data

The recording and evaluation of PIV images has been described in the previous two chapters. Investigations employing the PIV technique usually result in a great number of images which must be further processed. If looking for statistical quantities the recorded data can easily amount to some gigabytes, which is now possible with today's computer hardware. Even more data per investigation are to be expected in future. Thus, it is quite obvious that a fast, reliable and fully automatic further processing of the PIV data is essential.

In principle post-processing of PIV data is characterized by the following steps.

Validation of the raw data. After automatic evaluation of the PIV recordings a certain number of obviously incorrectly determined velocity vectors (so-called outliers) can usually be found by visual inspection of the raw data. In order to detect these incorrect data the raw flow field data have to be validated. For this purpose special algorithms have to be developed, which must work automatically.

Replacement of incorrect data. For most post-processing algorithms (e.g. calculation of vector operators) it is required to have complete data fields as is quite naturally the case for numerically obtained data. Such algorithms will not work if gaps (data drop-outs) are present in the experimental data. Thus, means to fill the gaps in the experimental data must be developed.

Data reduction. It is quite difficult to inspect several hundred velocity vector maps and to describe their fluid mechanical features. Usually techniques like averaging (in order to extract the information about the mean flow and its fluctuations), conditional sampling (in order to distinguish between periodic and nonperiodic parts of the flow), and vector field operators (e.g. vorticity, divergence in order to detect structures in the flow) are applied.

Analysis of the information. At present this is the most challenging task for the user of the PIV technique. PIV being the first technique to offer information about complete instantaneous velocity vector fields allows new insights in old and new problems of fluid mechanics. New tools for analysis such as proper orthogonal decomposition (POD) [153] or neural networks [43] are applied to PIV data.

Presentation and animation of the information. A number of software packages – commercially available as well as in-house developed – are obtainable for the graphical presentation of the PIV field data. It is also very important to support the easier understanding of a human observer of the main features of the flow field. This can be done by contour plotting, color coding, etc. Animation of the PIV data is very useful for better understanding in the case of time series of PIV recordings or 3D data.

In the following sections those steps of post-processing with special requirements due to the PIV technique will be explained in more detail.

6.1 Data validation

Some of the problems associated with PIV raw data after automatic evaluation can be seen in figure 6.1. It shows the instantaneous flow field above a NACA 0012 airfoil at a free stream Mach number Ma = 0.75. The vector of the average flow velocity (= 344 m/s) as calculated for this PIV recording has been subtracted from each individual velocity vector in order to enhance details of the flow field. The supersonic flow regime above the leading edge of the airfoil and the terminating shock with its strong velocity gradient can clearly be detected.

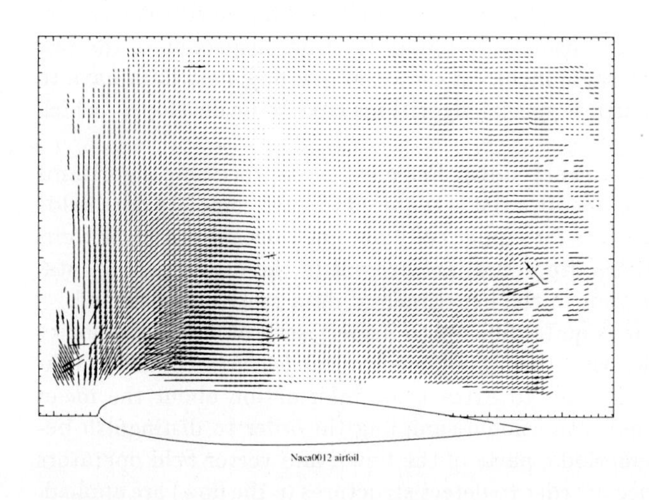

Fig. 6.1. Velocity vector map (raw data) of instantaneous flow field $(U - \overline{U}, V)$ above NACA 0012 airfoil (Ma= 0.75, $\alpha = 5°$, $l_c = 20\,\mathrm{cm}$, $\tau = 4\,\mu\mathrm{s}$, $\overline{U} = 344\,\mathrm{m/s}$)

For more details on the experiments see chapter 8. Typical features of incorrect velocity vectors, which can be detected in figure 6.1, are that:

- their magnitude and direction differ considerably from their surrounding neighbors,
- they very often appear at the edges of the data field (near the surface of the model, at the edges of drop-out areas, at the edges of the illuminated area),

– in most cases, they appear as single incorrect vectors.

From this description we can conclude that it is most likely that during the automatic evaluation procedure a correlation peak has been detected which is due to noise or artifacts (model surface, noise of different sources, etc.) and not due to the correlation of matched image pairs.

All these outliers can be detected by visual inspection. Dealing only with a small number of PIV recordings these erroneous velocity vectors may be deleted interactively. This is no longer possible if a great number of recordings has to be evaluated. However, for the further processing of the flow field data it is absolutely necessary to eliminate all such erroneous data. All subsequent operations involving differential operators on the raw vector data still including such outliers would enhance and smear out these errors locally and could thus mask data of good quality. Differential operators are, for example, the divergence, the vorticity operator, or the calculation of differences between numerical and experimental flow field data. In contrast, the application of operators utilizing averaging processes over a great number of data, e.g. mean value, variance, degree of turbulence, etc., are less severely affected by a few incorrect data values. It follows that all PIV data should generally be checked for erroneous data. Because of the great amount of data this can only be performed by means of an automatic algorithm. The leading idea for handling questionable data should be:

– The algorithm must ensure with a high level of confidence that no questionable data are stored in the final PIV data set.
– Questionable data should be rejected, if it cannot be decided by application of the algorithm whether data are valid or not.

As a consequence of the application of the validation algorithm the number of PIV data obtained from a given recording will be reduced by $\approx$0.1–1.5 % depending on the quality of the PIV recording and the type of flow to be investigated. The problem of filling gaps where no valid data were found on the flow field map (by means of interpolation or extrapolation) should only be attacked after having performed the data validation. It should again be emphasized that this procedure prevents information arising from incorrect data being spread into areas with data of good quality. For the same reason no smoothing of data should be carried out before having validated the data.

Different techniques for data validation have been described in the literature [41, 61, 70, 125, 151]. However, up to now no general solution can be offered for the problem of data validation in PIV. In our applications we use several algorithms, which have been developed and tested utilizing real PIV recordings obtained in different experimental situations and for different types of flows [123]. Some of these modules are alternatives, some of them may be applied successively. Two of these modules, the *global histogram operator* and the *dynamic mean value operator*, will be discussed in detail in the following.

Some definitions will be made for the following discussion of the different data validation algorithms. The instantaneous velocity vector field (U, V) has been sampled ("interrogated") on positions which form a regular grid in the flow field. In our case the grid, a part of which is shown in figure 6.2, consists of $I \times J$ grid points in the X and Y directions with constant distance $\Delta X_{\text{step}}, \Delta Y_{\text{step}}$ between neighboring grid points in both directions. The two-dimensional velocity vector at the position i, j $(i = 1 \cdots I, j = 1 \cdots J)$ is denoted $\boldsymbol{U}_{2\text{D}}(i, j)$. In the following discussion the relation between the central velocity vector $\boldsymbol{U}_{2\text{D}}(i, j)$ and one of its nearest neighbors $\boldsymbol{U}_{2\text{D}}(n)$ is considered. The nearest neighbors are labeled by $n, (n = 1, ..., N)$. Usually, N is chosen to be eight. The distance d between the central velocity vector $\boldsymbol{U}_{2\text{D}}(i, j)$ and its nearest neighbors $\boldsymbol{U}_{2\text{D}}(n)$ is either $\Delta X_{\text{step}}, \Delta Y_{\text{step}}$, or $\sqrt{\Delta X^2_{\text{step}} + \Delta Y^2_{\text{step}}}$, depending on its position on the grid. The magnitude of the vector difference between the central velocity vector $\boldsymbol{U}_{2\text{D}}(i, j)$ and $\boldsymbol{U}_{2\text{D}}(n)$ is $|\boldsymbol{U}_{\text{diff},n}| = |\boldsymbol{U}_{2\text{D}}(n) - \boldsymbol{U}_{2\text{D}}(i, j)|$.

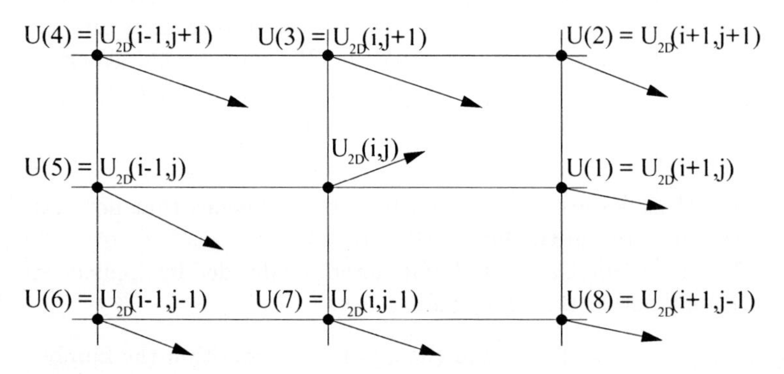

Fig. 6.2. Sketch of data grid with notation of vectors

For the demonstration of the performance of the different algorithms for data validation the flow field shown in figure 6.3 is utilized. It represents the lower left part of the flow field already shown in figure 6.1.

6.1.1 Global histogram operator

Principle. It is assumed that for real flow fields the vector difference between the neighboring velocity vectors is smaller than a certain threshold ϵ_{thresh}:

$$|\boldsymbol{U}_{\text{diff},n}| = |\boldsymbol{U}_{2\text{D}}(n) - \boldsymbol{U}_{2\text{D}}(i, j)| < \epsilon_{\text{thresh}} \quad \text{with} \quad \epsilon_{\text{thresh}} > 0 \,. \tag{6.1}$$

This is certainly true as long as the length scale of the flow is much greater than the distance d between the position of neighboring vectors. Consequently all correct velocity vectors must lie within a continuous area in the (u, v) plane for flow fields without discontinuities. In a first step this criterion is used to discriminate against incorrect data.

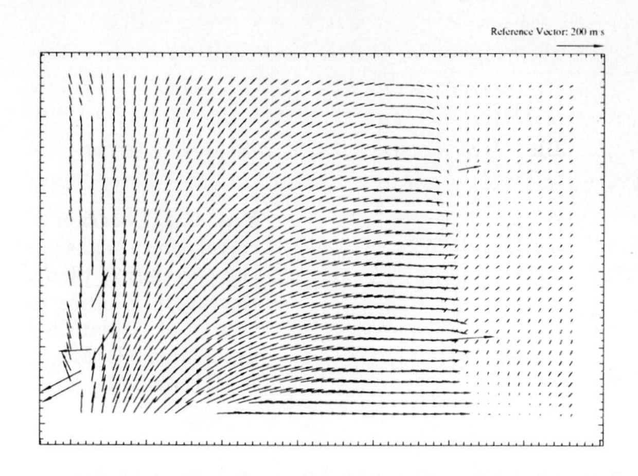

Fig. 6.3. Enlarged area (lower left part) of figure 6.1 for the demonstration of the data validation algorithm

Procedure. The velocity plane is divided into small cells in order to be able to calculate a 2D histogram of the velocity vectors. For each cell the number of velocity vectors as derived from the automatic evaluation of the PIV recording is summed.

An equivalent presentation is given in figure 6.4 where a point has been plotted in the correlation plane at the location of the highest correlation peak for each interrogation window. (This saves computer time as compared to working in the velocity plane.)

Two separated areas of accumulated correlation peaks (velocity vectors) can be detected, one circular region (I) and a second region (II) with greater scattering of the peak's locations (i.e. of the velocity vectors). Now a rectangular box can automatically be calculated, circumscribing the area with the highest accumulation of displacement peaks (velocity vectors). In the next step of the validation procedure the displacement peaks detected for each interrogation window will be checked individually. All displacement peaks (velocity vectors) lying outside the rectangle will be marked and rejected.

Discussion. Most outliers due to noise in the correlation plane can be rejected by application of this simple algorithm. Also, if auto-correlation is employed, data originating from the vicinity of the origin of the correlation plane (zero order central peak) can be eliminated by defining another rectangle of rejection around the zero order peak (not shown in figure 6.4). Noise originating from the recording process, which is concentrated in certain spectral bands (noise from A/C sources, imperfections of the optics for interrogation, structures of the CCD-camera, etc.) can be detected and eliminated because in most cases its response is located in areas of the correlation plane which are not connected to those areas where the correct velocity vectors accumulate.

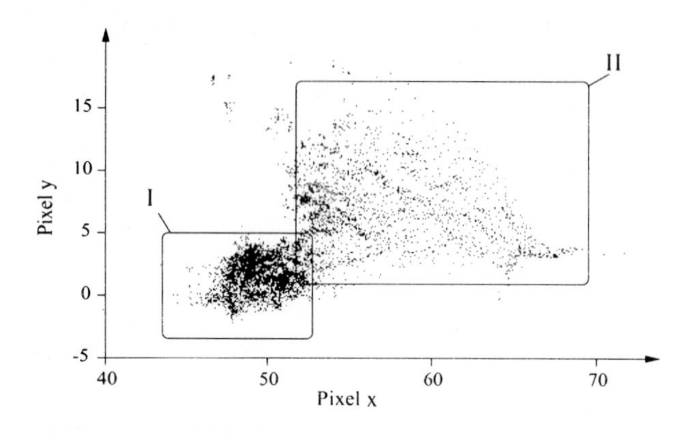

Fig. 6.4. Location of correlation peaks in the correlation plane. Rectangle indicates area of plausible data, area (I) Ma < 1, (II) Ma > 1

Figure 6.4 also demonstrates that there may be situations with two or more areas in the correlation (or velocity) plane, where the peaks (or velocity vectors) accumulate. This is the case if discontinuities are present in the flow field, e.g. for transonic flow fields, if shocks are embedded in the flow field. In figure 6.4 the area marked (I) is due to the subsonic part of the flow field at the right side of figure 6.3, whereas the area (II) is due to the supersonic part of the flow field just above the leading edge of the airfoil. Summarizing, one can say that the global histogram operator employs physical arguments (upper and lower limit of possible flow velocities) to remove all data, which physically cannot exist in the flow field. Moreover, the inspection of the global velocity histogram gives useful information about the quality of the PIV evaluation (number of incorrect data due to noise, dynamic range of the flow field, maximum utilization of the optimal range for PIV evaluation by selecting the proper time delay between the light pulses for illumination, etc.). If at this stage of the validation process a velocity vector is rejected, it will automatically be checked whether for this grid position the automatic evaluation procedure has detected more than one peak in the correlation plane (see section 5.4.4). If this should be the case, these data are checked as well, whether they fulfill the criteria explained above. It is obvious that the maximum number of peaks admitted during the automatic evaluation for a fixed grid position should be limited to two or three, as otherwise – with a great number of peaks for selection – the chance would be rather high to pick up peaks due to noise, which, however, accidentally fulfill the criterion for selection.

6.1.2 Dynamic mean value operator

Principle. Mostly, mean value tests are described in the literature for data validation for PIV. These algorithms check each velocity vector individually by comparing its magnitude $|\boldsymbol{U}_{2D}(i,j)|$ with the average value over its nearest neighbours $\boldsymbol{\mu_U}(i,j)$. Typically, a 3×3 neighborhood with eight nearest neighbors is selected. The velocity vector to be validated will be rejected if the absolute difference between its magnitude and the average over its neighbors is above a certain threshold ϵ_{thresh}. The test can be modified by applying it not only to the magnitude but also to the U and V components of the vector, or by utilizing a larger number of neighbors for comparison.

However, the application of this test to transonic flows has shown some problems if shocks are present in the flow field (discontinuity of the flow velocity along a line). Thus, the algorithm had to be improved for flows with velocity gradients by locally varying the threshold level ϵ for validation.

Procedure. The following expression is calculated for $N = 8$ neighbors:

$$\boldsymbol{\mu_U}(i,j) = \frac{1}{N} \sum_{n=1}^{N} \boldsymbol{U}_{2D}(n) \; . \tag{6.2}$$

The averaged magnitude of the vector difference between the average vector and its 8 neighbors is also calculated:

$$\sigma_{U}^{2}(i,j) = \frac{1}{N} \sum_{n=1}^{N} \left(\boldsymbol{\mu_U}(i,j) - \boldsymbol{U}_{2D}(n)\right)^2 \; . \tag{6.3}$$

The criterion for data validation is:

$$|\boldsymbol{\mu_U}(i,j) - \boldsymbol{U}_{2D}(i,j)| < \epsilon_{\text{thresh}} \tag{6.4}$$

where $\epsilon_{\text{thresh}} = C_1 + C_2\, \sigma_U(i,j)$ with $C_1, C_2 =$ constants.

Problems will arise in the case of drop-outs within the data field or at the edges of the data field, i.e. if less than $N = 8$ neighbors are available for comparison. Satisfying results were obtained when the drop-out areas were filled with the mean value of the whole data field and lines and rows were added at each edge of the data field by doubling the outermost lines and rows. However, these artificially generated data were only kept during the application of the local mean value test or other validation tests.

Discussion. The effect of the application of the dynamic mean value operator is demonstrated in figure 6.5, where all velocity vectors identified as “incorrect” are marked by thick vector symbols. Figure 6.5 clearly shows the power of the algorithm described above as those velocity data, which are incorrect – as they are obviously different in magnitude and direction from their neighbors – have been consequently marked, whereas the application of this algorithm does not lead to difficulties with the strong velocity gradients across the shock, i.e. no correct data are marked as “incorrect”. Thus, it has

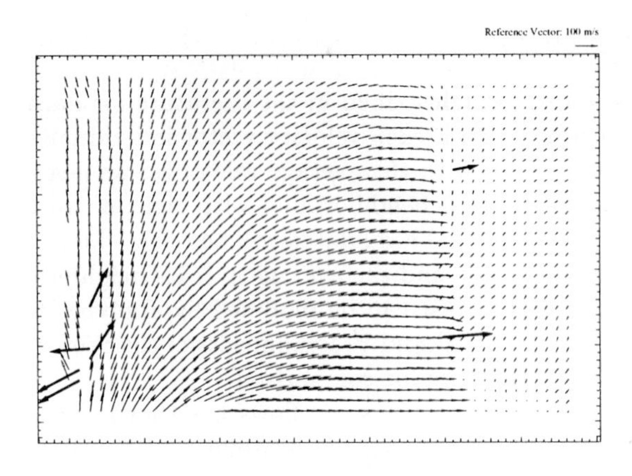

Fig. 6.5. Application of the dynamic mean value operator on the velocity vector map of figure 6.3. Vectors identified as "incorrect" are marked by thick vector symbols

been shown that this algorithm is able to handle flows with strong velocity gradients by the introduction of the locally varying threshold level for validation. The constants C_1 , C_2 have to be determined once experimentally and can then be utilized for the whole series of PIV recordings taken for the same type of flow.

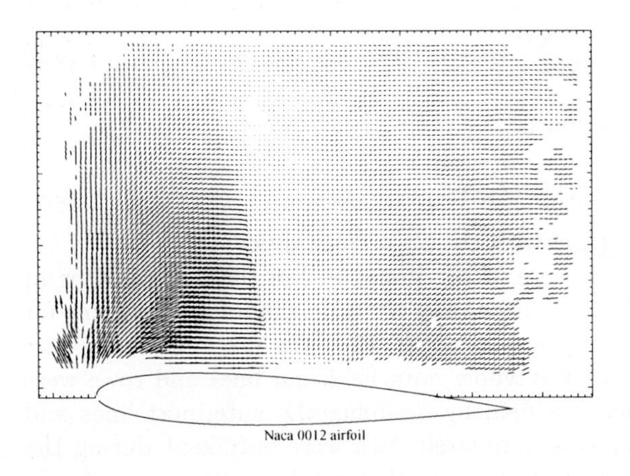

Fig. 6.6. Velocity vector map of the instantaneous flow field above a NACA 0012 airfoil of figure 6.1 after clean-up

6.1.3 Median test

A further validation method has been proposed by Westerweel [125]. The median filtering, which is frequently utilized in image processing to remove binary noise, is applied to the PIV raw data. Median filtering simply speaking means that all neighboring velocity vectors $\boldsymbol{U}_{2\mathrm{D}}(n)$ are ordered linearly either with respect to the magnitude of the velocity vector, or their U and V components. The central value in this order (that is either the fourth or fifth for eight neighbors) is the median value. The velocity vector under inspection $\boldsymbol{U}_{2D}(i,j)$ is considered valid if

$$|\boldsymbol{U}_{2D}(\text{median}) - \boldsymbol{U}_{2D}(i,j)| < \epsilon_{\text{thresh}} \,.$$

Simulations have shown the power of median testing for data validation [125]. However, more tests and applications of the different validation algorithms in different types of flows are required before an optimal validation algorithm can be recommended.

6.1.4 Implementation of data validation algorithms

Therefore, at present we perform data validation by application of the global histogram operator and by subsequent application of the dynamic mean value operator. The result of the automatic application of both operators on the PIV recording of figure 6.1 is shown in figure 6.6. All velocity data, which are incorrect – as they are obviously different in magnitude and direction from their neighbors – have been deleted. No difficulties appeared, when validating the flow field in front and behind the shock. According to future applications of the PIV technique to other types of flows the data validation algorithms have to be improved further. A successful procedure should be to collect as much a priori information about the flow field to be investigated as possible and to express this knowledge in the form of fluid mechanical or image processing operators. The first simple fluid mechanical operators have already been developed [119].

6.2 Replacement schemes

After having validated all PIV data it is possible to fill in missing data using, for instance, bilinear interpolation. According to WESTERWEEL [125] the probability that there is another spurious vector in the direct neighborhood of a spurious vector is given by a binominal distribution. For instance if the data contains 5% spurious vectors, more than 80% of the data can be recovered by a straight bilinear interpolation from the four valid neighboring vectors. (Incidentally, the bilinear interpolation also fulfills continuity.) The remaining missing data can be estimated by using some sort of weighted average of the surrounding data, such as the adaptive Gaussian window technique proposed by AGÜI & JIMÉNEZ [158].

Some post-processing methods also require smoothing of the data. The reason is that the experimental data are affected by noise in contrast to numerical data. A simple convolution of the data with a 2×2, 3×3 or larger smoothing kernel (with equal weights) is generally sufficient for this purpose. By choosing the kernel size to have spatial dimensions smaller than the effective interrogation window size, additional lowpass filtering of the velocity field can be minimized.

6.3 Vector field operators

In many fluid mechanical applications the velocity information by itself is of secondary interest in the physical description, which is principally due to the lack of simultaneous pressure and density field measurements. In general the pressure, density and velocity fields are required to completely recover all terms in the Navier–Stokes equation:

$$\rho \frac{\mathrm{D}\boldsymbol{U}}{\mathrm{D}t} = -\nabla p + \mu \nabla^2 \boldsymbol{U} + \boldsymbol{F} \tag{6.5}$$

where $\boldsymbol{F}$ represents the contribution of the body forces such as gravity. Clearly, the task of obtaining all of these field quantities simultaneously is unrealistic at present. However, the velocity field obtained by PIV can be used to estimate other fluid mechanically relevant quantities by means of differentiation or integration. Of the differential quantities the vorticity field is of special interest because this quantity, unlike the velocity, is independent of the frame of reference. In particular, if it is resolved temporally, the vorticity field can be much more useful in the study of flow phenomena than the velocity field by itself, especially in highly vortical flows such as turbulent boundary layers, wake vortices and complex vortical flows. For incompressible flow ($\nabla \cdot \boldsymbol{U} = 0$) the Navier–Stokes equation can actually be rewritten in terms of the vorticity, that is the vorticity equation:

$$\frac{\partial \boldsymbol{\omega}}{\partial t} + \boldsymbol{U} \cdot \nabla \boldsymbol{\omega} = \boldsymbol{\omega} \cdot \nabla \boldsymbol{U} + \nu \nabla^2 \boldsymbol{\omega} \tag{6.6}$$

which expresses the rate of change of vorticity of a fluid element (for simplicity, $\boldsymbol{F} = 0$). Although the pressure term has been eliminated from this expression, the estimation of the last term, $\nabla^2 \boldsymbol{\omega}$, is difficult from actual PIV data. Because of its frequent use in fluid mechanical descriptions the estimation of vorticity from PIV data will serve as an example for the available differentiation schemes given in the following sections.

Integral quantities can also be obtained from the velocity field. The instantaneous velocity field obtained by PIV can also be integrated yielding either single values through path integrals or another field quantity such as the stream function. Analogous to the vorticity field, the circulation which is obtained through path integration is also of special interest in the study of vortex dynamics, mainly because it is also independent of the reference frame. Other PIV applications may require the calculation of mass flow rates in a control volume type of analysis. The later sections of this chapter will be devoted to the aspects of integration.

6.4 Estimation of differential quantities

Before addressing the actual calculation schemes available for the differentiation of the velocity field data, it should be determined which terms can actually be calculated from the planar velocity field. Standard PIV data provide only the two components[1] of the three-dimensional vector field while more advanced PIV methods like stereoscopic PIV provide three-component velocity data. Unless several light sheet planes are recorded simultaneously the PIV method can only provide a single plane of velocity data thereby excluding all possibilities of calculating gradients normal to the light sheet. In order to see which differential terms actually can be calculated, the full velocity gradient tensor or deformation tensor, $\mathrm{d}\boldsymbol{U}/\mathrm{d}\boldsymbol{X}$, will be given first:

$$\frac{\mathrm{d}\boldsymbol{U}}{\mathrm{d}\boldsymbol{X}} = \begin{bmatrix} \frac{\partial U}{\partial X} & \frac{\partial V}{\partial X} & \frac{\partial W}{\partial X} \\ \frac{\partial U}{\partial Y} & \frac{\partial V}{\partial Y} & \frac{\partial W}{\partial Y} \\ \frac{\partial U}{\partial Z} & \frac{\partial V}{\partial Z} & \frac{\partial W}{\partial Z} \end{bmatrix} \tag{6.7}$$

This deformation tensor can be decomposed into a symmetric part and an antisymmetric part:

$$\frac{\mathrm{d}\boldsymbol{U}}{\mathrm{d}\boldsymbol{X}} = \begin{bmatrix} \frac{\partial U}{\partial X} & \frac{1}{2}\left(\frac{\partial V}{\partial X}+\frac{\partial U}{\partial Y}\right) & \frac{1}{2}\left(\frac{\partial W}{\partial X}+\frac{\partial U}{\partial Z}\right) \\ \frac{1}{2}\left(\frac{\partial U}{\partial Y}+\frac{\partial V}{\partial X}\right) & \frac{\partial V}{\partial Y} & \frac{1}{2}\left(\frac{\partial W}{\partial Y}+\frac{\partial V}{\partial Z}\right) \\ \frac{1}{2}\left(\frac{\partial U}{\partial Z}+\frac{\partial W}{\partial X}\right) & \frac{1}{2}\left(\frac{\partial V}{\partial Z}+\frac{\partial W}{\partial Y}\right) & \frac{\partial W}{\partial Z} \end{bmatrix} \tag{6.8}$$

$$+ \begin{bmatrix} 0 & \frac{1}{2}\left(\frac{\partial V}{\partial X}-\frac{\partial U}{\partial Y}\right) & \frac{1}{2}\left(\frac{\partial W}{\partial X}-\frac{\partial U}{\partial Z}\right) \\ \frac{1}{2}\left(\frac{\partial U}{\partial Y}-\frac{\partial V}{\partial X}\right) & 0 & \frac{1}{2}\left(\frac{\partial W}{\partial Y}-\frac{\partial V}{\partial Z}\right) \\ \frac{1}{2}\left(\frac{\partial U}{\partial Z}-\frac{\partial W}{\partial X}\right) & \frac{1}{2}\left(\frac{\partial V}{\partial Z}-\frac{\partial W}{\partial Y}\right) & 0 \end{bmatrix} \tag{6.9}$$

A substitution of the strain and vorticity components yields:

$$\frac{\mathrm{d}\boldsymbol{U}}{\mathrm{d}\boldsymbol{X}} = \begin{bmatrix} \epsilon_{XX} & \frac{1}{2}\epsilon_{XY} & \frac{1}{2}\epsilon_{XZ} \\ \frac{1}{2}\epsilon_{YX} & \epsilon_{YY} & \frac{1}{2}\epsilon_{YZ} \\ \frac{1}{2}\epsilon_{ZX} & \frac{1}{2}\epsilon_{ZY} & \epsilon_{ZZ} \end{bmatrix} + \begin{bmatrix} 0 & \frac{1}{2}\omega_Z & -\frac{1}{2}\omega_X \\ -\frac{1}{2}\omega_Z & 0 & \frac{1}{2}\omega_Y \\ -\frac{1}{2}\omega_X & \frac{1}{2}\omega_Y & 0 \end{bmatrix} \tag{6.10}$$

Thus the symmetric tensor represents the strain tensor with the elongational strains on the diagonal and the shearing strains on the off-diagonal, whereas the antisymmetric part contains only the vorticity components.

Given that standard PIV data provides only the U and V velocity components and that this data can only be differentiated in the X and Y directions,

[1] We ignore the fact that standard PIV only yields a two-dimensional projection of the three-dimensional vector.

only a few terms of the deformation tensor, $\mathrm{d}\boldsymbol{U}/\mathrm{d}\boldsymbol{X}$, can be estimated with PIV:

$$\omega_Z = \frac{\partial V}{\partial X} - \frac{\partial U}{\partial Y} \tag{6.11}$$

$$\epsilon_{XY} = \frac{\partial U}{\partial Y} + \frac{\partial V}{\partial X} \tag{6.12}$$

$$\eta = \epsilon_{XX} + \epsilon_{YY} = \frac{\partial U}{\partial X} + \frac{\partial V}{\partial Y} \tag{6.13}$$

Therefore only the vorticity component normal to the light sheet can be determined, along with the in-plane shearing and extensional strains. In this regard it is interesting to note that the additional availability of the third velocity component, W, by more advanced PIV methods, does not yield any additional strains or vorticity components.

Assuming incompressibility, that is, $\nabla \cdot \boldsymbol{U} = 0$, the sum of the in-plane extensional strains in equation (6.13) can be used to estimate the out-of-plane strain ϵ_{ZZ}:

$$\epsilon_{ZZ} = \frac{\partial W}{\partial Z} = -\frac{\partial U}{\partial X} - \frac{\partial V}{\partial Y} = -\eta \tag{6.14}$$

However it should be kept in mind that the quantity η only indicates the presence of out-of-plane flow; it does not recover the out-of-plane velocity.

6.4.1 Standard differentiation schemes

Since PIV provides the velocity vector field sampled on a two-dimensional, evenly spaced grid, finite differencing has to be employed in the estimation of the spatial derivatives of the velocity gradient tensor, $\mathrm{d}\boldsymbol{U}/\mathrm{d}\boldsymbol{X}$. Moreover, each of the velocity data, U_i, is disturbed by noise, that is, a measurement uncertainty, ϵ_U. Although the error analysis used for the estimation of the uncertainty in the differentials assumes the measurement uncertainty of each quantity to be decoupled from its neighbors, this is not always the case. For instance, if the PIV image is oversampled, that is, the interrogation interval (sample points) is smaller than the interrogation area dimensions ($\Delta X < \Delta X_0$ and/or $\Delta Y < \Delta Y_0$), the recovered velocity estimates are not independent because the neighboring interrogation areas partly sample the same particles. At low image densities, $\mathcal{N}$, this problem worsens especially in regions of high displacement gradients (see also figure 5.33 in section 5.5.7). For simplicity the differentiation schemes described next assume the measurement uncertainties to be independent of their neighbors.

Table 6.1 lists a number of finite difference schemes to obtain estimates for the first derivate, $\mathrm{d}f/\mathrm{d}x$, of a function $f(x)$ sampled at discrete locations $f_i = f(x_i)$. The "accuracy" in this table reflects the truncation error associated with derivation of each operator by means of Taylor series expansion. The actual uncertainty in the differential estimate due to the uncertainty in

Table 6.1. First order differential operators for data spaced at uniform ΔX intervals along the X-axis

Operator	Implementation	Accuracy	Uncertainty
Forward difference	$\left(\frac{\mathrm{d}f}{\mathrm{d}x}\right)_{i+1/2} \approx \frac{f_{i+1} - f_i}{\Delta X}$	$\mathrm{O}(\Delta X)$	$\approx 1.41 \frac{\epsilon_U}{\Delta X}$
Backward difference	$\left(\frac{\mathrm{d}f}{\mathrm{d}x}\right)_{i-1/2} \approx \frac{f_i - f_{i-1}}{\Delta X}$	$\mathrm{O}(\Delta X)$	$\approx 1.41 \frac{\epsilon_U}{\Delta X}$
Center difference	$\left(\frac{\mathrm{d}f}{\mathrm{d}x}\right)_i \approx \frac{f_{i+1} - f_{i-1}}{2\Delta X}$	$\mathrm{O}(\Delta X^2)$	$\approx 0.7 \frac{\epsilon_U}{\Delta X}$
Richardson extrapol.	$\left(\frac{\mathrm{d}f}{\mathrm{d}x}\right)_i \approx \frac{f_{i-2} - 8f_{i-1} + 8f_{i+1} - f_{i+2}}{12\Delta X}$	$\mathrm{O}(\Delta X^3)$	$\approx 0.95 \frac{\epsilon_U}{\Delta X}$
Least squares	$\left(\frac{\mathrm{d}f}{\mathrm{d}x}\right)_i \approx \frac{2f_{i+2} + f_{i+1} - f_{i-1} - 2f_{i-2}}{10\Delta X}$	$\mathrm{O}(\Delta X^2)$	$\approx 1.0 \frac{\epsilon_U}{\Delta X}$

the velocity estimates ϵ_U can be obtained using standard error propagation methods assuming the individual data to be independent of each other.

The difference between the Richardson extrapolation scheme and the least squares approach is that the former is designed to minimize the truncation error while the latter attempts to reduce the effect of the random errors, that is, the measurement uncertainty, ϵ_U. The least squares approach therefore seems to be the most suitable method for PIV data. In particular, for oversampled velocity data where neighboring data are no longer uncorrelated, the Richardson extrapolation scheme along with the less sophisticated finite difference schemes will perform poorly with respect to the least-squares approach. On the other hand, the least-squares approach has a tendency to smooth the estimate of the differential because the outer data $f_{i\pm2}$ are more weighted than the inner data $f_{i\pm1}$.

The effect of oversampling on the estimation of the differential quantities is demonstrated in figure 6.7 for vorticity fields computed from the same velocity data at different mesh spacings. Since the data is taken from a laminar vortex pair the vorticity contours are expected to be smooth (data from [154]). For a 50% interrogation window overlap all schemes produce reasonable results since neighboring data are only weakly correlated. The estimate obtained from the forward difference scheme is the most noisy because the data entering in the formula are correlated (by 50% overlap) which is not the case for the center difference scheme.

By doubling the interrogation window overlap, such as in figure 6.8, much noisier vorticity fields are obtained which has two related causes: first the grid spacing, $\Delta X, \Delta Y$ is reduced by a factor of two while the measurement uncertainty for the velocity, ϵ_U, stays the same. As a result the vorticity measurement uncertainty is doubled. Secondly, all or part of the data used in the differentiation scheme will be correlated because of the increased overlap. For instance velocity gradient induced bias errors will be similar in neighboring points which in turn results in a biased estimate of the vorticity. Thus, the estimation of differential quantities from the velocity field has to be optimized with respect to the grid spacing. A coarser grid yields less noisy estimates of the gradient quantity, but also results in a reduced spatial resolution. In the following section alternative differentiation schemes are introduced which perform well even in oversampled data.

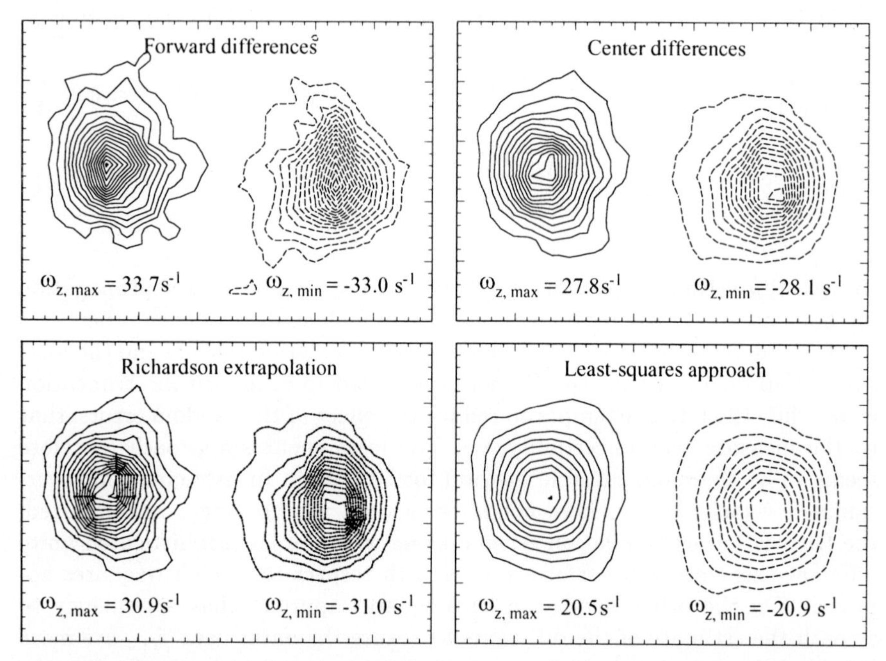

Fig. 6.7. Vorticity field estimates obtained from twice oversampled PIV data, e.g. the interrogation window overlap is 50%. The vortex pair is known to be laminar and thus should have smooth vorticity contours

6.4.2 Alternative differentiation schemes

The finite differencing formulae given in table 6.1 have been derived for functions of one variable, that is, they are applied in one dimension at a time. Almost by definition the velocity data obtained by PIV is provided on a two-dimensional grid which also holds for the differential quantities obtained from it. As a consequence the use of one-dimensional finite difference schemes for the estimation of the two-dimensional differential field quantities seems inadequate. Using the estimation of the out-of-plane vorticity component, ω_Z,

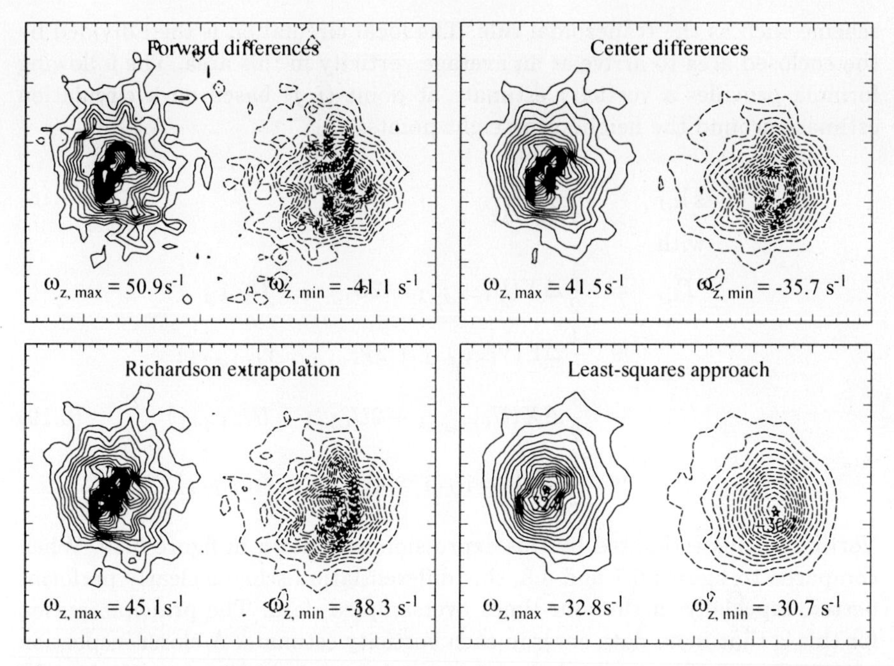

Fig. 6.8. Vorticity field estimates obtained from four times oversampled PIV data, e.g. the interrogation window overlap is 75%

as an example several alternative approaches to the problem of differential estimation will be given.

By definition the vorticity is related to the circulation by Stokes theorem

$$\Gamma = \oint \boldsymbol{U} \cdot \mathrm{d}\boldsymbol{l} = \int (\nabla \times \boldsymbol{U}) \cdot \mathrm{d}\boldsymbol{S} = \int \boldsymbol{\omega} \cdot \mathrm{d}\boldsymbol{S} \tag{6.15}$$

where $\boldsymbol{l}$ describes the path of integration around a surface $\boldsymbol{S}$. The vorticity for a fluid element is found by reducing the surface $\boldsymbol{S}$, and with it the path $\boldsymbol{l}$, to zero:

$$\hat{\boldsymbol{n}} \cdot \boldsymbol{\omega} = \hat{\boldsymbol{n}} \cdot \nabla \times \boldsymbol{U} = \lim_{S \to 0} \frac{1}{S} \oint \boldsymbol{U} \cdot \mathrm{d}\boldsymbol{l} \tag{6.16}$$

where the unit vector $\hat{\boldsymbol{n}}$ is normal to the surface S. Stokes theorem can also be applied to the (X, Y)-gridded PIV velocity data:

$$(\overline{\omega}_Z)_{i,j} = \frac{1}{A} \Gamma_{i,j} = \frac{1}{A} \oint_{\boldsymbol{l}(X,Y)} (U, V) \cdot \mathrm{d}\boldsymbol{l} \tag{6.17}$$

where $(\overline{\omega}_Z)_{i,j}$ reflects the average vorticity within in the enclosed area. In practice equation (6.17) is implemented by choosing a small rectangular contour (figure 6.9, for instance two mesh points wide and two mesh points high) around which the circulation is calculated using a standard integration scheme such as the trapezoidal rule. The local circulation is then divided by the enclosed area to arrive at an average vorticity in this area. The following formula provides a vorticity estimate at point (i, j) based on a circulation estimate around the neighboring eight points:

$$(\omega_Z)_{i,j} \;\widehat{=}\; \frac{\Gamma_{i,j}}{4 \Delta X \Delta Y} \tag{6.18}$$

with

$$\begin{aligned} \Gamma_{i,j} &= \frac{1}{2} \Delta X (U_{i-1,j-1} + 2U_{i,j-1} + U_{i+1,j-1}) \\ &+ \frac{1}{2} \Delta Y (V_{i+1,j-1} + 2V_{i+1,j} + V_{i+1,j+1}) \\ &- \frac{1}{2} \Delta X (U_{i+1,j+1} + 2U_{i,j+1} + U_{i-1,j+1}) \\ &- \frac{1}{2} \Delta Y (V_{i-1,j+1} + 2V_{i-1,j} + V_{i-1,j-1}) \,. \end{aligned} \tag{6.19}$$

Vorticity fields estimated by this expression are shown in figure 6.10. When compared to figures 6.7 and 6.8, this differentiation scheme clearly performs better, especially in the four-times oversampled data. The principle reason for this is that more data enter in each vorticity estimate. A closer inspection of equation (6.18) reveals that the expression is equivalent to applying the center difference scheme (table 6.1) to a smoothed (3×3 kernel) velocity field [7]. While the vorticity estimation by one-dimensional finite differences (table 6.1) requires only 4 to 8 velocity data values this expression utilizes 12 data values. The uncertainty in the vorticity estimate, assuming uncorrelated velocity data, then reduces to $\epsilon_\omega \approx 0.61 \epsilon_U / \Delta X$ compared to $\epsilon_\omega \approx \epsilon_U / \Delta X$ for center differences or $\epsilon_\omega \approx 1.34 \epsilon_U / \Delta X$ for the Richardson extrapolation method. Also the effects due to data oversampling are not as significant as with some of the simpler one-dimensional differentiation schemes because no differences of directly adjoining data are used.

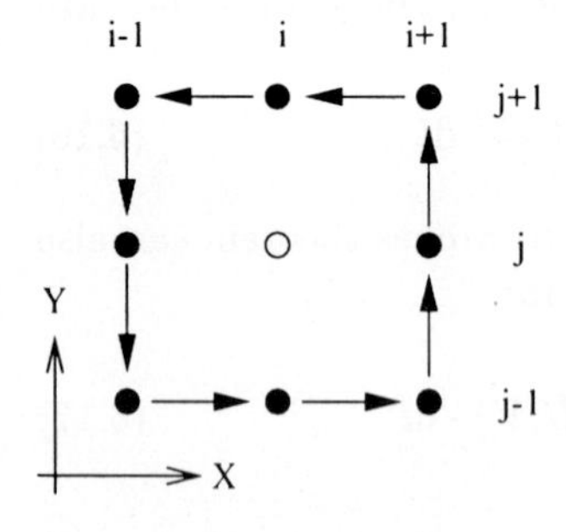

Fig. 6.9. Contour for the circulation calculation used in the estimation of the vorticity at point (i, j)

A similar approach may be used in the estimation of the shear strain and the out-of-plane strain.

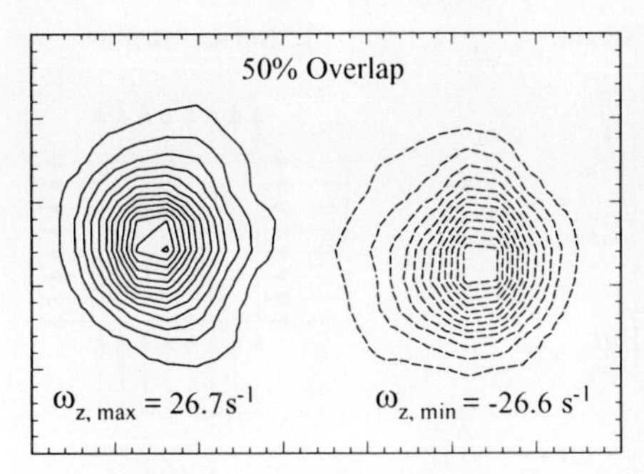

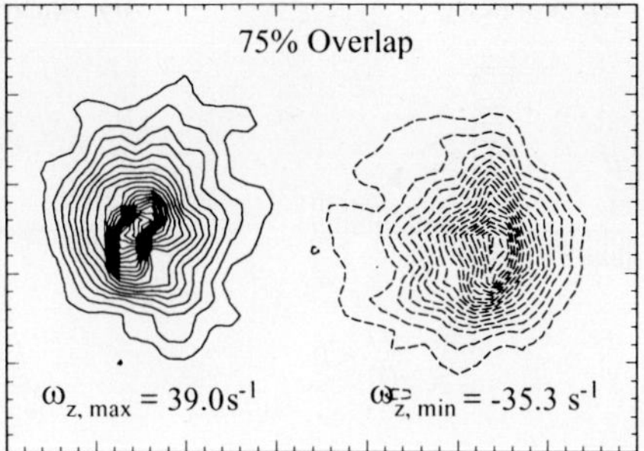

Fig. 6.10. Vorticity field estimates obtained from PIV velocity fields by the circulation method: (left) the velocity field is twice oversampled, (right) four times oversampled. The contours of this laminar vortex pair are known to be smooth such that the nonuniformities are due to measurement noise

$$
\begin{aligned}
(\epsilon_{xy})_{i,j} = \left(\frac{\partial U}{\partial Y} + \frac{\partial V}{\partial X}\right)_{i,j} \quad &\widehat{=} \quad -\frac{U_{i-1,j-1} + 2U_{i,j-1} + U_{i+1,j-1}}{8\Delta Y} \\
&+ \quad \frac{U_{i+1,j+1} + 2U_{i,j+1} + U_{i-1,j+1}}{8\Delta Y} \\
&- \quad \frac{V_{i-1,j+1} + 2V_{i-1,j} + V_{i-1,j-1}}{8\Delta X} \\
&+ \quad \frac{V_{i+1,j-1} + 2V_{i+1,j} + V_{i+1,j+1}}{8\Delta X}
\end{aligned}
\tag{6.20}
$$

$$
\begin{aligned}
-(\epsilon_{zz})_{i,j} = \left(\frac{\partial U}{\partial X} + \frac{\partial V}{\partial Y}\right)_{i,j} \quad &\widehat{=} \quad \frac{V_{i-1,j-1} + 2V_{i,j-1} + V_{i+1,j-1}}{8\Delta Y} \\
&- \quad \frac{V_{i+1,j+1} + 2V_{i,j+1} + V_{i-1,j+1}}{8\Delta Y} \\
&+ \quad \frac{U_{i+1,j-1} + 2U_{i+1,j} + U_{i+1,j+1}}{8\Delta X} \\
&- \quad \frac{U_{i-1,j+1} + 2U_{i-1,j} + U_{i-1,j-1}}{8\Delta X}
\end{aligned}
\tag{6.21}
$$

For the out-of-plane or normal strain an analogy to the vorticity/circulation relation can be given: in place of the circulation the net flow across the boundaries of the contour is calculated. However, no such analogy exists for the shear strain, ϵ_{XY}. Figure 6.11 graphically illustrates the three differential estimation schemes described in this section.

Aside from the above mentioned techniques for differential estimation in PIV velocity data, the literature has suggested alternative methods. Earlier

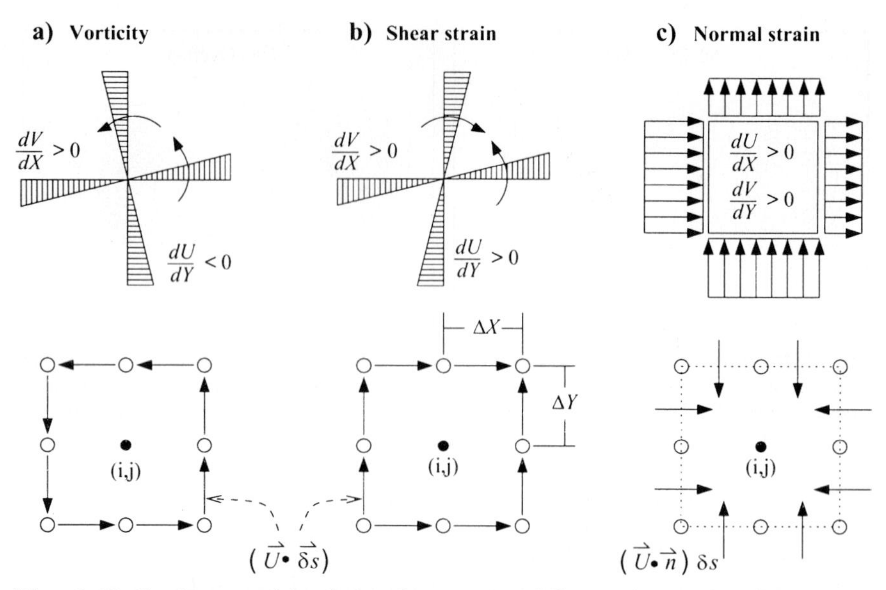

Fig. 6.11. Implementation of the three major differential quantities obtainable with planar PIV data. The deformation of the fluid element is given on top while the bottom shows the path of integration

it was noted that the estimation uncertainty, ϵ_Δ, for the same differentiation scheme is directly proportional to the grid spacing $(\Delta X, \Delta Y)$, that is, $\epsilon_\Delta \approx \epsilon_U/\Delta X$. Once the interrogation window overlap exceeds 50%, the velocity data entering in the differentiation are increasingly correlated (i.e. biased) and cause the differential estimates to be biased. For these reasons LOURENÇO & KROTHAPALLI [121] suggested the use of an adaptive scheme for the computation of vorticity. The method is based on Richardson extrapolation and is aimed at minimizing the total error in the vorticity estimate by combining vorticity estimates at several different grid spacings. Even better results can be obtained by also including a least squares second order polynomial approximation in the differentiation scheme. The extension of this principally one-dimensional approach to differential estimation would certainly also be possible.

The task of differential estimation from velocity field data can also be studied from a two-dimensional signal processing point of view as suggested by LECUONA et al. [120]. Linear filter theory is used to derive and optimize a variety of one- and two-dimensional derivating filters whose performance is tested on noisy PIV data. Vorticity estimates from one such filter (filter f) are shown in figure 6.12. Compared to the circulation method equation (6.18) this differential filter is more susceptible to the side-effects of oversampling described before because it is designed to perform well at higher spatial frequencies.

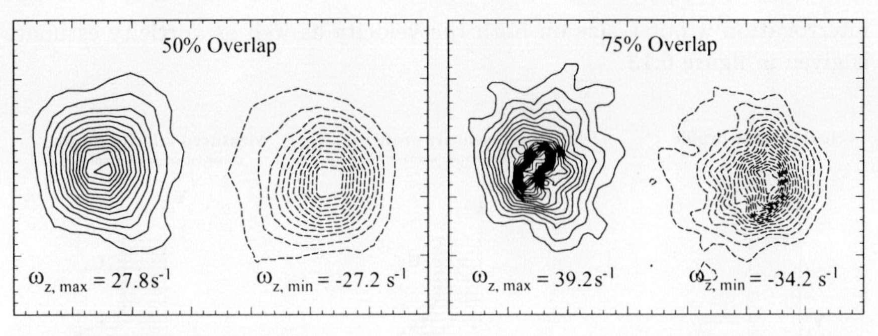

Fig. 6.12. Vorticity field estimates obtained from PIV velocity fields using a linear, two-dimensional filter: (left) the velocity field is twice oversampled, (right) four times oversampled

6.4.3 Uncertainties and errors in differential estimation

As already noted in the previous section a variety of factors enter in the uncertainty of a differential estimate.

Uncertainty in velocity: Each PIV velocity estimate $U_{i,j}$ is associated with a measurement uncertainty ϵ_U whose magnitude depends on a wide variety of aspects such as interrogation window size, particle image density, displacement gradients, etc. (see section 5.5). Since differential estimates from the velocity data require the computation of local differences on neighboring data the noise increases inversely proportional to the local difference, $U_b - U_a$, as the spacing between the data $\Delta X = |X_a - X_b|$ is reduced. That is, the estimation uncertainty in the differential, ϵ_Δ, scales with $\epsilon_U / \Delta X$.

Oversampled velocity data: It is common practice to oversample a PIV recording during interrogation at least twice to fulfill the Nyquist sampling theorem as well as to bring out small-scale features in the flow. Because of this oversampling, neighboring velocity data are estimated partially from the same particle images and therefore are correlated with each other. Because of this, neighboring data are likely to be biased to a similar degree, especially in regions containing high velocity gradients and/or low seeding densities (see section 5.5.7). This localized velocity bias then causes the differential estimate to be biased as well. The oversampling effects can be observed very well by comparing figures 6.7 and 6.8.

Interrogation window size: The size of the interrogation window in the object plane ($\Delta X_0 \times \Delta Y_0$) defines the spatial resolution in the recovered velocity data, provided the sampling positions fulfill the Nyquist criterion ($\Delta X \leq \Delta X_0/2$ and $\Delta Y \leq \Delta Y_0/2$). The spatial resolution in the velocity field in turn limits the obtainable spatial resolution of the differential estimate. Depending on the utilized differentiation scheme the spatial resolution will be reduced to some degree due to smoothing effects. The effect of the interrogation window size on both the velocity as well as vorticity estimate is given in figure 6.13.

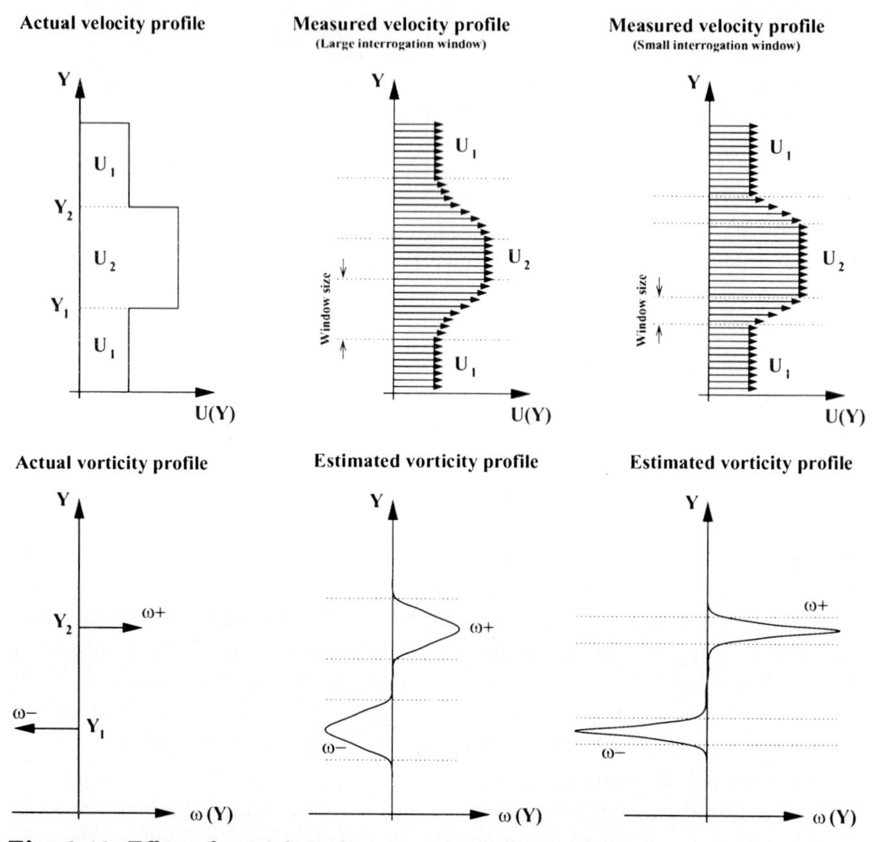

Fig. 6.13. Effect of spatial resolution on vorticity estimation

Curvature effects: The standard PIV method only is a first order approximation to the true particle image displacement. Because it generally relies on only two illumination pulses, effects due to acceleration and curvature are lost. In regions of rotation (i.e. velocity gradients) this straight line approximation underestimates the actual particle image displacement and thereby the local velocity. Differential estimates will then have a tendency to be biased to lower magnitudes as well. By reducing the illumination pulse delay, Δt, this effect can be reduced at the cost of increased noise in the differential estimate due to the velocity measurement uncertainty, ϵ_U, itself.

6.5 Estimation of integral quantities

6.5.1 Path integrals – circulation

By definition, the vorticity integrated over an area A equals the circulation, Γ. With Stokes theorem equation (6.15) this operation reduces to a line integral of the dot product between the local velocity vector, $\boldsymbol{U}$, and the incremental path element vector $\mathrm{d}\boldsymbol{X}$ where the integration path is defined by the boundary, C, of the enclosed area A:

$$\Gamma = \int_A \omega \,\mathrm{d}A \tag{6.22}$$

$$= \oint_C \boldsymbol{U} \cdot \mathrm{d}\boldsymbol{X} \tag{6.23}$$

For two-component velocity data constrained to the XY plane with $\boldsymbol{U} = (U, V)$, the above reduces to:

$$\Gamma = \iint_{A(X,Y)} \omega_Z \,\mathrm{d}X\,\mathrm{d}Y \tag{6.24}$$

$$= \oint_{C(X,Y)} \boldsymbol{U}(\boldsymbol{X},\boldsymbol{Y}) \cdot \mathrm{d}\boldsymbol{X} \tag{6.25}$$

$$= \oint_{C(X,Y)} U\,\mathrm{d}X + V\,\mathrm{d}Y \tag{6.26}$$

Given the path of integration, the evaluation of equation (6.26) is straightforward using integration schemes such as the trapezoidal approximation or Simpson's rule. To determine the circulation of clearly defined, nearly round vortical structures, a circular integration path centered at the position of maximum vorticity is generally sufficient. By plotting the circulation with respect to integration path radius an asymptotic convergence toward the value of the structure's circulation can be observed (provided no other vortices are included by the integration contour). This convergence coincides with a decay of vorticity away from the vortex core.

For more complex vortical structures the assignment of a reasonable integration path is not as simple. For vortical structures the ideal integration path would be defined by a dividing stream line which separates it from other vortical structures. However, the computation of the required stream function from the unsteady velocity data is nontrivial and often nonunique (see section 6.5.3). Since the circulation actually is an area-integral of vorticity an integration along a constant-vorticity contour near zero will evaluate to a value close to the vortex's actual circulation. This approach was chosen for instance in the evaluation of the Karman vortex street data shown in figure 6.14. Although this approach is very robust, the difficulty often lies in retrieving the desired closed contour from the vorticity data. Once the contour is available a bi-linear interpolation of the velocity on to the contour is sufficient for the evaluation of equation (6.26).

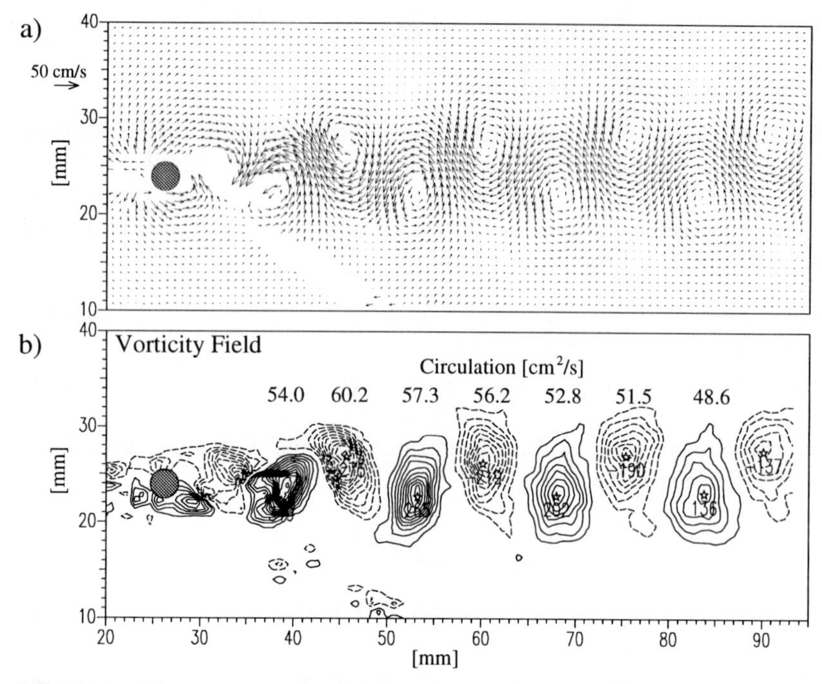

Fig. 6.14. The contours of the vorticity field are used as integration paths for the circulation whose magnitude is given above the individual vortices (data courtesy of SCHRÖDER [148])

6.5.2 Path integrals – mass flow

In some applications the rate mass or volume across a control surface, CS, is of interest and is expressed as a surface integral:

$$\dot{\boldsymbol{M}} = \frac{\mathrm{d}m}{\mathrm{d}t} = \iint_{CS} \rho(\boldsymbol{U} \cdot \hat{\boldsymbol{n}})\, \mathrm{d}S \tag{6.27}$$

For two-dimensional data constrained to the xy plane, the surface reduces to a path integral similar to equation (6.26):

$$\dot{\boldsymbol{M}}_{XY} = \frac{\mathrm{d}m_{XY}}{\mathrm{d}t} = \oint_C \rho(U\, \mathrm{d}Y - V\, \mathrm{d}X) \tag{6.28}$$

The units of $\dot{\boldsymbol{M}}_{XY}$ are mass flow per unit depth and if $\rho \equiv 1$ then equation (6.28) represents a volume flow rate per unit depth. With regard to its numerical implementation similar integration schemes as for the estimation of the circulation (section 6.5.1) can be used.

In cases where three-dimensional velocity data is available in a plane, the actual mass (or volume) flow rate across this surface or portions thereof can be determined using an area integral:

$$\dot{M} = \iint_{A(X,Y)} W \, \mathrm{d}X \, \mathrm{d}Y \tag{6.29}$$

where W is the velocity component normal to the light sheet. In this case the approximation of the integral is more complicated than for the previously described line integrals.

6.5.3 Area integrals

The following integration schemes are based on the assumption that the integrand, that is, the flow field, is two-dimensional as well as incompressible. In this case potential theory relates the velocity field, $\boldsymbol{U} = (U(X,Y), V(X,Y))$, to the stream function, $\boldsymbol{\Psi}$, and potential function, $\boldsymbol{\Phi}$:

$$U = \frac{\partial \boldsymbol{\Psi}}{\partial Y} = \frac{\partial \boldsymbol{\Phi}}{\partial X} \tag{6.30}$$

$$V = -\frac{\partial \boldsymbol{\Psi}}{\partial X} = \frac{\partial \boldsymbol{\Phi}}{\partial Y} \tag{6.31}$$

which can be integrated over the domain (i.e. XY plane) to:

$$\boldsymbol{\Psi} = \int_Y U \, \mathrm{d}Y - \int_X V \, \mathrm{d}X \tag{6.32}$$

$$\boldsymbol{\Phi} = \int_X U \, \mathrm{d}X + \int_Y V \, \mathrm{d}Y \tag{6.33}$$

Although these purely kinematic conditions will work reasonably well for the flows studied with PIV, the inherent problem is that, depending on the chosen frame of reference, nonunique solutions to $\boldsymbol{\Psi}$ and $\boldsymbol{\Phi}$ are obtained. This is due to the fact that equation (6.32) and equation (6.33) are a simplification of the Poisson equation:

$$\nabla^2 \boldsymbol{\Psi} = -\omega_Z \tag{6.34}$$

to a Laplace equation, $\nabla^2 \boldsymbol{\Psi} = 0$ using the assumption of irrotationality (i.e. $\omega_Z = 0$). The integration of equation (6.34) is rather difficult because the integrand can only be approximated from the velocity field data (see section 6.4). Further, the boundary conditions along the edges of the field of view need to be defined prior to integrating equation (6.34).

Figure 6.15 shows the result of integrating equation (6.32) on an actual flow, a vortex pair, which can be assumed to be nearly two-dimensional, but not irrotational. Depending on the choice of the reference frame two entirely different results are obtained. For instance, if the vortex propagation speed is subtracted, the bounding streamline of the Kelvin oval, that is, the body of fluid moving with the vortex pair, can be approximated. This is not the case for streamlines computed in a laboratory-fixed reference frame (figure 6.15 left).

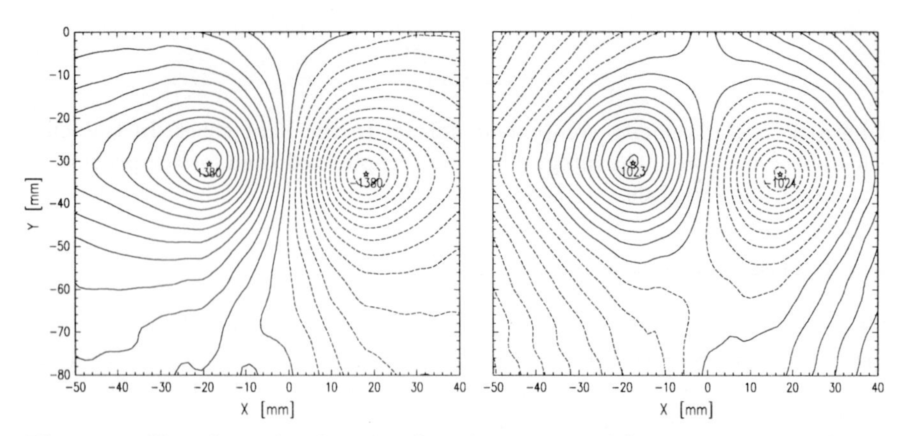

Fig. 6.15. Two-dimensional stream function computed from vortex pair velocity data in a laboratory-fixed reference frame (left) and in a reference frame moving 20 mm/s upward with the vortex pair (right).

Since equation (6.32) and equation (6.33) are path-independent integrals, the numerical integration of the velocity field can be freely chosen. In this case an integration scheme as presented by IMAICHI & OHMI [164] is used. The trapezoid approximation is used to integrate between two neighboring points. To start the integration a starting point, P_o, is chosen, preferably near the middle of the velocity field since errors in the individual velocity data are propagated through integration. As illustrated in figure 6.16 there are two principal integration methods: a column-major integration or a row-major integration. In the first case, the integration proceeds in opposite horizontal directions away from the starting point, P_o, producing new values of the integral for each node on the horizontal. These new estimates are then used as initial values for the integration in opposite directions along the vertical columns, producing estimates of the integral throughout the domain. A second estimate of the integral can then be obtained by reversing the order of the integration scheme, that is, by starting the integration off in opposite directions along the vertical line containing the starting point. The two results are then arithmetically averaged together.

Since the described integration scheme tends to propagate disturbances due to noisy or erroneous data "down-stream" of its occurrence, more sophisticated integration schemes, such as a multigrid approach, could be used. In this case the integral is first computed on highly smoothed and subsampled versions of the flow field and successively updated as the sampling mesh is refined.

6.5.4 Pressure field estimation

If the flow field under investigation is nearly two-dimensional, steady (i.e. $\mathrm{d}U/\mathrm{d}t = 0$) as well as incompressible (i.e. $\mathrm{d}\rho/\mathrm{d}t = 0$) the pressure field can

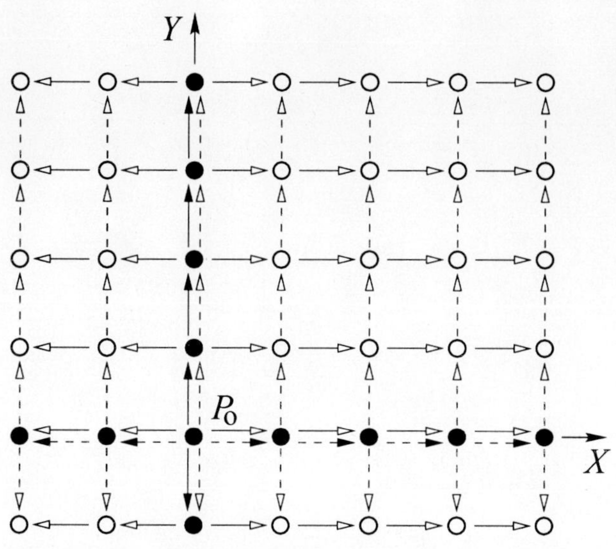

Fig. 6.16. Integration routes used to integrate two-dimensional stream and potential functions as well as pressure. Two paths of integration follow either the dashed or solid arrows (after [164])

be estimated through the numerical integration of the steady Navier–Stokes equations in two-dimensional form [164]:

$$U\frac{\partial U}{\partial X} + V\frac{\partial U}{\partial Y} = -\frac{1}{\rho}\frac{\partial p}{\partial X} + \nu\left(\frac{\partial^2 U}{\partial X^2} + \frac{\partial^2 U}{\partial Y^2}\right), \tag{6.35}$$

$$U\frac{\partial V}{\partial X} + V\frac{\partial V}{\partial Y} = -\frac{1}{\rho}\frac{\partial p}{\partial Y} + \nu\left(\frac{\partial^2 V}{\partial X^2} + \frac{\partial^2 V}{\partial Y^2}\right). \tag{6.36}$$

To obtain the pressure field, the pressure gradients $\partial p/\partial X$ and $\partial p/\partial Y$ are approximated using finite difference approximations of the velocity gradients. The pressure gradients are then iterated starting from a point near the center with an initial value of p_0 using an integration scheme similar to figure 6.16.

7. Three-component PIV measurements on planar domains

In spite of all its advantages, the PIV method underlies some shortcomings that make further developments on the basis of instrumentation necessary. One of these disadvantages is the fact that the "classical" PIV method is only capable of recording the projection of the velocity vector into the plane of the light sheet; the out-of-plane velocity component is lost while the in-plane components are affected by an unrecoverable error due to the perspective transformation as described in section 2.4.3. For highly three-dimensional flows this can lead to substantial measurement errors of the local velocity vector. This error increases as the distance to the principal axis of the imaging optics increases. Thus it is often advantageous to select a large viewing distance in comparison to the imaged area to keep the projection error to a minimum. This is easily achieved using long focal length lenses.

Nevertheless, an increasing number of PIV applications require the additional knowledge of the out-of-plane velocity component.

A variety of approaches capable of recovering the complete set of velocity components have been described in the literature [116, 32]. The most straightforward, but not necessarily easily implemented, method is additional PIV recording from a different viewing axis using a second camera, which can be generally referred to as stereoscopic PIV recording [112, 117, 108]. Stereoscopic recording can also be achieved with a single camera by placing a set of mirrors in front of the recording lens [101]. Holographic PIV recording is another approach capable of recovering the third displacement component [159, 106]. If sufficient illumination power is present, the light sheet can be expanded into a thick slab such that three-dimensional PIV measurements are possible throughout a volume [102]. A completely different approach, referred to as *dual-plane* PIV, is implemented by offsetting the light sheet a small amount between the recordings to obtain a third PIV recording [114]. Based on measuring the change of the respective correlation peak height from one recording to the next, the out-of-plane displacement component can be estimated. A similar approach was utilized for the analysis of PIV image sequences obtained in a scanning light sheet set-up [103].

Of all the existing three-dimensional PIV methods, holography is capable of the highest measurement precision but, at its present stage of development, is not well suited for experiments, where set-up time, optical access and observation distances are important factors.

For application to low-speed liquid flows the "dual-plane" extension to standard PIV may be the most easily implemented as it only requires an additional third illumination pulse and a slight out-of-plane displacement of the light sheet. The measurement precision of the third component does however depend on a continuous and known shape of the light sheet's intensity profile among other factors many of which are still under investigation. Unless special measures are taken, the intensity profiles of the commonly utilized frequency doubled Nd:YAG lasers are not adequate for providing out-of-plane information with reasonable precision. Also, the recording of three separate images in short succession (microsecond range) is not easily tackled with existing PIV equipment; high-speed video or cinematography is necessary. Therefore, the dual-plane technique, which is described in section 7.2, has not yet successfully been applied to flows of medium or high speed.

The approach described next – followed by many researchers – is that of stereoscopic imaging for which much knowledge has already been accumulated [112, 101, 165].

7.1 Stereo PIV

In the following a stereoscopic approach will be described, which has been developed for application in an industrial wind tunnel environment. The adaptation of this approach to applications in liquid flows can easily be performed by changing the angles between the lens plane and sensor plane according to the refraction of the air-water interfaces. A detailed description of the adaptation of stereo PIV to liquid flows is given in [112, 113].

Since the measurement precision of the out-of-plane component increases as the opening angle between the two cameras reaches 90 degrees, it is not always possible to mount the cameras on to a common base, when using large focal length lenses, and much less to provide a symmetric arrangement. Therefore a general description for asymmetric recording and associated calibration has been developed.

Another problem that arises when large focal length imaging lenses are used is that their limited angular aperture restricts the distance between the lenses in a translation imaging approach (figure 7.1a). Designed for use with a fixed format sensor centered on the optical axis of the lens, most lenses are not only limited in their optical aperture but are also characterized by a strong decrease in the modulation transfer function (MTF) toward the edges of the field of view. To adequately image small particles a good MTF at small f numbers ($f_{\#} < 4$) is a stringent requirement. Since lens systems with an oblique principal axis are practically nonexistent, a departure from the translation imaging method of figure 7.1a is unavoidable. As the best MTF is generally present near the lens principal axis the alternative angular displacement method (figure 7.1b) aligns the lens with the principal viewing direction. The additional requirement for small f-numbers is associated with a very small depth of field which only can be accommodated by additionally tilting the back plane according to the Scheimpflug criterion in which the image plane, lens plane and object plane for each of the cameras intersect in a common line [113, 32]. This has the side-effect of introducing a strong perspective distortion. In essence the magnification factor is no longer constant across the field of view and requires an additional means of calibration.

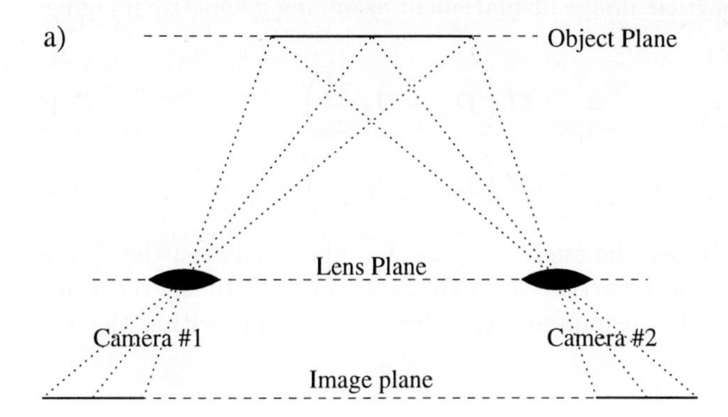

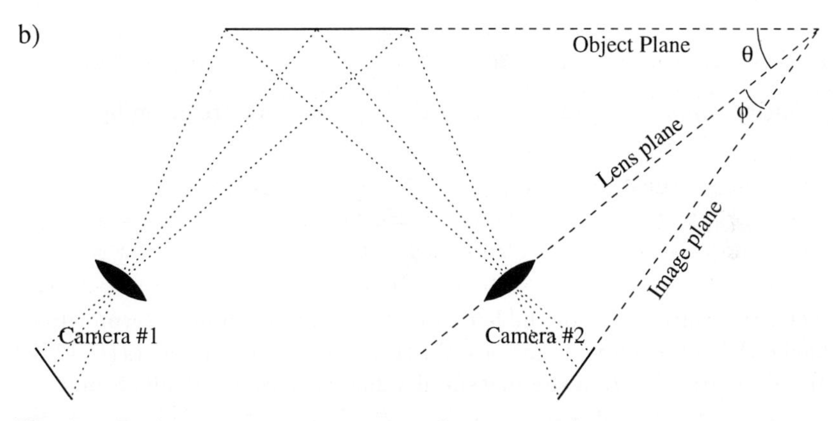

Fig. 7.1. Basic stereoscopic imaging configurations: a) lens translation method, b) angular lens displacement with tilted back plane (Scheimpflug condition)

In the following sections the generalized, that is, nonsymmetric, description for stereoscopic PIV imaging is given first and is followed by a methodology for calibrating for the perspective distortion. The feasibility of this approach is demonstrated in an experiment with an unsteady flow field, which is described in section 8.3.

7.1.1 Reconstruction geometry

This section describes the geometry necessary to reconstruct the three-dimensional displacement field from the two projected, planar displacement fields. Past descriptions of stereoscopic PIV imaging systems assume a symmetric configuration [90, 107, 112, 113, 117]. In the present case the two cameras may be placed in any desirable configuration provided the viewing axes are not collinear.

In section 2.4.3 we determined the basic equations- equation (2.19) and equation (2.20)- for particle image displacement assuming geometric imaging:

$$x'_i - x_i = -M\left(D_X + D_Z \frac{x'_i}{z_0}\right) \tag{7.1}$$

$$y'_i - y_i = -M\left(D_Y + D_Z \frac{y'_i}{z_0}\right) \tag{7.2}$$

In the following we will use the angle α in the XZ plane between the Z axis and the ray from the tracer particle through the lens center to the recording plane (see figure 2.28). Correspondingly, β defines the angle within the YZ plane.

$$\tan\alpha = \frac{x'_i}{z_0}$$

$$\tan\beta = \frac{y'_i}{z_0}$$

The velocity components measured by the right camera are given by:

$$U_\mathrm{r} = -\frac{x'_i - x_i}{M\Delta t}$$

$$V_\mathrm{r} = -\frac{y'_i - y_i}{M\Delta t}$$

The velocity components for the left camera U_l and V_l can be determined accordingly. Using the above equations, the three velocity components (U, V, W) can be reconstructed from the measured values. For $\alpha, \beta \geq 0$ we obtain:

$$U = \frac{U_\mathrm{r}\tan\alpha_\mathrm{l} + U_\mathrm{l}\tan\alpha_\mathrm{r}}{\tan\alpha_\mathrm{r} + \tan\alpha_\mathrm{l}} \tag{7.3}$$

$$V = \frac{V_\mathrm{r}\tan\beta_\mathrm{l} + V_\mathrm{l}\tan\beta_\mathrm{r}}{\tan\beta_\mathrm{r} + \tan\beta_\mathrm{l}} \tag{7.4}$$

$$W = \frac{U_\mathrm{l} - U_\mathrm{r}}{\tan\alpha_\mathrm{r} + \tan\alpha_\mathrm{l}} \tag{7.5}$$

$$= \frac{V_\mathrm{l} - V_\mathrm{r}}{\tan\beta_\mathrm{r} + \tan\beta_\mathrm{l}} \tag{7.6}$$

These formulae are general and apply to any imaging geometry. However, the denominators can approach zero as the viewing axes become collinear in either of their two-dimensional projections. For example, in the set-up described in section 8.3 the cameras are approximately positioned in the same vertical position as the field of view which makes the angles β_r, β_l and their tangents $\tan\beta_r$ and $\tan\beta_l$ very small. Clearly, W can only be estimated with higher accuracy using equation (7.5), while V has to be rewritten using equation (7.5) which does not include $\tan\beta_r$ and $\tan\beta_l$ in the denominator:

$$V = \frac{V_r + V_l}{2} + \frac{W}{2}(\tan\beta_l - \tan\beta_r) \tag{7.7}$$

$$V = \frac{V_r + V_l}{2} + \frac{U_l - U_r}{2}\left(\frac{\tan\beta_l - \tan\beta_r}{\tan\alpha_r + \tan\alpha_l}\right) \tag{7.8}$$

If $\tan\beta_r$ and $\tan\beta_l$ are very small, then V is given as the arithmetic mean of V_r and V_l with the out-of-plane component W having no effect.

To use the above reconstruction, the displacement data set must first be converted from the image plane to true displacements in the global coordinate system taking account of all magnification issues. All of this is taken care of in the image dewarping and calibration procedure described next.

7.1.2 Image dewarping

In order to determine the local magnification factor the mapping between the image (x, y) and the object plane (X, Y) has to be determined. Geometric back projection which is based on geometric optics could be used; however, it requires exact knowledge of the imaging parameters such as the lens focal length, f, the angles between the various planes, θ, ϕ (see figure 7.1b), the actual position of the lens plane (which is not simple to determine) and the nominal magnification factor, M_o (the magnification along the principal optical axis):

$$X = \frac{f\,x\,\sin\phi}{M_0 \sin\theta(x\,\sin\phi + f\,M_0)}$$

$$Y = \frac{f\,y}{x\,\sin\phi + f\,M_0}$$

These approximate expressions, given in [111], do not incorporate nonlinearities such as lens distortions and are sensitive to small variations in each of the parameters.

A more robust approach is the second order warping approach employed by other researchers [20, 111]:

$$X = a_1 x^2 + a_2 y^2 + a_3 xy + a_4 x + a_5 y + a_6 \quad (7.9)$$
$$Y = b_1 x^2 + b_2 y^2 + b_3 xy + b_4 x + b_5 y + b_6 \quad (7.10)$$

The above equations do not constitute a mapping based on the geometry at hand. Nevertheless the twelve unknown parameters can be easily determined using a least squares approach if at least six image-object point pairs are given. The advantage of this approach is that the imaging parameters such as focal length, magnification factor, etc., never need to be determined. Also lens distortions or other image nonlinearities can be accounted for by the higher order terms.

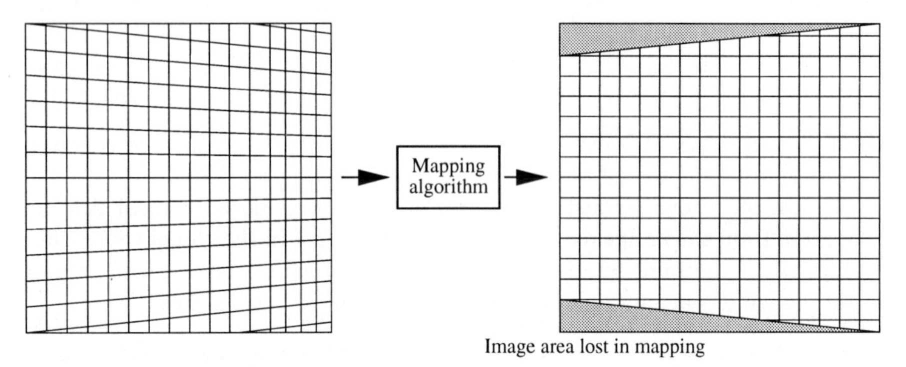

Fig. 7.2. The backprojection algorithm has to map the recorded image on the left to the reconstructed image on the right.

For the reconstruction of the images we implemented the projection equations based on perspective projection as put forth in [173, 16]. Using homogeneous coordinates the perspective projection is expressed by:

$$\begin{bmatrix} w_o X \\ w_o Y \\ w_o \end{bmatrix} = \begin{bmatrix} a_{11} & a_{12} & a_{13} \\ a_{21} & a_{22} & a_{23} \\ a_{31} & a_{32} & a_{33} \end{bmatrix} \cdot \begin{bmatrix} w_i x \\ w_i y \\ w_i \end{bmatrix} \quad (7.11)$$

where w_o and w_i are constants and $a_{33} = 1$. When rewritten in standard coordinates the following two nonlinear expressions are obtained:

$$X = \frac{a_{11} x + a_{12} y + a_{13}}{a_{31} x + a_{32} y + 1} \quad (7.12)$$

$$Y = \frac{a_{21} x + a_{22} y + a_{23}}{a_{31} x + a_{32} y + 1} \quad (7.13)$$

The principle property of the perspective projection is that it maps a rectangle onto a general four-sided polygon. In other words this mapping preserves only the straightness of lines. By setting a_{31} and a_{32} equal to zero the perspective transformation equation (7.11) reduces to the more frequently used affine transformation which can only map a rectangle onto a parallelogram.

To account for geometric distortions due to imperfect imaging optics (pincushion and barrel distortion) equation (7.12) can be extended to a higher order:

$$
\begin{aligned}
X &= \frac{a_{11}x + a_{12}y + a_{13} + a_{14}x^2 + a_{15}y^2 + a_{16}xy}{a_{31}x + a_{32}y + a_{33} + a_{34}x^2 + a_{35}y^2 + a_{36}xy} && (7.14)\\
Y &= \frac{a_{21}x + a_{22}y + a_{23} + a_{24}x^2 + a_{25}y^2 + a_{26}xy}{a_{31}x + a_{32}y + a_{33} + a_{34}x^2 + a_{35}y^2 + a_{36}xy} && (7.15)\\
a_{33} &= 1
\end{aligned}
$$

The determination of the unknowns in equation (7.12) and especially in equation (7.14) by means of a linear least squares method is not as simple as for the second order warping approach given in equation (7.9) because the equations no longer constitute linear polynomials but rather are ratios of two polynomials of the same order. Strongly deviating or erroneous point pairs cause a standard least squares method to rapidly diverge from the "true" best match. To find the best match to the eight or seventeen unknowns, a nonlinear least squares method such as the Levenberg–Marquart method [21] is required.

The Levenberg–Marquart method is implemented by first solving for the unknowns in the first order projection equations of equation (7.12) and using these as initial estimates for the solution of the higher order unknowns in equation (7.14).

7.1.3 Calibration procedure

The calibration grid, a thin overhead transparency with black line rulings at a 5 mm square spacing, is clamped between two 2 mm glass plates and placed inside the light sheet. Calibration consists of a three step process:

Step 1: After inversion (positive to negative), each of the two calibration grid images is cross-correlated with a + correlation mask; the correlation peaks indicate the position of the line crossings and are fitted with the typical three-point (Gaussian) estimator used to obtain subpixel shift information. The same approach to grid registration was already implemented by [105] although no image enhancement is required in this case.

Step 2: The coordinates of the detected line crossing are automatically reconstructed into a rectangular grid. Each of its nodes is assigned with corresponding coordinates in physical space. At this stage care has to be taken that the assigned physical coordinates match between the two camera views.

Step 3: The previously mentioned Levenberg–Marquart nonlinear least squares fit based on equation (7.14) is applied to each of the two image-object point pair sets yielding a set of reconstruction coefficients for each camera view.

By choosing an adequate multiplication factor for the points in physical space, a set of coefficients can be calculated which map a given image to a new but back projected image with a constant magnification factor. In the application given in section 8.3, the 1018×1008 pixel input images are mapped to 1450×1200 pixel output images with a constant magnification factor of 10 pixel/mm. In the original image the magnification factor varied between 8.95 and 8.43 pixel/mm vertical and between 7.54 and 6.74 pixel/mm horizontal. In this case the choice of a magnification factor which results in a significantly larger output image is primarily a practical one: the particle image density is sufficiently high to reduce the interrogation window to be less than 32×32 pixel. Since the interrogation algortihm is based on radix-2 sized interrogation windows (due to the use of FFT's in the correlation) a reduction to a 16×16 pixel interrogation window is not feasible due to a significantly higher measurement uncertainty and a drop in the validation rate. By enlarging the image by approximately 20% the spatial resolution can also be increased by 20% as the same 32×32 pixel window samples a smaller area.

The residuals of the nonlinear least squares fit, given in table 7.1, indicate that the recovered line crossing could be assigned to within $\pm 25\,\mu m = \pm 0.25$ pixel. This value is on the same order as the registration of the line crossings by means of cross-correlation. Table 7.1 further shows that the present set-up does not require a higher degree of fit which is a clear indication that the imaging lenses are nearly free of geometric distortions.

Table 7.1. Residuals of the nonlinear least squares fit to the projection equations (7.12) and (7.14). The corresponding magnification factor is 10 pixel/ mm

Degree of fit	Deviations	Camera 1		Camera 2	
		x	y	x	y
1st order	Mean	$\pm 22\,\mu m$	$\pm 11\,\mu m$	$\pm 21\,\mu m$	$\pm 12\,\mu m$
	Maximum	$83\,\mu m$	$62\,\mu m$	$69\,\mu m$	$80\,\mu m$
2nd order	Mean	$\pm 20\,\mu m$	$\pm 10\,\mu m$	$\pm 18\,\mu m$	$\pm 11\,\mu m$
	Maximum	$77\,\mu m$	$59\,\mu m$	$70\,\mu m$	$82\,\mu m$

The back projection is implemented by stepping through the output image in subpixel increments (typically 4×4 subpixels for each pixel) and accumulating the intensity contributions from each of the corresponding positions in the input image. The more subpixels are chosen the smoother the output image will become at the cost of increased computational time.

Although the mismatch between the two back projected images is less than ± 0.5 pixel across the viewing area, this method of image reconstruction is not ideal because it requires the calibration grid to be perfectly aligned within the light sheet. If this is not the case, the image reconstruction will still yield nicely overlapping dewarped images of the calibration grid, while the correct light sheet position may differ. To test for this effect, back projected particle image recordings of the same time instant but different viewing directions were cross-correlated with each other. A second degree least squares fit to the recovered displacements filtered out all small-scale variations due to the measurement uncertainty inherent in the correlation method and due to the Poisson distribution of the particles within the light sheet which may cause groups of particles to be concentrated toward one side of the light sheet. The final displacement field, shown in figure 7.3, shows deviations in excess of 10 pixels (1 mm) toward the edges of the field even though the reconstructed calibration grids overlapped to better than 1 pixel (0.1 mm, table 7.1).

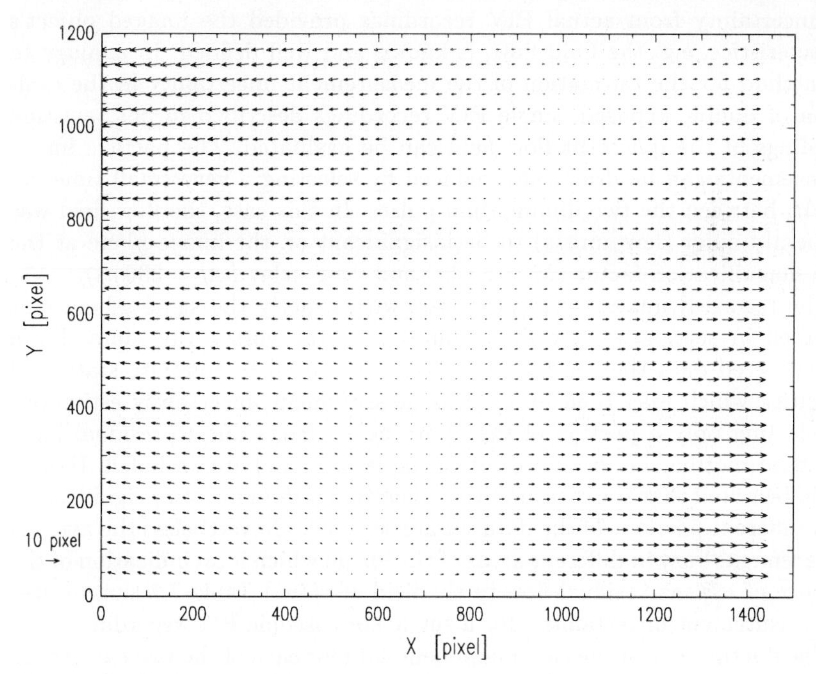

Fig. 7.3. Residual misalignment between the two viewing directions due to a slight misalignment of the calibration grid within the light sheet. A 6 pixel horizontal misalignment has already been removed

This suggests that the calibration grid was not perfectly parallel with the light sheet but slightly rotated along both the X-axis and Y-axis near the center of the images. In fact a rotation of the calibration grid of less than 0.6° is already sufficient to produce the measured horizontal misalignment of 10 pixels. Unless this misalignment is accounted for, the interrogation areas will be mismatched by up to 1 mm which corresponds to a mismatch of one third of a interrogation window. For the data given in section 8.3 the interrogation windows were offset by 6 pixel between the two viewing directions to partially account for the misalignment. For future applications a more reliable calibration scheme which ensures an exact alignment of the calibration grid with the light sheet is required.

7.1.4 Error estimation

In light of the large variety of relevant parameters an estimation of the measurement uncertainty for a complex device such as this stereoscopic PIV system is not trivial. In general two approaches can be used: statistical approaches based on a simulated system through the use of numerically generated PIV recordings, that is, Monte Carlo simulations, such as described in section 5.5. The second method, which is applied here, is to directly estimate the uncertainty from actual PIV recordings provided the imaged object's characteristics, e.g. the fluid flow, is known and well defined. In analogy to the method for the estimation of the measurement uncertainty in the evaluation of double exposed, single PIV recordings described in [99], existing recordings of the quiescent flow field can be evaluated. The particle image displacements can be drastically reduced by selecting a very small time delay, Δt, between the two illumination pulses. In this case, the flow field was sufficiently quiet (few mm/s) to be insignificant on the image plane at the given magnification factor ($M \approx 1:14$) and time delay ($\Delta t = 200\,\mu$s).

The recorded images were processed with exactly the same schemes as described in section 8.3 for the evaluation of the vortex ring data. From the recovered data the central 50×50 mm was used to compute statistical quantities which may then be used as measurement uncertainty estimates. By selecting only a portion of the entire field, effects due to residual large scale fluid motion (i.e. from a draft in the laboratory) are excluded. Here it should be noted that the measurement uncertainty does not vary significantly when different portions of the data set are selected. (Nevertheless the average displacements vary in different areas of the image which is an indication of the presence of a weak residual flow in the field of view.) Table 7.2 summarizes the measurement uncertainties for a single stereoscopic PIV recording.

The fluctuations of the two-component data for each of the two cameras as well as the reconstructed displacement field is given. Overall the uncertainties match their predicted behavior, that is, the in-plane components (σ_U, σ_V) are roughly a factor $1/\sqrt{2}$ smaller than the fluctuations obtained from a single recording which corresponds to the estimate predicted by standard error propagation. The out-of-plane fluctuation is predicted to be a factor $1/\tan\alpha$ larger than the in-plane components which, for the present angle of $\alpha = 35°$, seems to underestimate the actual value. If the displacement uncertainty is expressed in terms of pixels, the data obtained for a single viewing axis falls directly in line with the Monte Carlo predictions given in [156] and restated at the bottom of table 7.2. The preprocessing of the images also included a binarization such that the Monte Carlo estimates for single-bit images apply.

Table 7.2. Measurement uncertainties for the present stereoscopic PIV system as obtained from processing of images with quiescent flow

	Fluctuations in [μm]			Fluctuations in [pixel]		
	RMS	Min.	Max.	RMS	Min.	Max.
Total						
$\sigma_{\Delta x}$	±6.30	-20.1	24.2	±0.063	-0.201	0.242
$\sigma_{\Delta y}$	±6.24	-20.9	24.1	±0.062	-0.209	0.241
$\sigma_{\Delta z}$	±10.3	-29.0	33.7	±0.103	-0.290	0.337
Camera #1						
$\sigma_{\Delta x_1}$	±11.3	-37.9	41.1	±0.113	-0.379	0.411
$\sigma_{\Delta y_1}$	±9.69	-33.8	30.5	±0.097	-0.338	0.305
Camera #2						
$\sigma_{\Delta x_2}$	±7.41	-30.8	28.0	±0.074	-0.308	0.280
$\sigma_{\Delta y_2}$	±8.19	-29.8	34.3	±0.082	-0.298	0.343
Monte Carlo simulation results						
$\sigma_{\Delta x}$, 8-bit	–	–	–	±0.060	-0.216	0.084
$\sigma_{\Delta x}$, 1-bit	–	–	–	±0.090	-0.291	0.225

The higher measurement uncertainty in the horizontal direction can be explained by the increased horizontal stretching of the original image due to the backprojection which transforms initially round images into ellipses. For particle images larger than the optimum diameter of $d_\tau \approx 2$ pixel the measurement uncertainty increases as the diameter increases (see section 5.5.2). In this case the particle image diameters were on the order of $d_\tau \approx 3$ pixel in the original images, whereas they measured more than 4 pixels after the backprojection. Another noteworthy observation is the increased level of uncertainty for camera #1 whose particle images were not as well focused as on camera #2.

Overall the measurement system in its present configuration (M, α, etc.) can be said to have a displacement measurement uncertainty of $\sigma_{D_X} = \sigma_{D_Y} \approx 6\,\mu$m for the in-plane components and $\sigma_{D_Z} \approx 10\,\mu$m for the out-of-plane component. Since the cameras are placed to either side of the light sheet the included angle is not constant across the field of view and hence the measurement uncertainty for the out-of-plane component is not constant as in the "classical" stereoscopic imaging arrangement. For the rather narrow field of view ($\approx 5°$) this variation is of no significance. For longer focal length lenses the even smaller field of view further reduces this effect.

7.1.5 Conclusion and outlook

A stereoscopic PIV camera system based on a pair of digital cameras in an angular imaging arrangement has been described. To achieve focused particle images across the entire field of view at small f-numbers, the camera back planes are tilted backwards satisfying the Scheimpflug imaging criterion. The accompanying projective distortion in the recorded images is accounted for prior to their evaluation using a nonlinear image dewarping algorithm based on projective geometry. The reconstruction method also serves as an accurate means of recovering the magnification factor. Live image display allows for an easy alignment of digital cameras to meet the Scheimpflug criterion. To avoid the introduction of systematic errors, care must be taken in the alignment of the calibration grid with the light sheet plane. The application of the camera system to an unsteady air flow (see section 8.3) indicated displacement measurement uncertainties on the order of 6 μm for the in-plane components and 10 μm for the out-of-plane component in a field of view covering 145×120 mm. The measurement precision of the out-of-plane component could be further increased by aligning the camera axes orthogonal to each other; in the present case the enclosed angle was roughly 70°.

7.2 Dual-plane PIV

In this section we will describe an approach to obtain information about the out-of-plane velocity components, which is based on the analysis of the height of the peak in the correlation plane. This value depends on the portion of paired particle images, which itself depends on the out-of-plane velocity component and on other parameters. To circumvent problems with other influences (e.g. background light, amount and size of images), images from an additional light sheet plane parallel to the first one were also captured for peak height normalization. This concept is referred to as "dual-plane" PIV [114] or as the "spatial-correlation" technique [104]. Experimental results of this technique are shown in section 8.4. Two problems arise when measuring highly three-dimensional flows with PIV.

First, it is obvious from the description of the recording process that an appropriate pulse separation Δt can only be selected for flows with a limited extent of the out-of-plane velocity component. This restriction is imposed because particles moving perpendicular to the light sheet will leave and enter the light sheet in between the two illumination pulses and thus will not correlate when evaluating the PIV recording. This leads to a decreased probability in detecting the particle image displacement, as has been shown by numerical simulation (see section 5.4). If the detection probability is increased by a strongly reduced Δt, an amplification of the measurement noise will be obtained as described in section 3.

Second, Second, allowing a significant velocity component perpendicular to the light sheet plane leads to an additional error because the camera lens reproduces the tracer particles by perspective projection and not by parallel projection. The only way to avoid this error when observing highly three-dimensional flows is to measure all three components of the velocity vectors (see section 2.4.3).

The most commonly used technique to measure instantaneous three-dimensional velocity fields in a plane is stereoscopic PIV, which has been described in the previous section. For this technique the achievable accuracy of the measurement of the out-of-plane velocity component is less than that of the in-plane component, but the out-of-plane measurement error can be reduced below 1% of full scale in best cases (see section 7.1). As already mentioned, specially developed calibration techniques and two recording cameras are necessary for the stereoscopic approach, which both need optical access to the test section. This makes the operation of stereoscopic PIV more difficult than that of dual-plane PIV, when dealing with low speed liquid flows.

7.2.1 Mode of operation

A three-dimensional representation of the measurement volume is shown in figure 7.4. The fluid element (left hand side) contains the particles, which were in the interrogation volume during the first exposure at $t = t_0$. Due to the out-of-plane motion of this fluid element it will partly leave the light sheet between $t = t_0$ and $t = t_0 + \Delta t$. Therefore, particles which have left the light sheet will not be imaged at $t = t_0 + \Delta t$. In dual-plane PIV a slightly displaced light sheet has to be generated additionally. If the displacement of this light sheet corresponds to the displacement of the fluid element the particles will be illuminated and imaged on a third frame at $t = t_0 + 2\Delta t$.

When dealing with a sufficient number of particles in the measurement volume, the number of particle image pairs inside the interrogation windows of two frames can be used to estimate the out-of-plane flow component. This number is proportional to the number of particles within the interrogation volume at t, decreased by the number of second images lost due to out-of-plane motion and by the number lost by in-plane motion. Using evaluation methods with a constant size and fixed location of the interrogation window, and assuming a constant particle density, the number of lost particle image pairs is proportional to the dark area shown in figure 7.4. Now, following the analysis for single-plane PIV (see section 3), we obtain the following results for the expectation of the image cross-correlations $E\{R_{\mathrm{II'}}(\boldsymbol{s} = \boldsymbol{d})\}$ and $E\{R_{\mathrm{I'I''}}(\boldsymbol{s} = \boldsymbol{d})\}$ between the pairs of interrogation windows $I(x, y)$,$I'(x, y)$ and $I'(x, y)$, $I''(x, y)$ respectively:

$$E\{R_{\mathrm{II'}}(\boldsymbol{s} = \boldsymbol{d})\} = C_{\mathrm{R}}\, R_\tau(\mathbf{0})\, F_{\mathrm{O}}(D_Z - Z' + Z)\, F_i(D_X, D_Y) \qquad (7.16)$$

$$E\{R_{\mathrm{I'I''}}(\boldsymbol{s} = \boldsymbol{d})\} = C_{\mathrm{R}}\, R_\tau(\mathbf{0})\, F_{\mathrm{O}}(D_Z - Z'' + Z')\, F_i(D_X, D_Y) \qquad (7.17)$$

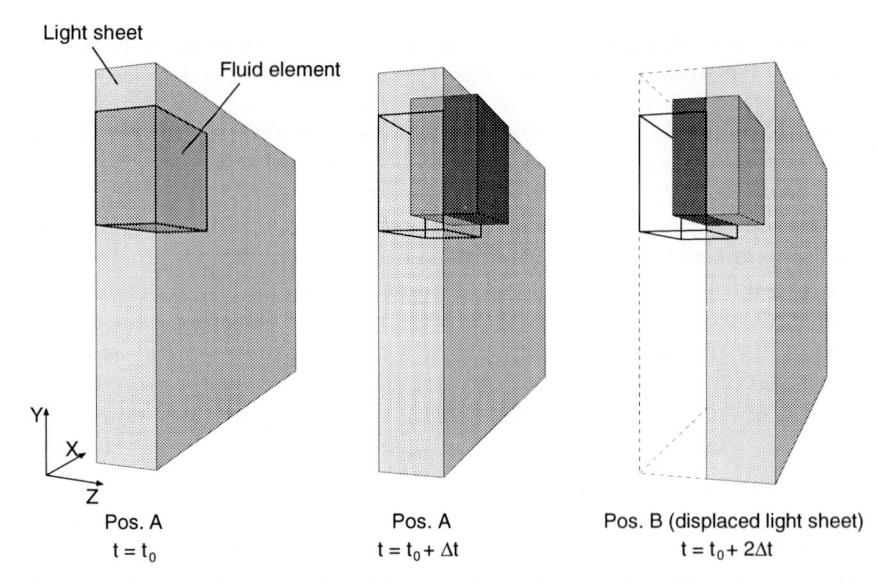

Fig. 7.4. Size and location of the interrogation volume and position of the particles, which were illuminated by the first light pulse, represented at the time of the first, second and third exposure at light sheet position (A), (A), and (B) respectively

where C_{R} is a constant factor (defined in section 3), R_τ is the image self-correlation and F_i is interpreted as the in-plane loss of correlation. Z, Z' and Z'' represent the position of the light sheet in the Z direction during the first, second and third exposure respectively. F_0, the loss-of-correlation due to out-of-plane motion, is the relevant term for the description of "dual-plane" evaluation. It is given by:

$$F_0(D_Z) = \frac{\int I_0(Z) I_0(Z + D_Z)\,\mathrm{d}Z}{\int I_0^2(Z)\,\mathrm{d}Z}\,. \tag{7.18}$$

In a practical situation we have to estimate the expectation of the correlation by the finite spatial average. Therefore, we will replace the (mathematically correct) expectation by its estimator: the computed correlation $R_{\mathrm{II}'}$ and $R_{\mathrm{I}'\mathrm{I}''}$ respectively (for brevity of notation we omit $\boldsymbol{s}$). Provided that R_τ and the light sheet intensity profile $I_Z(Z)$ are known, it is in principle possible to determine $|D_Z - Z' - Z|$ from $R_{\mathrm{II}'}$ only. However, in a practical situation the evaluation of $|D_Z - Z' - Z|$ leaves the sign of the out-of-plane motion undetermined. Dual-plane PIV avoids these complications by taking the ratio of $R_{\mathrm{II}'}$ and $R_{\mathrm{I}'\mathrm{I}''}$, which yields a general expression that relates D_Z for an arbitrary light-sheet intensity profile $I_Z(Z)$ and given shifts $Z'' - Z'$ and $Z' - Z$:

$$\frac{R_{\mathrm{II}'}}{R_{\mathrm{I}'\mathrm{I}''}} = \frac{F_0(D_Z - Z' + Z)}{F_0(D_Z - Z'' + Z')} \tag{7.19}$$

It can be seen from the above equation, that taking the ratio of the correlation estimators also compensates for the loss-of-correlation due to in-plane-motion. Obviously, the ability to solve D_Z in equation (7.19) depends on the accurate knowledge of the light-sheet profile $I_Z(Z)$. Most CW-lasers offer a sufficient beam-pointing stability, and their intensity profiles can easily be determined to find the exact relation between the out-of-plane velocity and the cross-correlation values. When using solid-state pulsed laser systems – for example a Nd:YAG laser – a variation of the pulse-to-pulse beam-pointing of up to 20% of the beam diameter has to be taken into account. If the spatial beam profile varies in time, the intensity profile should be determined simultaneously, e.g. by using a beam splitter and a CCD sensor.

In the remainder of this section we will deal with two particular solutions of equation (7.19), namely the solutions for (1) $Z = Z' < Z''$ (which is the situation for the investigation given in section 8.4), and (2) for $Z = Z'' < Z'$ (which is the situation if flows with bidirectional out-of-plane components have to be observed). To keep the subsequent analysis simple we consider a light sheet with a top-hat shaped intensity profile and a width ΔZ_0 i.e.

$$I_0(Z) = \begin{cases} I_Z & \text{if} \quad |Z| \leq \Delta Z_0/2 \\ 0 & \text{elsewhere} \end{cases}$$

and

$$F_0(Z) = \begin{cases} 1 - |Z|/\Delta Z_0 & \text{if} \quad |Z| \leq \Delta Z_0 \\ 0 & \text{elsewhere}\,. \end{cases} \tag{7.20}$$

For $Z = Z' < Z''$, the solution of equation (7.19) for the out-of-plane velocity $W = D_Z/\Delta t$, with $F_0(Z)$ given by equation (7.20), yields:

$$W = \frac{\Delta Z}{\Delta t} \cdot \begin{cases} \dfrac{R_{I'I''} - O_Z R_{II'}}{R_{II'} - R_{I'I''}} & \text{for} \quad -\Delta Z \cdot O_Z \leq D_Z \leq 0 \\[2ex] \dfrac{R_{I'I''} - O_Z R_{II'}}{R_{II'} + R_{I'I''}} & \text{for} \quad 0 \leq D_Z \leq Z'' - Z' \\[2ex] \dfrac{R_{I'I''} + (2 - O_Z) R_{II'}}{R_{II'} - R_{I'I''}} & \text{for} \quad Z'' - Z' \leq D_Z \leq \Delta Z \end{cases} \tag{7.21}$$

with $O_Z = 1 - (Z'' - Z')/\Delta Z_0$. For the experiments described in section 8.4 we only considered the solution for $0 < D_Z < Z'' - Z'$.

For a situation where the out-of-plane fluid velocity can be both positive or negative, a reverse displacement of the light sheet positions between first and second, and second and third exposure has to be taken into account. If the absolute value of displacement is the same for both directions equation (7.19) has to be solved for $Z = Z'' < Z'$. The solution, with $F_O(Z)$ again given by equation (7.20), differs slightly with respect to the solution given by equation (7.21), i.e.

$$W = \frac{\Delta Z}{\Delta t} \frac{R_{\mathrm{I'I''}} - R_{\mathrm{II'}}}{R_{\mathrm{II'}} + R_{\mathrm{I'I''}}} O_Z \tag{7.22}$$

for $-(Z' - Z'') < D_Z < Z' - Z''$. Note that the range of D_Z for this solution is symmetric with respect to $D_Z = 0$. The following simplifications are implied in the above formulas:

First, a top-hat intensity profile of the light sheets in the Z direction has been assumed instead of a Gaussian distribution, which would be more adequate for CW lasers. This leads to the fact that $F_0(Z)$, which is the normalized correlation of the intensity distributions in the Z direction of two successive pulsed light sheets and is therefore also a Gaussian function, is approximated by a triangle function.

Second, the effect of the variation of the displacement within the interrogated cell, and the fraction of second images lost by in-plane motion is assumed to be identical for both correlations. This is only a rough approximation as long as the second and third frames are not captured at the same time, which would require a more sophisticated image separation technique.

Third, the fluctuating noise component is neglected. Its effect on the measurement accuracy can be reduced by averaging results over neighboring interrogation cells. However, this has to be balanced against a decrease in spatial resolution.

7.2.2 Conclusions

In this section, we described the approach of using information rendered by the correlation function to estimate the out-of-plane components of velocity fields. Among other limitations, a larger number of particle images per interrogation window is required for this technique than for conventional PIV. The existence of a practically achievable maximum of the image density therefore results in a lower spatial resolution and/or accuracy compared to the in-plane measurement. However, we found the results of this approach and the ease of operation of the described technique to be encouraging. For investigations of low speed liquid flows, the operation of dual-plane PIV is easier than that of stereoscopic PIV, since the only calibration necessary is the determination of thickness and overlap of the light sheets. Furthermore, only one camera is needed. By changing the light sheet position after each exposure this technique can also be applied to flows with out-of-plane components in the positive and negative Z direction.

8. Examples of applications

In this chapter some of the applications of DLR's mobile PIV system will be described. For each experiment the most important parameters of the flow field under investigation, of the illumination and recording set-up will be given. If not stated otherwise, the evaluation has been carried out by means of cross-correlation methods, i.e. in the case of photographic recording the negatives have been scanned and digitized prior to evaluation.

8.1 Aerodynamics

8.1.1 Grid turbulence

Table 8.1. PIV recording parameters for grid turbulence

Flow geometry	$U_\infty = 10\,\mathrm{m/s}$ parallel to light sheet
Maximum in-plane velocity	$U_{\mathrm{max}} \approx 10.1\,\mathrm{m/s}$
Field of view	$160 \times 120\,\mathrm{mm}^2$
Interrogation volume	$2.1 \times 2.1 \times 2\,\mathrm{mm}^3$ $(H \times W \times D)$
Dynamic spatial range	DSR $\approx 57:1$ ($\approx 35:1$[a])
Dynamic velocity range	DVR $\approx 500:1$ ($\approx 1000:1$[a])
Observation distance	$z_0 \approx 1.5\,\mathrm{m}$
Recording method	single frame/double exposure
Ambiguity removal	image shifting/rotating mirror
Recording medium	35mm film, ASA 3200, 100 lps/mm
Recording lens	$f = 100\,\mathrm{mm}$ $f_\# = 2.8$
Illumination	Nd:YAG laser[b] 70 mJ/pulse
Pulse delay	$\Delta t = 225\,\mu\mathrm{s}$
Seeding material	oil droplets ($d_\mathrm{p} \approx 1\,\mu\mathrm{m}$)

[a] For optical evaluation
[b] Frequency doubled

The accuracy attainable with the PIV technique in best cases can be assessed by investigating laminar or weakly turbulent flow fields. Thus, PIV recordings have been taken in the DLR low turbulence wind tunnel (TUG), which is of an Eiffel type. Screens in the settling chamber and a high contraction ratio of 15:1 lead to a low turbulence level in the test section (cross section $0.3 \times 1.5\,\mathrm{m}^2$). The experiments were performed on the center line of the wind tunnel. Different grids can be installed near the end of the nozzle of the wind tunnel in order to generate turbulence. The basic turbulence level in the test section of the TUG of Tu = 0.06% (measured by means of a hot wire) can be increased by this modification. The flow was seeded in the settling chamber upstream of the screens used to reduce the turbulence of the wind tunnel flow.

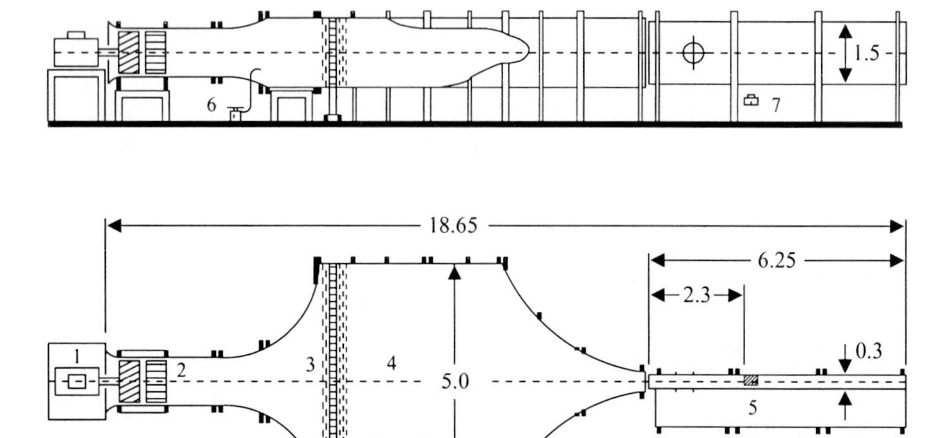

Fig. 8.1. Low turbulence wind tunnel

The PIV recordings were taken with a photographic camera and a high speed rotating mirror to increase the velocity resolution (see section 4.3, figure 4.12). Figure 8.2 shows the instantaneous flow field behind a grid [131] with a mesh size of 19 mm and a rod diameter of 1.5 mm. The distance between grid and observation area was 1.3 m (i.e. 160 rod diameters, slowly decaying grid turbulence).

The mean flow velocity in the wind tunnel was $\approx 10\,\mathrm{m/s}$. The field of view was $16 \times 12\,\mathrm{cm}^2$ and is subdivided into 115×78 interrogation spots for digital evaluation. In order to enhance the visual impression of the spatial structures in the flow field, a reference velocity U_{ref} is subtracted from the U component of each local vector. In this case U_{ref} was set equal to the mean flow velocity. Thus, the instantaneous field of the fluctuating velocity component ($U - U_{\mathrm{ref}}$, V) is obtained directly.

The degree of turbulence, as measured by means of a hot wire, was Tu = 0.46%. The small turbulent structures present in the flow field can be clearly

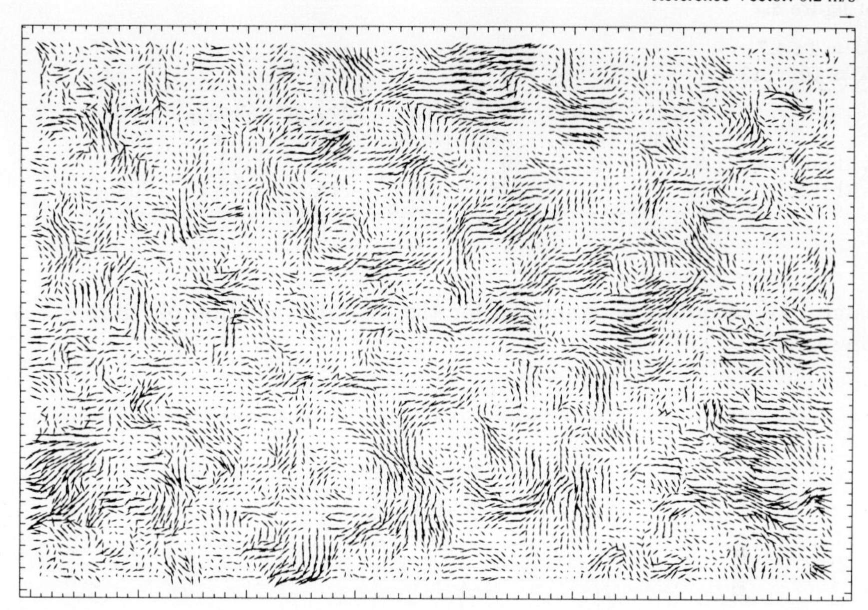

Fig. 8.2. Instantaneous flow field of the fluctuating velocity component behind a grid at $U_{\text{inf}} = 10\,\text{m/s}$ and Tu = 0.46%; $U_{\text{shift}} - 5.8\,\text{m/s}$. Result of digital evaluation

seen on the vector map. The measurement was performed by applying a "negative" image shift of $U_{\text{shift}} = -5.8\,\text{m/s}$ to the mean flow of $U_\infty = 10\,\text{m/s}$. The measurement parameters associated with this experiment are given in table 8.2. It lists the main parameters for the previously described experiment with and without the use of image shifting (I.S.). These parameters are the minimum and maximum velocity, U_{min} and U_{max}, the shift velocity, U_{shift}, and the resulting particle image distances, Δx_{shift}, Δx_{min}, Δx_{max}, in the film plane calculated with magnification, M, and the pulse separation, Δt. Table 8.2 shows how the application of image shifting (I.S.) increases the velocity resolution. Typical values for the number N of particles per interrogation volume are also given.

By application of image shifting the time delay between the two illumination pulses was increased from $80\,\mu$s to $225\,\mu$s, without changing the average particle image displacement. In the case of optical evaluation, image shifting is required in order to be able to adjust the average displacement of the tracer particles to be in the range of $\approx 200\,\mu$m for optimal evaluation. As already mentioned in section 5.3.2, data of optically evaluated recordings contain lower noise levels. However, this has to be balanced against the larger size of the interrogation spot which reduces the spatial resolution of the measurement. This can be seen in figure 8.3 where the fully optical evaluation of the

Table 8.2. Image recording parameters associated with the instantaneous flow field of figure 8.2

$M = 1:4.7$
$N \approx 10\ldots13$

	U_{min} [m/s]	U_{max} [m/s]	Δt [μs]	U_{shift} [m/s]	Δx_{shift} [μm]	Δx_{min} [μm]	Δx_{max} [μm]
without I.S.	9.9	10.1	80	0	0	198	202
with I.S.			225	-5.8	-278	196	206

same PIV recording as employed for the digital evaluation of figure 8.2 is presented.

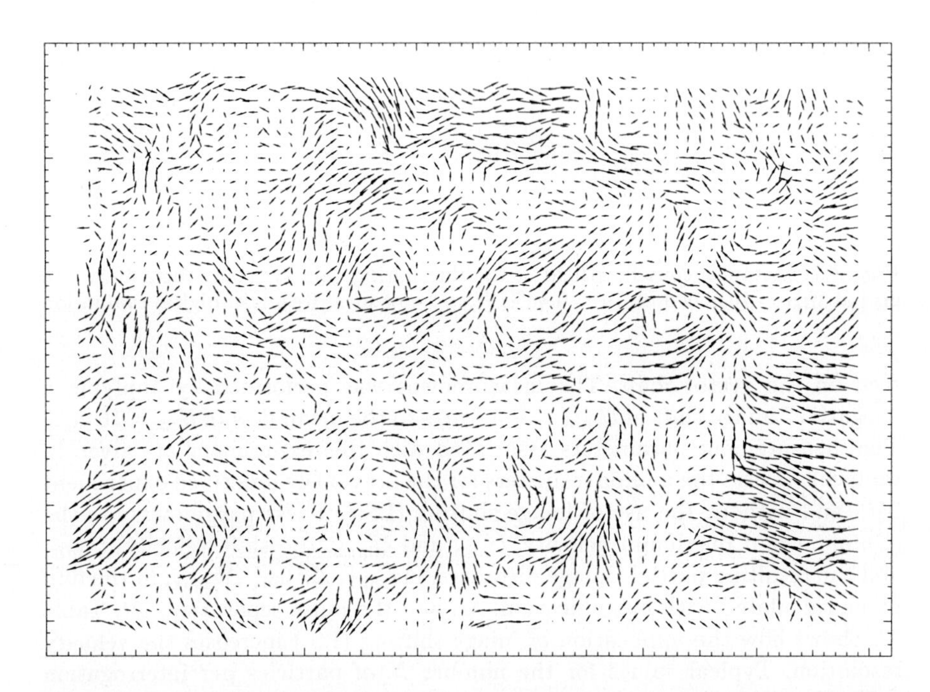

Fig. 8.3. Instantaneous flow field of the fluctuating velocity component behind a grid at $U_{inf} = 10$ m/s and Tu = 0.46%; $U_{shift} - 5.8$ m/s. Result of optical evaluation

8.1.2 Boundary layer instabilities

Table 8.3. PIV recording parameters for boundary layer instabilities.

Flow geometry	parallel to light sheet and plate
Maximum in-plane velocity	$U_{\mathrm{max}} \approx 12\,\mathrm{m/s}$
Field of view	$70 \times 70\,\mathrm{mm}^2$
Interrogation volume	$1.9 \times 1.9 \times 0.5\,\mathrm{mm}^3$ $(H \times W \times D)$
Dynamic spatial range	DSR $\approx 31 : 1$
Dynamic velocity range	DVR $\approx 137 : 1$
Observation distance	$Z_0 \approx 0.6\,\mathrm{m}$
Recording method	dual frame/single exposure
Ambiguity removal	frame separation (frame-straddling)
Recording medium	full frame interline transfer CCD
Recording lens	$f = 60\,\mathrm{mm}$ $f_{\#} = 2.8$
Illumination	Nd:YAG laser[a] 320 mJ/pulse
Pulse delay	$\Delta t = 80\,\mu\mathrm{s}$
Seeding material	oil droplets ($d_{\mathrm{p}} \approx 1\,\mu\mathrm{m}$)

[a] Frequency doubled

In the case of periodic flows the conditional sampling technique can be utilized in order to record instantaneous velocity vector maps always at the same phase angle. The excitation of the periodic process and the recording sequence must be phase locked. As an example for the application of conditional sampling the investigation of instabilities in a boundary layer will be described.

The transitional process in a boundary layer is determined by a mechanism of generation and interaction of various instabilities. Small oscillations may cause primary instability – two-dimensional waves, the so-called Tollmien-Schlichting (TS) waves. The growth of such TS waves leads to a streamwise periodic modulation of the basic flow, which gets sensitive to three-dimensional, spanwise periodic disturbances. These disturbances are amplified and lead to a three-dimensional distortion of the TS waves and farther downstream to the generation of three-dimensional Λ vortices. The extension of the knowledge about this mechanism enables the prediction and control of transitions as required for applications in fluid mechanical engineering.

In order to study the behavior of instabilities, quantitative data of velocity fields with known initial conditions have been acquired in a flat plate boundary layer, also in the TUG wind tunnel (see figure 8.1). In order to get reproducible and constant conditions for the development of the instabilities it is necessary to know the initial amplitude of the velocity fluctuations at the beginning of the observation area [153]. In the experiment of KÄHLER and WIEGEL this is achieved by introducing controlled disturbances by means of a device for acoustic excitation which consists of a single, spanwise slot for the controlled input of two-dimensional disturbances and 40 separate slots (positioned spanwise as well) for the input of controlled three-dimensional disturbances. The velocity at the outer edge of the boundary layer was about $U = 12\,\text{m/s}$. The average free stream turbulence level was $\text{Tu} = 0.065\%$. The light sheet (thickness $\delta_Z = 0.5\,\text{mm}$ in the observation area) was oriented parallel to the plate. Its height above the plate could be varied but was usually 0.5 mm in the experiment. The observation area was $70 \times 70\,\text{mm}^2$. The digital recording technique was employed.

By applying different input signals to the acoustic excitation it was possible to excite different transition types. We mention here the *fundamental type*, the *subharmonic type* and the *oblique type*. Figure 8.4 presents the phase locked field of the instantaneous velocity fluctuations $(U - U_{\text{mean}}, V)$ obtained by exciting the *oblique type* for two different disturbances. The Λ-vortices exhibit in an aligned pattern. The spanwise wavelength of these Λ-vortices (here $\approx 20\,\text{mm}$) matches with the wavelength of the controlled input of the 3D-waves.

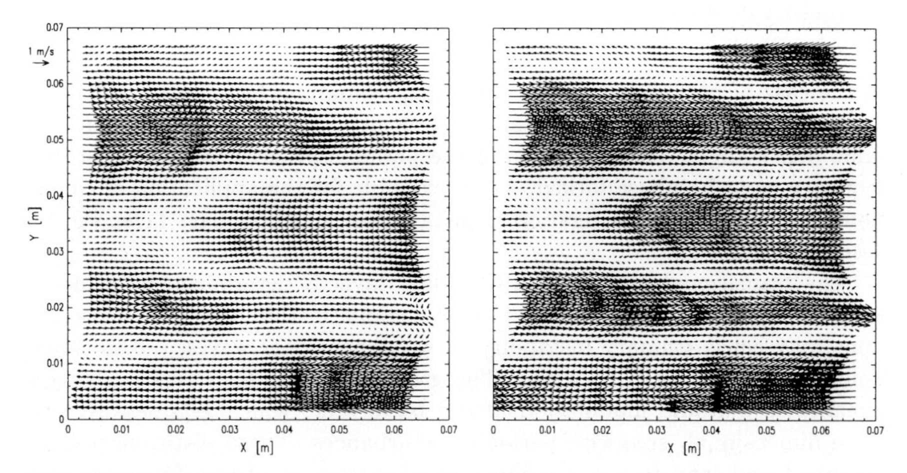

Fig. 8.4. Field of instantaneous velocity fluctuations of boundary layer instabilities above a flat plate for two different amplitudes of the input signal

The direction of the flow is from left to right. The mean velocity U_{mean} (calculated by averaging over all velocity vectors in the recording) has been subtracted from all velocity vectors in order to show the fluctuating components of the velocity vector field.

8.1.3 Turbulent boundary layer

Table 8.4. PIV recording parameters for turbulent boundary layer over a flat plate with zero pressure gradient

Flow geometry	parallel to light sheet
Maximum in-plane velocity	$U_\infty = 10.3, 14.9, 19.8\,\mathrm{m/s}$
Field of view	$30 \times 30\,\mathrm{mm}^2$
Interrogation volume	$2.0 \times 2.0 \times 1.0\,\mathrm{mm}^3$ $(H \times W \times D)$
	$2.0 \times 1.0 \times 1.0\,\mathrm{mm}^3$ $(H \times W \times D)$
	$2.0 \times 0.5 \times 1.0\,\mathrm{mm}^3$ $(H \times W \times D)$
	$1.0 \times 1.0 \times 1.0\,\mathrm{mm}^3$ $(H \times W \times D)$
Dynamic spatial range	DSR $\approx 31:1$
Dynamic velocity range	DVR $\approx 44:1$
Observation distance	$z_0 \approx 1.5\,\mathrm{m}$
Recording method	dual frame/single exposure
Ambiguity removal	frame separation (frame-straddling)
Recording medium	full frame interline transfer CCD
Recording lens	$f = 180\,\mathrm{mm}$ $f_\# = 2.8$
Illumination	Nd:YAG laser[a] 70 mJ/pulse
Pulse delay	$\Delta t = 7 - 20\,\mu\mathrm{s}$
Seeding material	oil droplets ($d_\mathrm{p} \approx 1\,\mu\mathrm{m}$)

[a] Frequency doubled

The following PIV application in a turbulent boundary layer at the wall of a flat plate illustrates two problems: to obtain PIV data close to a wall and to recover PIV data even in a flow with gradients (due to the velocity profile of the boundary layer).

In the present series of experiments the measurement position was 2.3 m downstream of a tipping region in the low-turbulence wind tunnel (see figure 8.1) at the DLR-Göttingen research center [128]. At this position the turbulent boundary layer thickness δ was of the order of 5 cm, of which the lower 3 cm was imaged using the high-resolution digital PIV recording system. At free stream velocities of 10.3, 14.9 and 19.8 m/s between 90 and 100 PIV image pairs were recorded. By removing a constant velocity profile of $U_\mathrm{ref} = 8\,\mathrm{m/s}$ from the PIV data set, the small scale structures in the boundary layer are highlighted as can be seen in figure 8.5. It is remarkable how close to the wall velocity data could be recovered.

In the first part of the evaluation the boundary layer profile and the RMS components of the velocity fluctuations were calculated as an average over all PIV recordings. These averaged quantities agree very well with the results from theory and pointwise velocity measurements as carried out by means of a hot wire. The nondimensional velocity profiles given in figure 8.6 start near the outer edge of the viscous sublayer ($y^+ \approx 10$) and extend well into

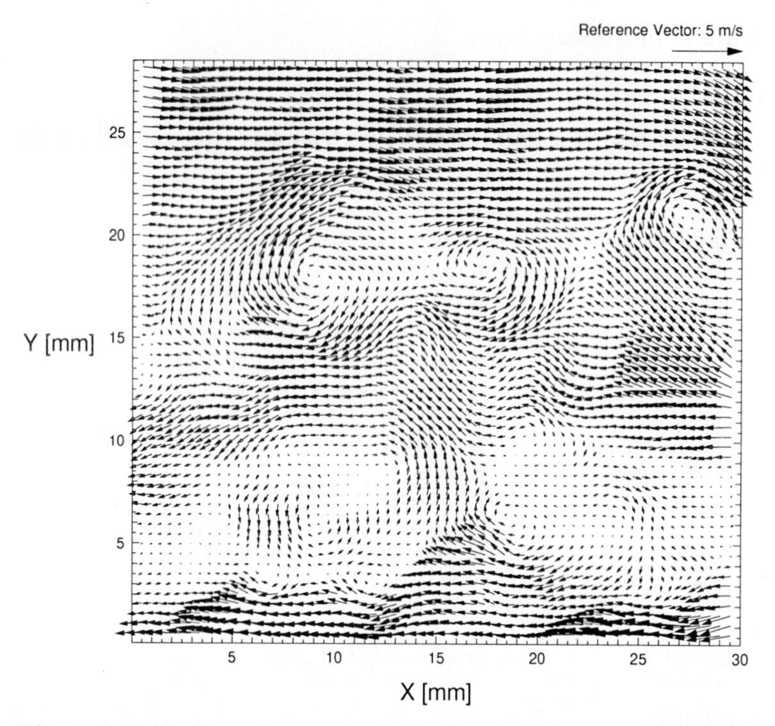

Fig. 8.5. Field of instantaneous velocity fluctuations in a fully turbulent boundary layer, $(U - U_{\mathrm{ref}}, V)$. Position of the wall at $Y = 0$.

the region where the large scale structures in the boundary layer cause a departure from the logarithmic profile ($y^+ \approx 200$).

As already mentioned, the strong velocity gradients within the interrogation areas close to the wall have mainly two effects.

First, due to the inhomogeneous displacement of paired particle images, the amplitude of the signal peak R_{D^+} is diminished. In addition, the diameter of the peak is broadened in the direction of shear. Therefore the velocity variation in the near wall region will decrease the likelihood of detection of the displacement peak.

Second, besides these experimental difficulties it has to be carefully checked whether the velocity vector assigned to the center of the interrogation window really represents the flow velocity at this location also in the presence of velocity gradients, as has been obtained by averaging over the interrogation window (see figure 6.13).

To investigate the effect of different interrogation area size on the number of outliers all PIV recordings were interrogated four times. The result can be seen in table 8.5 (for more details see [128]). The number of outliers in the 64×32 window is smaller compared to the other cases, because the number

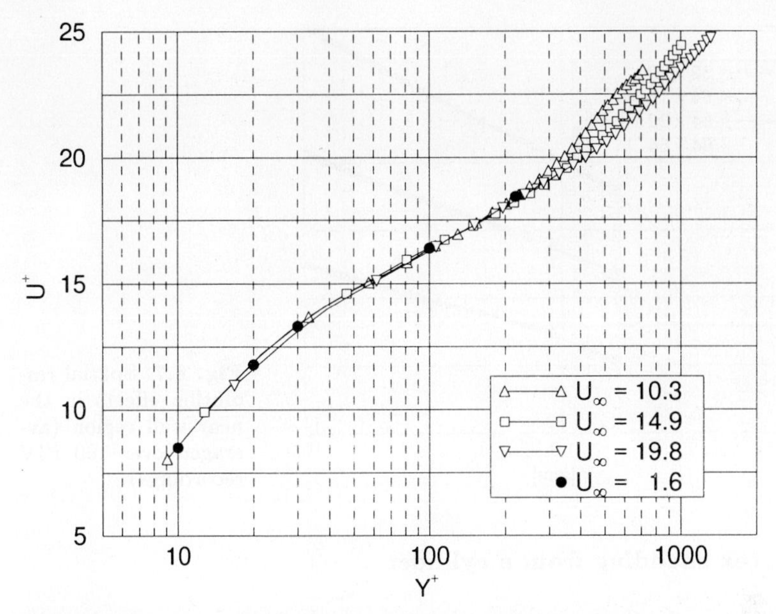

Fig. 8.6. Mean velocity profiles, scaled with inner variables (averaged over 100 PIV recordings)

Table 8.5. Number of outliers as a function of the interrogation area size, shape and free stream velocities

$\Delta x_0 \times \Delta y_0$ [pixel]	outliers [%] [10.3 m/s]	outliers [%] [14.9 m/s]	outliers [%] [19.8 m/s]
32 × 32	1.07	1.00	1.26
64 × 16	0.72	0.58	1.03
64 × 32	0.20	0.21	0.30
64 × 64	0.14	0.19	0.17

of particles is two times larger. The fraction of outliers is only of the order of 1% in the worst case, which clearly shows the reliability of the measurement technique.

Figure 8.7 represents the semilogarithmic mean velocity profiles as a function of the distance from the wall. For wall distances $y \geq 2$ mm the mean velocity is independent of the size of the interrogation window for three different measurements ($U_\infty = 10.3, 14.9$ and 19.8).

However, for $0 < y < 2$ mm the curves do not coincide due to the different averaging. The extension of the interrogation windows in the y direction is mainly responsible for this. Rectangular windows (extending parallel to the wall) show better performance as compared to square windows.

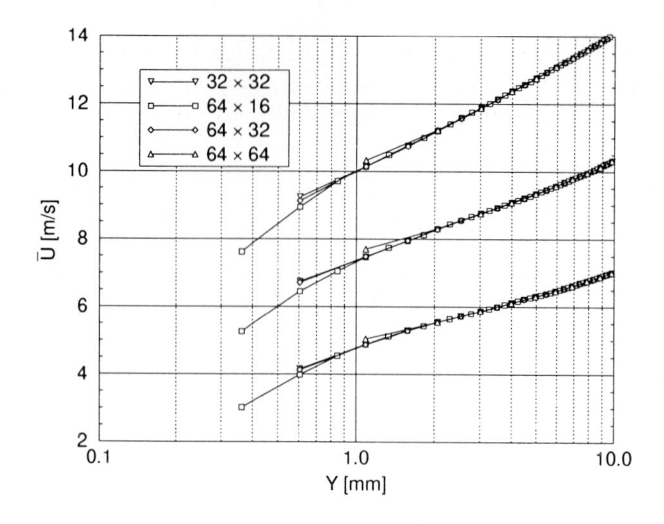

Fig. 8.7. Spatial resolution effects in the near wall region (averaged over 100 PIV recordings)

8.1.4 Vortex shedding from a cylinder

Table 8.6. PIV recording parameters for vortex shedding from cylinder

Flow geometry	$U_\infty = 0.8\,\mathrm{m/s}$ parallel to light sheet
Maximum in-plane velocity	$U_{\max} \approx 1.0\,\mathrm{m/s}$
Field of view	$105 \times 71\,\mathrm{mm}^2$
Interrogation volume	$1.8 \times 1.8 \times 1.5\,\mathrm{mm}^3$ $(H \times W \times D)$
Dynamic spatial range	DSR $\approx 57:1$
Dynamic velocity range	DVR $\approx 40:1$
Observation distance	$z_0 \approx 0.3\,\mathrm{m}$
Recording method	single frame/double exposure
Ambiguity removal	none
Recording medium	35mm film, ASA 3200, 100 lps/mm
Recording lens	$f = 100\,\mathrm{mm}$ $f_{\#}2.8$
Illumination	Nd:YAG laser[a] @ 70 mJ/pulse
Pulse delay	$\Delta t = 600 - 1000\,\mu\mathrm{s}$
Seeding material	oil droplets ($d_\mathrm{p} \approx 1\,\mu\mathrm{m}$)

[a] Frequency doubled

The shedding of vortices from a cylinder has been the subject of a number of investigations in the past. The regular, periodically alternating two-dimensional arrangement of vortices in a Kármán vortex street is well known. Recently, the three-dimensional structure of the wake behind the cylinder became of interest and was studied by means of flow visualization and hot-wire anemometry. Discontinuities in the relation between Strouhal number and Reynolds number were observed for Reynolds numbers $40 < \mathrm{Re} < 160$ depending on the boundary conditions. Finally, different modes with oblique shedding of the vortices at different angles between the center line of the vortex and the axis of the cylinder have been identified. Correlation methods had to be applied in order to obtain information about the spatial structure of the flow field.

However, with PIV it is immediately possible to "see" even the instantaneous spatial structures in the flow. For this purpose investigations by means of photographic PIV have been carried out in the low speed wind tunnel of MPI Göttingen at a mean flow velocity of = 0.8 m/s [148]. As an example figure 8.8 shows details of the instantaneous flow field of vortices shedding from a cylinder of diameter 2.5 mm at Re = 130. The cylinder is arranged vertically at 1.6 diameters behind the plane of the light sheet (and of the plane of the vector map as seen in figure 8.8) and attached to the wall of the wind tunnel (lower horizontal line of plot).

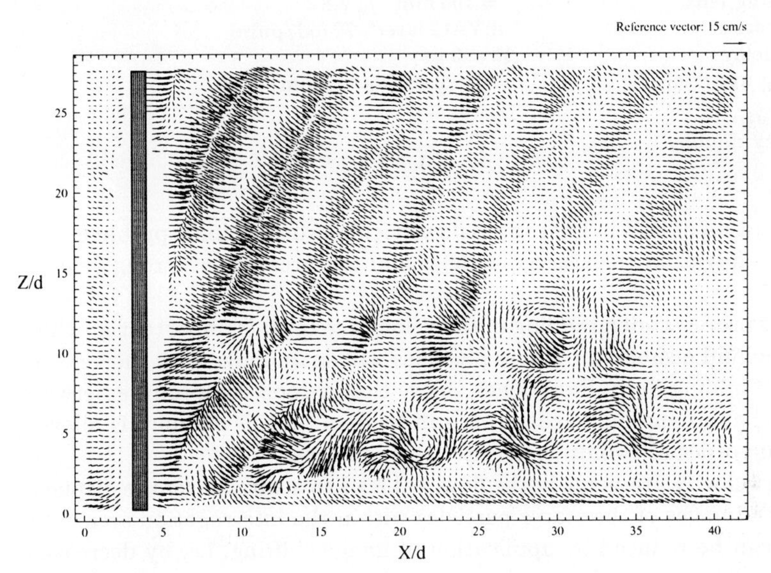

Fig. 8.8. Field of instantaneous velocity fluctuations in the wake of a cylinder at Re = 130, $(U - U_{\text{ref}}, V)$.

Figure 8.8 presents a cut through one side of the vortex street. It must be taken into account in the interpretation of the spatial structures shown in figure 8.8 that again the reference velocity U_{ref} obtained by averaging over all data has been subtracted. The end cell of the vortex, which forms the connection between the boundary layer and that vortex representing "mode n=3", can be clearly detected. The different angles at which the vortex sheds from the cylinder causes the branch like structure of the instantaneous velocity field. Details about the Kármán ratio, circulation and dissipation are also given in [175].

8.1.5 Transonic flow above an airfoil

Table 8.7. PIV recording parameters for transonic flow above a NACA0012 airfoil

Flow geometry	$Ma = 0.75$ parallel to light sheet
Maximum in-plane velocity	$U_{\max} \approx 520\,\mathrm{m/s}$
Field of view	$300 \times 200\,\mathrm{mm}^2$
Interrogation volume	$5.6 \times 5.6 \times 1\,\mathrm{mm}^3$ $(H \times W \times D)$
Dynamic spatial range	DSR $\approx 57:1$
Dynamic velocity range	DVR $\approx 150:1$
Observation distance	$z_0 \approx 1\,\mathrm{m}$
Recording method	single frame/double exposure
Ambiguity removal	image shifting/rotating mirror
Recording medium	35mm film, ASA 3200, 100 lps/mm
Recording lens	$f = 100\,\mathrm{mm}$ $f_{\#}2.8$
Illumination	Nd:YAG laser[a] 70 mJ/pulse
Pulse delay	$\Delta t = 3\,\mu\mathrm{s}$
Seeding material	oil droplets ($d_\mathrm{p} \approx 1\,\mu\mathrm{m}$)

[a] Frequency doubled

The application of PIV in high speed flows yields two additional problems: the resulting behavior of the tracer particles and the presence of strong velocity gradients.

For the proper understanding of the velocity maps it is important at which distance behind a shock the tracer particles will again move with the velocity of the surrounding fluid. Experience shows that a good compromise between particle behavior and light scattering can be found if this distance is allowed to be of the order of two or three interrogation areas.

Strong velocity gradients in the flow will lead to a variance of the displacement of the images of the tracer particles within the interrogation area. This influence can be reduced by application of image shifting, i.e. by decreasing the temporal separation between the two illumination pulses and increasing the displacement between the images of the tracer particles by image shifting to the optimum for evaluation. This is especially important if auto-correlation and optical evaluation methods are applied as in this case it is required to be able to adjust the displacement of the tracer particles to the range for optimal evaluation, i.e. $\approx 200\,\mu\mathrm{m}$.

In the case of optical evaluation methods image shifting helps also to solve the problem of large variations of the displacements of the tracer particle images within the PIV recording. As described earlier, a successful evaluation is achieved for a range of particle image displacements of $150\,\mu\mathrm{m} \leq d_{\mathrm{opt}} \leq 250\,\mu\mathrm{m}$. The upper and lower limits for this optimal particle image displacement is determined by the flow to be investigated and can be adapted to the optimal range of displacement on the recording medium by applying the image shifting technique, and adding an additional shift in the direction of the mean flow. Less data drop-out can be expected by this means.

Strong velocity gradients are present in flow fields containing shocks as are present in transonic wind tunnels. Figure 8.9 shows such an instantaneous flow field – again above a NACA 0012 airfoil with a chord length of $C_l = 20$ cm – at $Ma_\infty = 0.75$ [144]. By subtracting the speed of sound from all velocity vectors the supersonic flow regime and the shock are clearly detectable. Due to the application of image shifting ($U_{\text{shift}} = 174$ m/s) the requirement according to equation (4.1) could be fulfilled with an optimum interrogation spot diameter of 0.7 mm even at the location of the shock. No data drop-out is found even in interrogation spots located in front and behind the shock (flow velocities from $U = 280$ m/s to 520 m/s). The associated parameters for the PIV recording are presented in table 8.8.

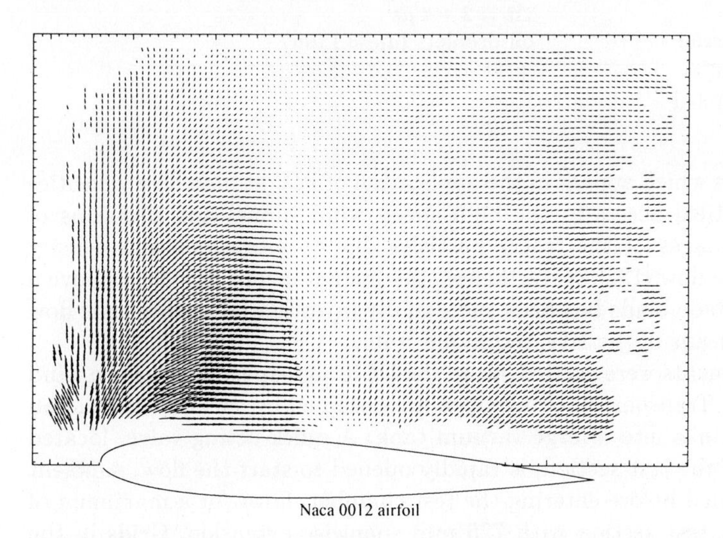

Fig. 8.9. Instantaneous flow field above a NACA 0012 airfoil at $\alpha = 5°$ and $Ma_\infty = 0.75$, $U_{\text{shift}} = 174$ m/s, $C_l = 20$ cm

Table 8.8. Image recording parameters associated with the instantaneous flow field of figure 8.9

$M = 1:6.7$ $N \approx 15$

	U_{min} [m/s]	U_{max} [m/s]	Δt [μs]	U_{shift} [m/s]	ΔX_{shift} [μm]	ΔX_{min} [μm]	ΔX_{max} [μm]
without I.S.	200	520	5	0	0	149	388
with I.S.			3	174	78	167	311

8.1.6 Cascade blade with cooling air ejection

Table 8.9. PIV recording parameters for cascade flow

Flow geometry	$Ma = 1.27$ parallel to light sheet
Maximum in-plane velocity	$U_{\max} \approx 400\,\mathrm{m/s}$
Field of view	$150 \times 100\,\mathrm{mm}^2$
Interrogation volume	$2.8 \times 2.8 \times 1\,\mathrm{mm}^3$ $(H \times W \times D)$
Dynamic spatial range	DSR $\approx 57:1$
Dynamic velocity range	DVR $\approx 100:1$
Observation distance	$z_0 \approx 1\,\mathrm{m}$
Recording method	single frame/double exposure
Ambiguity removal	image shifting/rotating mirror
Recording medium	35mm film, ASA 3200, 100 lps/mm
Recording lens	$f = 100\,\mathrm{mm}$ $f_{\#}2.8$
Illumination	Nd:YAG laser[a] 70 mJ/pulse
Pulse delay	$\Delta t = 2 - 4\,\mu\mathrm{s}$
Seeding material	oil droplets ($d_\mathrm{p} \approx 1\,\mu\mathrm{m}$)

[a] Frequency doubled

Other problems which appear in the application of PIV at high flow velocities are limited optical access in the high speed wind tunnel and problems of focusing the images of the tracer particles due to vibrations and density gradients in the flow [132]. Nevertheless, the instantaneous flow field above a model of a cascade blade has been successfully investigated at transonic flow velocities by means of the PIV technique [146].

The experiments were carried out in the DLR high-speed blow-down wind tunnel (HKG). Transonic flow velocities are obtained by sucking air from an atmospheric intake into a large vacuum tank. A quick-acting valve, located downstream of the test section, is rapidly opened to start the flow. Ambient air, which is dried before entering the test chamber, flows for a maximum of 20 s through a test section with 725 mm spanwise extension. Grids in the settling chamber and a high contraction ratio lead to a low turbulence level in the test section. The aim of this investigation was to study the effect of the ejection of cooling air on the wake behind a model of a cascade blade. Due to a specially adapted wind tunnel wall above and below the model and an adjustable tailboard above the model, the flow field of a real turbine blade could be simulated in a realistic manner. The PIV recordings were taken with the photographical PIV recording system utilizing the high speed rotating mirror for image shifting at a time delay between the two laser pulses of $2 - 4\,\mu$s. Figure 8.10 presents the instantaneous flow velocity field at the trailing edge of the plate (thickness 2 cm) for a cooling mass flow rate of 1.4 % at a free stream Mach number Ma = 1.27. Expansion waves and terminating shocks can be easily seen. No data were obtained in the area above the model as the laser light was blocked off by the model. Data drop-out was also found in the area directly downstream of the model. The reason is mainly that the size of the interrogation area could not be further decreased at evaluation. This would have been necessary in order to satisfactorily resolve the strong velocity gradients close to the trailing edge of the model. Without ejection of cooling air, the wake behind the plate can be characterized as a vortex street. With ejection of air this is no longer true: two separate thin shear layers can be detected in the presentation of the instantaneous vorticity shown in figure 8.11.

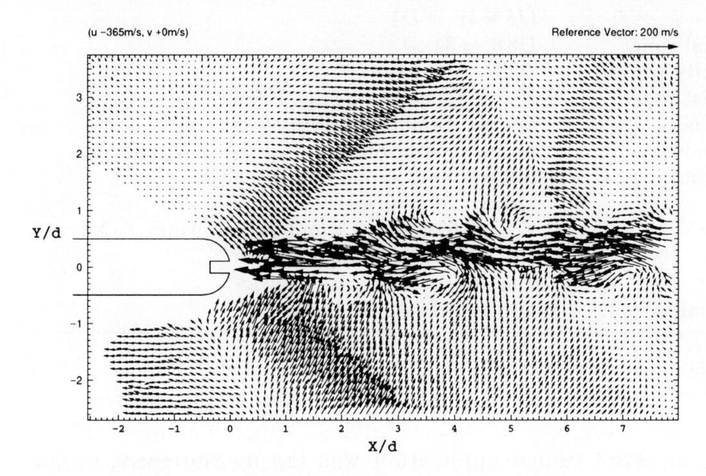

Fig. 8.10. Flow velocity field behind a cascade blade at Ma = 1.27 and a cooling mass flow rate of 1.4%.

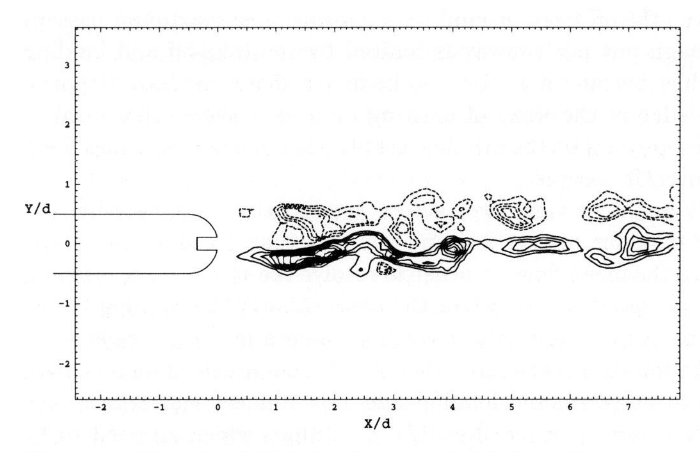

Fig. 8.11. Vorticity field behind a cascade blade at Ma = 1.27 and a cooling mass flow rate of 1.4%.

8.1.7 Aircraft wake

Table 8.10. PIV recording parameters for study of flow downstream of a transport aircraft model

Flow geometry	$U_\infty = 60\,\mathrm{m/s}$ normal to light sheet
Maximum in-plane velocity	$U_{\mathrm{max}} \approx 30\,\mathrm{m/s}$
Field of view	$220 \times 220\,\mathrm{mm}^2$ and $310 \times 310\,\mathrm{mm}^2$
Interrogation volume	$7 \times 7 \times 3\,\mathrm{mm}^3$ and $9.8 \times 9.8 \times 3\,\mathrm{mm}^3$ $(H \times W \times D)$
Dynamic spatial range	DSR $\approx 31:1$
Dynamic velocity range	DVR $\approx 30:1$
Observation distance	$z_{0,1} \approx 8.0\,\mathrm{m}$ and $z_{0,2} \approx 3.5\,\mathrm{m}$
Recording method	double frame/single exposure
Ambiguity removal	frame separation (frame-straddling)
Recording medium	full frame interline transfer CCD (1008 × 1018 pixel)
Recording lens	$f_1 = 300\,\mathrm{mm}$ $f_{\#}2.8$ and $f_2 = 100\,\mathrm{mm}$ $f_{\#}2.8$
Illumination	Nd:YAG laser[a] 320 mJ/pulse
Pulse delay	$\Delta t_1 = 20\,\mu\mathrm{s}$ and $\Delta t_2 = 30\,\mu\mathrm{s}$
Seeding material	oil droplets ($d_{\mathrm{p}} \approx 1\,\mu\mathrm{m}$)

[a] Frequency doubled

Another demanding wind tunnel application was the measurement of the wake vortices of a lifting wing in the 6 m by 8 m test section of the Deutsch–Niederländischer Windkanal (DNW).

The study of the wake vortices has recently been of increased interest because their life-time and strength dictates at which intervals commercial aircraft can safely take off from or land on a runway. The maximum passenger/payload through-put per runway is limited by the take-off and landing intervals which has become a serious problem for many modern airports. A detailed knowledge of the wake of existing aircraft is therefore crucial in understanding the physics of the trailing vortex phenomenon and may help improve future aircraft designs.

From the PIV point of view two factors are especially noteworthy: the high observation distance of $z \approx 8\,\mathrm{m}$ and the fact that the light sheet was oriented normal to the mean flow. In order to resolve the stream-wise oriented wake vortices it was necessary to orient the observation plane normal to the free stream which meant placing the recording camera into the free stream, downstream of the model wing (figure 8.12). Aside from the need for a camera mount resistant to aerodynamic loading and flow-induced vibrations, this viewing geometry results in a set of critical conditions which all need to be simultaneously satisfied: due to the high free stream velocity (60 m/s) the residence time of the tracer particles within the light sheet is very short. This meant that the time delay between the two light pulses, Δt, had to

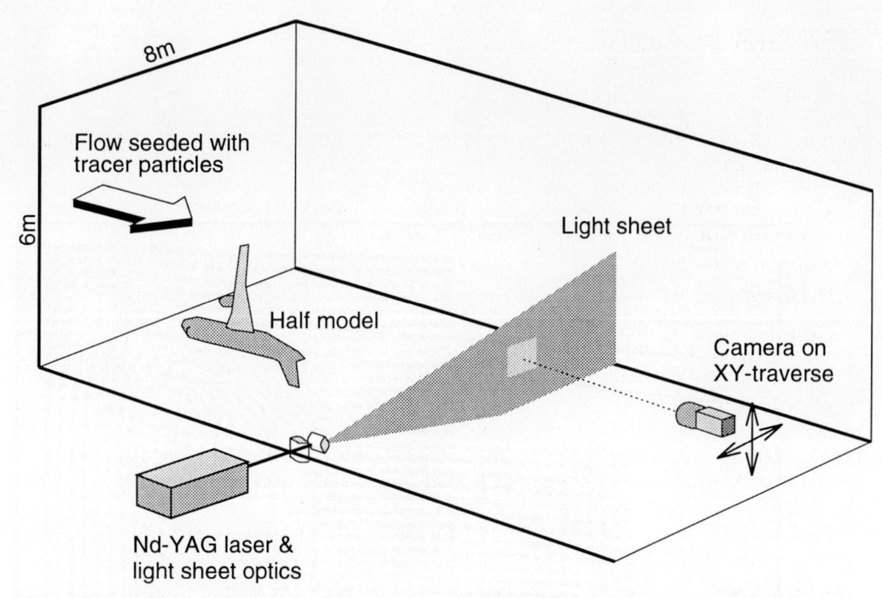

Fig. 8.12. PIV recording setup for the investigation of the flow downstream of a transport aircraft model in landing configuration

be kept short to maintain a sufficient match of particles between the two sequential recordings. At the same time the light sheet thickness could also not be increased significantly beyond 2 mm because the light scattered off the small seeding particles (1 μm oil droplets) could barely expose the CCD sensor, even at the available energy levels of up to 320 mJ per pulse. By choosing a rather small observation area of about 20 cm square, a maximum particle image displacement of up to 3 pixels could be maintained which still provided sufficient dynamic range in the velocity measurements. As the investigated flow field covered a much bigger area, the camera was mounted on an XY-in-flow traversing system allowing the flow field to be sampled at different positions. A composite at three different sampling positions is shown in figure 8.13. In this case the displayed vector maps are not instantaneous velocity maps but rather averages of up to 50 individual recordings, that is, time-averaged flow-fields.

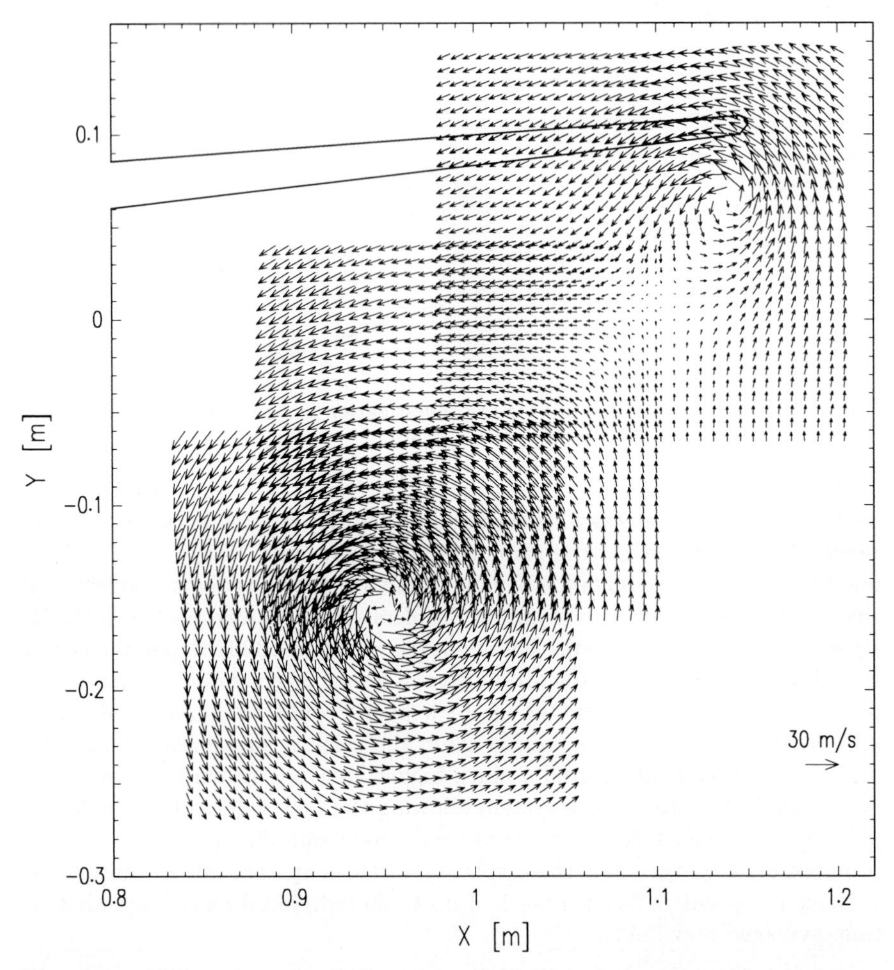

Fig. 8.13. Time-averaged velocity map of the flow downstream of a transport aircraft model. The segments were obtained by traversing the camera on an XY-translation stage

8.1.8 Flow field above a pitching airfoil

Table 8.11. PIV recording parameters for pitching NACA0012 airfoil

Flow geometry	$U_\infty = 28\,\mathrm{m/s}$ parallel to light sheet
Maximum in-plane velocity	$U_{\mathrm{max}} \approx 50\,\mathrm{m/s}$
Field of view	$300 \times 200\,\mathrm{mm}^2$
Interrogation volume	$5.6 \times 5.6 \times 1\,\mathrm{mm}^3$ $(H \times W \times D)$
Dynamic spatial range	DSR $\approx 57:1$
Dynamic velocity range	DVR $\approx 200:1$
Observation distance	$z_0 \approx 0.8\,\mathrm{m}$
Recording method	single frame/double exposure
Ambiguity removal	image shifting/rotating mirror
Recording medium	35mm film, ASA 3200, 100 lps/mm
Recording lens	$f = 60\,\mathrm{mm}$ $f_{\#}2.8$
Illumination	Nd:YAG laser[a] 70 mJ/pulse
Pulse delay	$\Delta t = 12\,\mu\mathrm{s}$
Seeding material	oil droplets ($d_{\mathrm{p}} \approx 1\,\mu\mathrm{m}$)

[a] Frequency doubled

The instantaneous flow field on moving models can only be measured by means of PIV. As an example for such investigations figure 8.14 shows the instantaneous flow field above a pitching NACA 0012 airfoil with a chord length of $C_l = 20\,\mathrm{cm}$ at an angle of attack of $\alpha = 24°$ in upstroke motion at a mean flow velocity of $U_\infty = 28\,\mathrm{m/s}$ [134]. A strong vortex with flow reversals (separation) can be observed above the airfoil and is frequently referred to as a "dynamic stall" vortex.

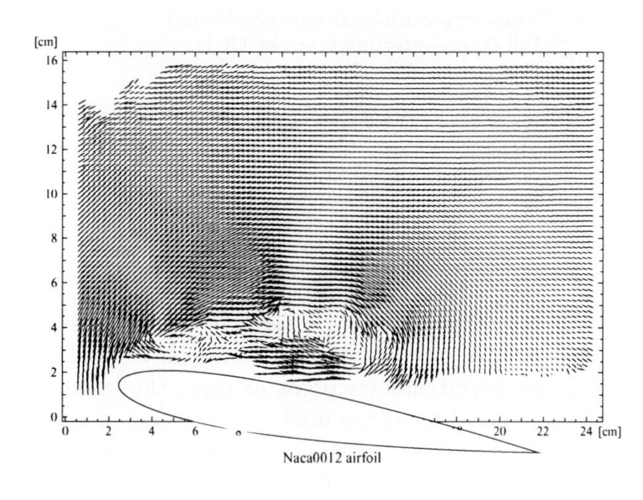

Fig. 8.14. Instantaneous flow field above NACA 0012 airfoil at $\alpha = +24°$ in upstroke motion and $U_\infty = 28\,\mathrm{m/s}$, $U_{\mathrm{shift}} = 128\,\mathrm{m/s}$, $C_l = 20\,\mathrm{cm}$

A single frame/double exposure recording technique was applied and flow reversal was present, so image shifting had to be utilized.

Table 8.12 shows how the application of image shifting (I.S.) solves the problem of ambiguity of the direction of the velocity vector.

Table 8.12. Parameters in recording of instantaneous flow field shown in figure 8.14

$M = 1:7.9$ $N \approx 30$

	U_{min} [m/s]	U_{max} [m/s]	Δt [μs]	U_{shift} [m/s]	ΔX_{shift} [μm]	ΔX_{min} [μm]	ΔX_{max} [μm]
without I.S.	-28	+50	56	0	0	-198	+354
with I.S.			12	128	194	+152	+270

8.1.9 Helicopter aerodynamics

Table 8.13. PIV recording parameters for orthogonal blade-vortex interaction measurements on helicopter rotor model

Flow geometry	$U_\infty = 15.7$ m/s normal to light sheet
Maximum in-plane velocity	$U_{max} \approx 30$ m/s
Field of view	65×65 mm^2
Interrogation volume	$2.1 \times 2.1 \times 2$ mm^3 $(H \times W \times D)$
Dynamic spatial range	DSR $\approx 31:1$
Dynamic velocity range	DVR $\approx 40:1$
Observation distance	$z_0 \approx 1.5$ m
Recording method	double frame/single exposure
Ambiguity removal	frame separation (frame-straddling)
Recording medium	full frame interline transfer CCD (1008 × 1018 pixel)
Recording lens	$f = 180$ mm $f_{\#}2.8$
Illumination	Nd:YAG laser[a] 70 mJ/pulse
Pulse delay	$\Delta t = 7 - 20$ μs
Seeding material	oil droplets ($d_p \approx 1$ μm)

[a] Frequency doubled

The application of PIV for the investigation of rotor aerodynamics is a challenging task because it involves a variety of complex technical as well as aerodynamical aspects. From the aerodynamics point of view, the increasing use of civil helicopters in urbanized areas has made the problem of noise emission increasingly important. Of the various noise sources on rotor craft, blade/vortex interactions (BVI) has been identified as a major source of impulsive noise, especially during vertical descent and landing. Generally, blade/vortex interactions occur when tip vortices generated through the lift of advancing rotor blades are struck by the following rotor blades. Here two types of BVI have to be distinguished: (1) orthogonal BVI occurs when the intersection angle between vortex and blade is around 90°, for which example data is given here, and (2) parallel BVI is present when the blade and vortex axis are close to parallel.

The aerodynamics of BVI can only be investigated by nonintrusive techniques such as PIV or laser Doppler anemometry. Both types of measurements were performed on the same rotor model to allow a direct comparison between the two methods [145]. The model rotor, located at the Department of Aerospace Engineering (ILR) of RWTH Aachen, consists of four hinged, fully active, blades (NACA-0015, square tips) with a diameter of 1 m. Driven by a 65 kW electric motor, the rotor was installed in the open test section of the ILR low speed wind tunnel. Forward flight was simulated by tilting the rotor plane $-3°$ into the free stream ($U_\infty = 15.7\,\mathrm{m/s}$) at a rotor speed of $f = 25\,\mathrm{Hz}$.

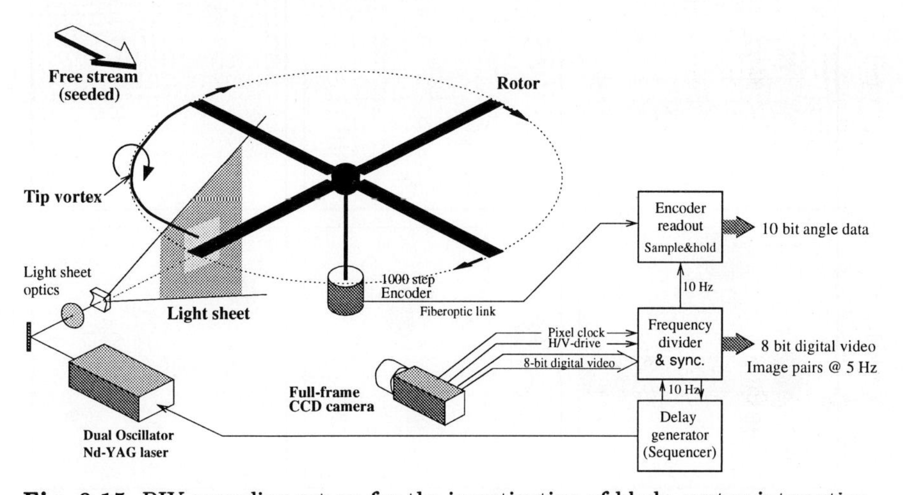

Fig. 8.15. PIV recording set-up for the investigation of blade–vortex interaction

In this configuration, the PIV measurement plane was located at an azimuthal angle of 90° on the advancing blade side as depicted in figure 8.15. Using the frame straddling technique, up to 200 digital image pairs of the flow around the blade's tip were taken for a given configuration with a field of view of $65 \times 65\,\mathrm{mm}^2$. The PIV acquisition was complicated by the fact that the mean flow direction was normal to the imaged plane requiring the use of a rather short pulse delay ($\Delta t = 7\,\mu\mathrm{s}$) in order to ensure sufficient frame-to-frame matching of particles. Also, since the camera could not be randomly triggered at a predefined rotor azimuth angle, a large number of images were randomly recorded along with the azimuthal position of the rotor. The data set was subsequently sorted to reconstruct the evolving flow field. A selection of three velocity and vorticity maps are displayed in figure 8.16 and clearly show the temporal evolution of the BVI flow field. In figure 8.16 (a) the rotor blade has just left the image plane and is moving away from the viewer. A strong tip vortex can be observed in the tip region while the vortex near the top edge is due to the passage of the preceding blade. The movement of the vortices with respect to each other can be observed in the following two frames (figure 8.16 b,c). In figure 8.16 (c) the vortex structure initially at the top of figure 8.16 (a) is moving into the rotor plane and will be struck by the next advancing blade giving rise to BVI. Remnants of such an interaction can be seen in the rotor plane on the right side of figure 8.16 (a). Computer animations have been compiled on the data to further illustrate the flow's evolution.

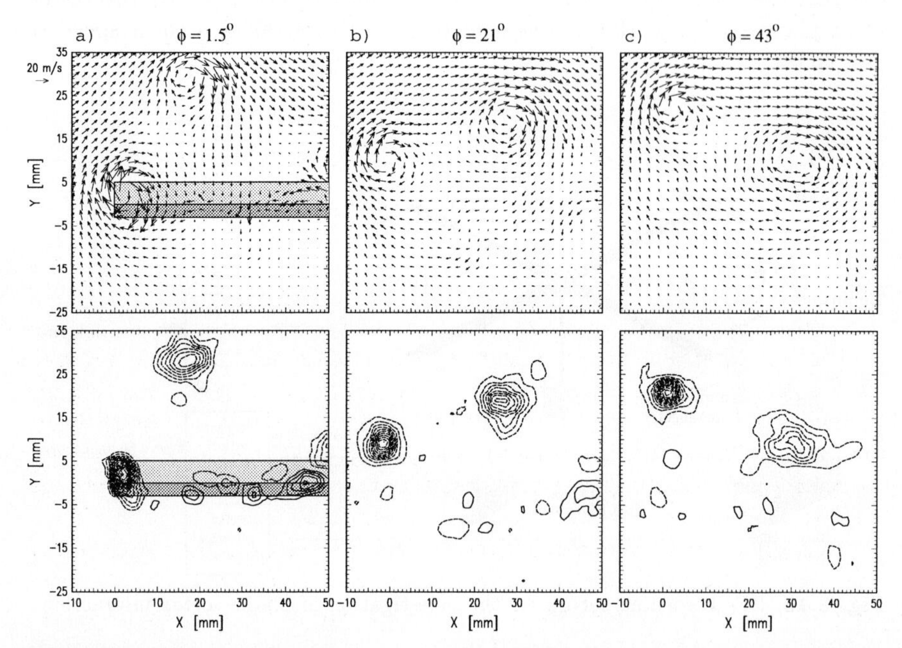

Fig. 8.16. Instantaneous velocity (top) and vorticity (bottom) maps of the blade–vortex interaction. The position of the blade is indicated on the left frame

8.2 Liquid flows

8.2.1 Vortex–free-surface interaction

Table 8.14. PIV recording parameters for vortex–free-surface interaction.

Flow geometry	Nearly two-dimensional flow aligned with the light sheet
Maximum in-plane velocity	$U_{max} \approx 10\,\mathrm{cm/s}$
Field of view	$103 \times 97\,\mathrm{mm}^2$
Interrogation volume	$6.4 \times 6.4 \times 1.5\,\mathrm{mm}^3$ $(H \times W \times D)$
Dynamic spatial range	DSR $\approx 16:1$
Dynamic velocity range	DVR $\approx 100:1$
Observation distance	$z_0 \approx 1.5\,\mathrm{m}$ (through glass/water)
Recording method	double frame/single exposure
Ambiguity removal	frame separation (frame-straddling)
Recording medium	frame transfer CCD (512 × 480 pixel)
Recording lens	$f = 50\,\mathrm{mm}$ $f_{\#}1.8$
Illumination	5W CW argon-ion laser, mechanical shutter
Pulse delay	$\Delta t = 10\,\mathrm{ms}$
Pulse duration	2 ms
Seeding material	silver-coated, glass spheres ($d_p \approx 10\,\mu\mathrm{m}$)

The present example was chosen to illustrate the possibility of PIV in providing time-resolved measurements, which is of importance in many fluid mechanical investigations. Time resolution is possible when the image frame rate exceeds the time scales present in the flow. For this investigation the frame rate of the utilized video equipment was 30 Hz (RS-170) resulting in an image-pair rate of 15 Hz using the frame-straddling approach, whereas the time-scales in the flow were longer than 1/10th of a second.

The vortex-pair flow under investigation is motivated in the context of understanding the fundamentals of the interaction of vortical structures with a free surface [155]. The vortex pair was generated by a pair of counter-rotating flaps whose sharp tips were located at $y \approx -10\,\mathrm{cm}$. Once the flaps were closed the separation vortex from each tip formed a symmetric vortex pair which propagated toward the surface within two seconds. The interesting interaction events typically took place in the following 2–5 seconds. This meant that on the order of 10 seconds worth of PIV data were needed to resolve a single interaction process. This translated to 300 separate PIV recordings, which were acquired using either a computer-based, real-time digitization and hard-disk array or a real-time, analog video disk. The latter was chosen in place of a video tape recorder because of its superior playback stability and reproducibility which is essential for the digitization of the video images prior to their evaluation. In spite of more limited storage capacity of the real-time digitization system its primary advantage is the immediate availablity on the computer for evaluation.

Figure 8.17 shows four selected instantaneous PIV velocity and corresponding vorticity fields from a 150 image-pair sequence. Since the flow was repeatable, it was imaged in several different planes in order to reconstruct the entire flow field. Using the circulation measurement schemes described in section 6.5.1 time-resolved circulation measurements of the vortex structures could be obtained from the velocity maps to study the vortex dynamics at the surface, namely, vortex reconnection and dissipation. Further details on these experiments can be found in Willert & Gharib [155].

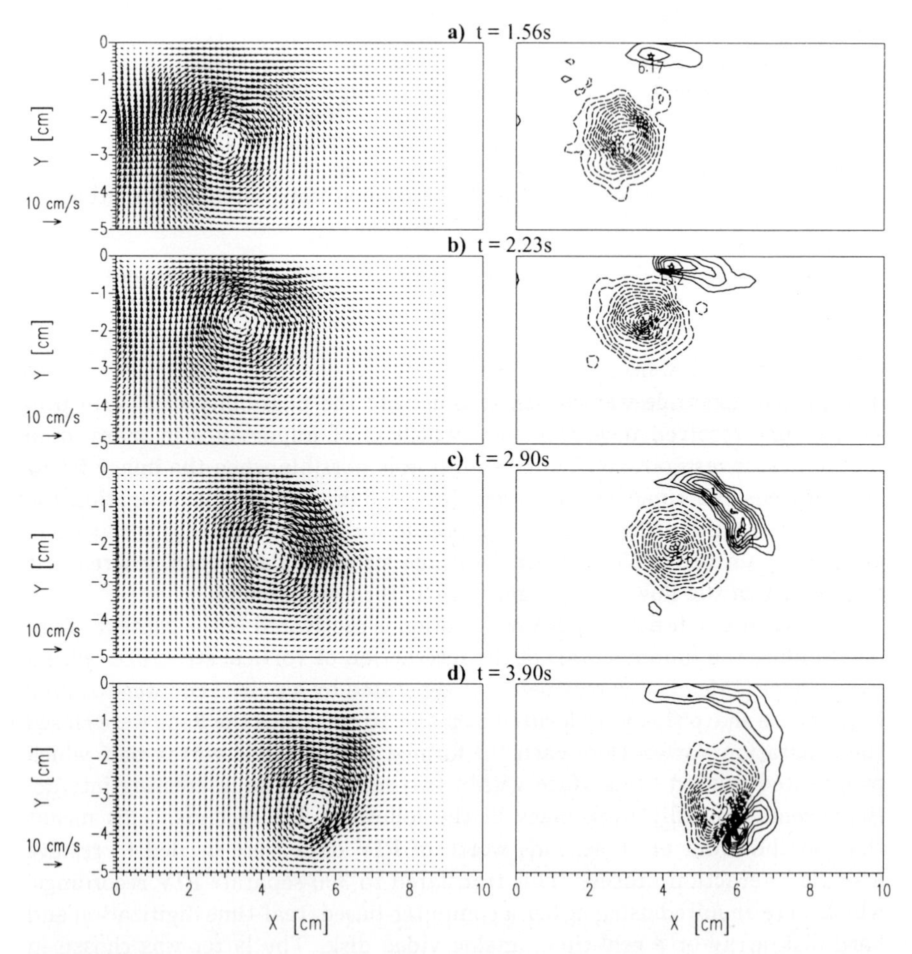

Fig. 8.17. Time resolved PIV measurements of the right core of a vortex pair impinging on a contaminated free surface (at $y = 0$). Shown in the right column are vorticity estimates computed from the velocity fields to the left

8.2.2 Study of thermal convection and Couette flow between two concentric spheres

Table 8.15. PIV recording parameters for thermal convection

Flow geometry	$U_\infty = 0.5\,\mathrm{cm/s}$ parallel to light sheet
Maximum in-plane velocity	$U_{\max} \approx 0.5\,\mathrm{cm/s}$
Field of view	$50 \times 40\,\mathrm{mm}^2$
Interrogation volume	$1.6 \times 1.6 \times 2\,\mathrm{mm}^3$ $(H \times W \times D)$
Dynamic spatial range	DSR $\approx 24:1$
Dynamic velocity range	DVR $\approx 200:1$
Observation distance	$z_0 \approx 1.5\,\mathrm{m}$
Recording method	dual frame/single exposure
Ambiguity removal	frame separation
Recording medium	full frame interline transfer CCD (782×582 pixel)
Recording lens	$f = 100\,\mathrm{mm}$ $f_{\#}2.8$ to 22
Illumination	continuous argon-Ion laser, 1 watt, internal shutter of camera
Pulse delay	$\Delta t = 40\,\mathrm{ms}$
Seeding material	glass particles ($d_\mathrm{p} \approx 10\,\mu\mathrm{m}$)

These experimental investigations of flows by means of PIV have been carried out by DLR in cooperation with the Center of Applied Space Technology and Microgravity (ZARM), University of Bremen, in order to complete their numerical simulations and LDV measurements [126]. The experimental set-up is shown in figure 8.18.

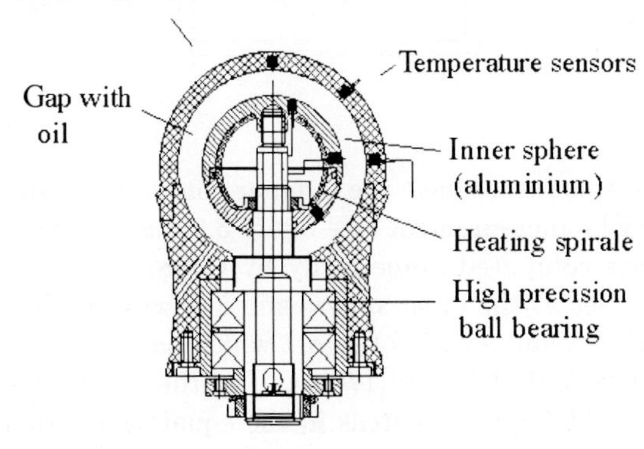

Fig. 8.18. The experimental apparatus to study the thermal convection and the Taylor flow

A fluid (silicone oils M20 and M3) seeded with $10\,\mu\mathrm{m}$ diameter glass particles with a volumetric mass near $1.05\,\mathrm{g}/cm^3$ and a refraction index of 1.55, is filled in the gap between two concentric spheres. The outer sphere is composed of two transparent acrylic glass hemispheres (refraction index of 1.491), with a radius of 40.0 mm, and the inner sphere is made out of aluminum with a radius of 26.7 mm. To minimize optical distortions because of the curvature of the model the outer sphere is included in a rectangular cavity filled with silicone oil to provide a plane liquid–air interface and reduce optical refraction.

To study the thermal convection flows, the inner sphere is heated homogeneously up to 45°C whereas the outer sphere is held at constant temperature. Six temperature sensors are installed on both spheres as indicated in figure 8.18. A 25 Hz CCD camera with an internal shutter (40 ms between each frame) was used in combination with a continuous argon-ion laser. This was possible because of the low velocity flow studied ($\approx$ 0.5 cm/s). A 100 mm Zeiss Makro Planar objective lens was used during the flow measurements with a $f_{\#}$ number of 2.8. For a magnification between 1/2 and 1/4, and a f-number of $f_{\#} = 11$, the particle image diameters are in the range between 22 and 18μm, that is to say between 2 and 3 pixels which give the lowest measurement uncertainty. For the particles utilized in this experiment, the gravitational velocity is found to be: $v_g = 2.9 \cdot 10^{-7}$m/s with M20 oil and $v_g = 3 \cdot 10^{-6}$m/s with M3 oil, disturbances that can be disregarded. The investigations were carried out in a meridional light sheet (figure 8.19) through the sphere's center.

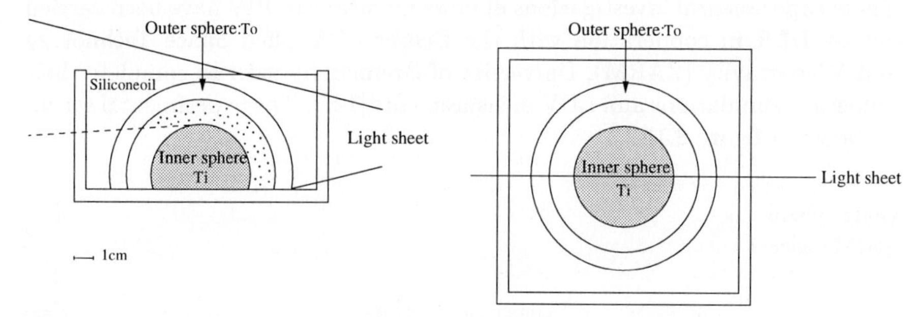

Fig. 8.19. The light sheet position

For small ΔT between the two spheres we have the laminar convective state, and the flow structures of the PIV measurement (see figure 8.20) are in good agreement with the streamlines computed numerically by GARG [174]. We have an upward flow of 0.1 cm/sec at the inner sphere and a downward flow of 0.05 cm/sec at the outer sphere, and a ratio of 2 to 1 which has also been predicted theoretically by MACK and HARDEE [176], and at the north pole we have a radial outward flow of 0.2 cm/s whereas in the equatorial region we have an area of zero velocity in the middle of the sphere as in the model of GARG.

By increasing the ΔT a time dependent pulsating ring vortex sets in at the north pole near the outer sphere (see figure 8.20). The maximal velocities increase up to 1 cm/s in the polar region and we have an upward flow of about 0.35 cm/s at the inner sphere and a downward flow of 0.1 cm/s at the outer sphere. The convective motion is dominant at the boundary regions and near to the pole in contrast to the vanishing velocities in a wide range of radial positions. Additionally there are small radial inward flows at the outer sphere boundary which is in agreement with the numerical simulations.

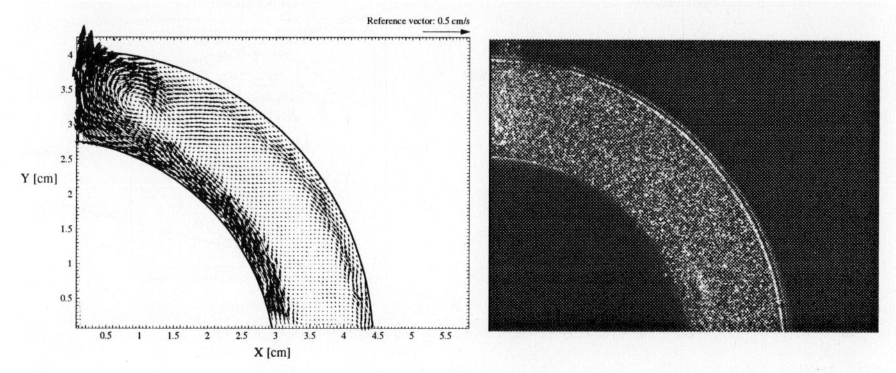

Fig. 8.20. Thermal convection velocity fields and flow picture with the two exposures

Table 8.16. PIV recording parameters for Couette flow

Flow geometry	$U_\infty = 10\,\mathrm{cm/s}$ parallel to light sheet
Maximum in-plane velocity	$U_{\max} \approx 10\,\mathrm{cm/s}$
Field of view	$50 \times 50\,\mathrm{mm}^2$
Interrogation volume	$1.6 \times 1.6 \times 2\,\mathrm{mm}^3$ $(H \times W \times D)$
Dynamic spatial range	DSR $\approx 31:1$
Dynamic velocity range	DVR $\approx 200:1$
Observation distance	$z_0 \approx 0.5\,\mathrm{m}$
Recording method	dual frame/single exposure
Ambiguity removal	frame separation (frame-straddling)
Recording medium	full frame interline transfer CCD (1008 × 1018 pixel)
Recording lens	$f = 60\,\mathrm{mm}$ $f_{\#}$ 2.8 to 22
Illumination	Nd:YAG laser[a] 70 mJ/pulse
Pulse delay	$\Delta t = 20\,\mathrm{ms}$
Seeding material	glass particles ($d_\mathrm{p} \approx 10\,\mu\mathrm{m}$)

[a] Frequency doubled

The Couette flow study required the use of another set-up: the velocity being contained between 5 cm/s and 10 cm/s therefore the previous delay used between each frame was too large. As a consequence, a pulsed Nd:YAG laser synchronized with a large format video camera was used allowing us to select appropriate pulse delays. The frame-straddling technique was employed for ambiguity removal. The study has been performed at 0.4 cm above the pole region. The rotation of the inner sphere was 250 revolutions per minute for the experiment and $\Delta T = 0$. The velocity vector maps are presented in figure 8.22.

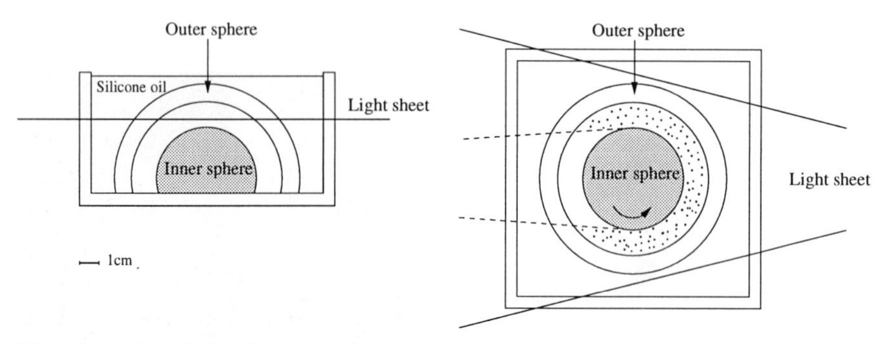

Fig. 8.21. The light sheet position: 0.4 cm up to the pole region

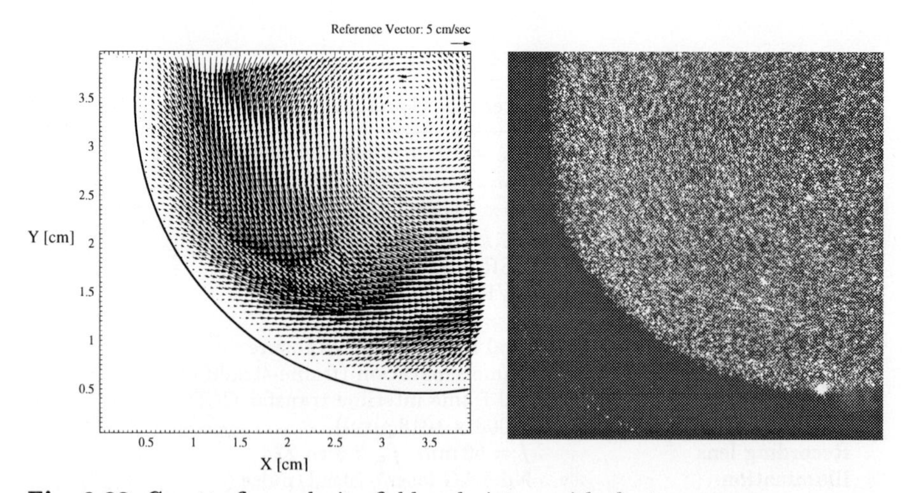

Fig. 8.22. Couette flow velocity field and picture with the two exposures

8.3 Stereo PIV applied to a vortex ring flow

The various methods of image reconstruction and calibration as described in section 7.1 were applied in the measurement of the unsteady vortex ring flow field. Figure 8.23 outlines a simply realized vortex ring generator with very reproducible flow characteristics. The vortex ring is generated by discharging a bank of electrolytic capacitors (60 000 μF) through a pair of loudspeakers which are mounted facing inward on to two sides of a wooden box. By forcing the loudspeaker membranes inward, air is impulsively forced out of a cylindrical, sharpened nozzle (inner diameter= 34.7 mm) on the top of the box. The shear layer formed at the tip of the nozzle then rolls up into a vortex ring and separates from the nozzle as the membranes move back to their equilibrium positions due to the decay in supply voltage. As long as the charging voltage is kept constant, the formation of the vortex ring will be very reproducible. The generator also has a seeding pipe with a check valve allowing the interior of the box and ultimately the core of the vortex ring to be seeded.

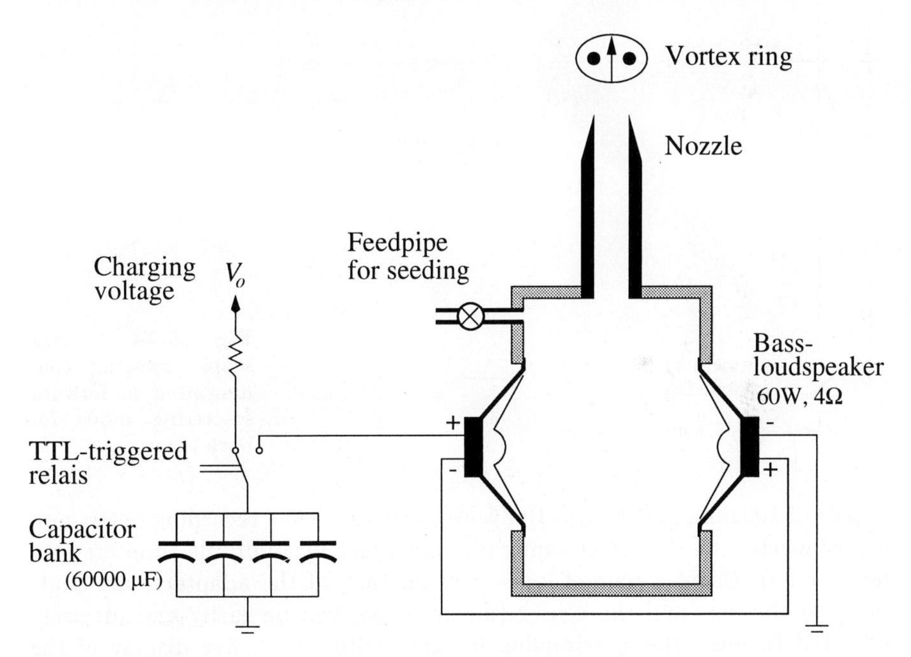

Fig. 8.23. Schematic of the vortex ring generator used to obtain an unsteady, yet reproducible flow field

8.3.1 Imaging configuration and hardware

A noteworthy feature of the imaging configuration outlined in figure 8.24 – which has also been used for the error estimation given in section 7.1 –

is that the cameras are positioned on both sides of the light sheet. This arrangement allows both cameras to make use of the much higher forward scattering properties of the small (1 μm) oil droplets used for seeding. The principal viewing axes are both around 35° from the light sheet normal such that the combined opening angle is approximately 70° near the center of the image.

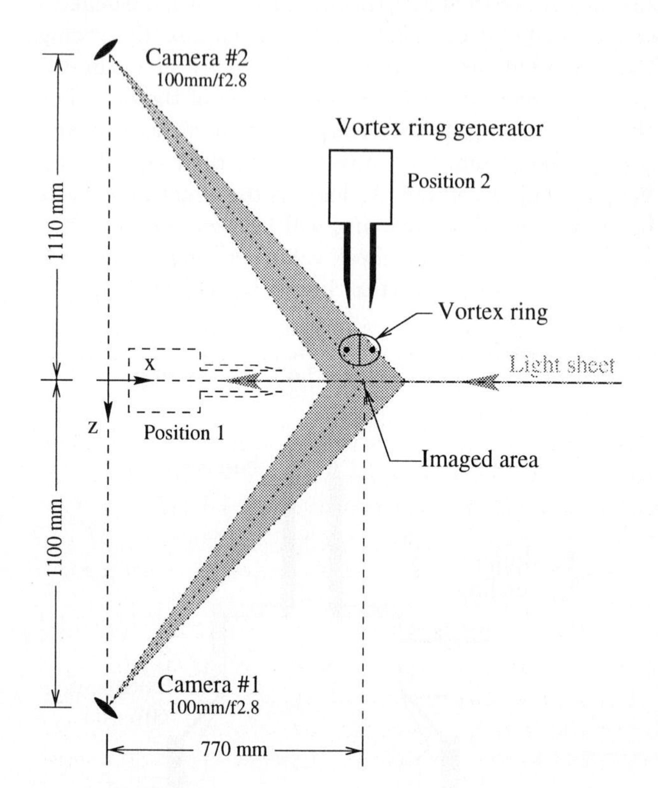

Fig. 8.24. Stereoscopic imaging configuration in forward scattering mode for both cameras

A pair of 100 mm, $f_{\#}$ 2.8 objective lenses constitute the recording optics and are connected to the CCD cameras using specially built tilt-adapters (see figure 8.25). Using a pair of set screws on each of the adapters, the angle between the lens and the sensor (image plane) can be easily and precisely adjusted to meet the Scheimpflug imaging criterion. A live display of the particle images at large f-numbers permit an accurate adjustment within minutes. In the arrangement shown in figure 8.24 this angle (ϕ in figure 7.1b) was measured to be approximately 2.7°. The field of view covered about 145 mm horizontally by 115 mm vertically across the center of the image. The edge loss due to the Scheimpflug imaging arrangement was about 5 mm vertical from side to side, but since both cameras were positioned nearly symmetrically, the field of view could be matched very well, thereby allowing three-dimensional PIV measurements across the entire sensor area. This is an advantage over the "classical" stereoscopic arrangement in which both cameras view from the same side of the light sheet because the nonoverlapping areas are of no use in the three-dimensional reconstruction.

Fig. 8.25. A specially built tilt-adapter between the lens and the sensor allows adjustment according to the Scheimpflug criterion

The cameras used for this experiment are based on a full frame interline transfer CCD sensor with a 1008H by 1018V pixel resolution. The light sheet was generated by a frequency doubled, double oscillator Nd-YAG laser with more than 300 mJ per pulse. Synchronization between the cameras and the laser was achieved by means of a multiple channel sequencer. Since one of the cameras was not capable of operating in a triggered mode it provided the master timing of the entire PIV recording system. The second was operated in an asynchronously triggered mode. Two separate personal computers with interface cards captured the image pairs from the cameras at a common image pair rate of 5 Hz. (In principle the use of a common PC for both cameras would have also been possible.) One of the computers provided the trigger pulse for the vortex generator as soon as the image acquisition was started. By adding a time delay (or by moving the vortex generator back and forth) the position of the vortex ring within the PIV recording could be adjusted.

The light sheet thickness was set at approximately 2.5 mm, while the pulse delay was varied within $300 \leq \Delta t \leq 500\,\mu$s with the vortex ring propagating in-line with the light sheet (position 1 in figure 8.24), and $\Delta t = 200\,\mu$s while propagating normal to the light sheet (position 2 in figure 8.24). With maximum velocities of 3.5 m/s this translated to maximum displacements of 0.7 mm for the vortex ring passage normal to the light sheet. Effectively, the loss of pairs was kept to less than 30% thereby ensuring a high data yield even in regions of high out-of-plane motion. The f-number was set to $f_{\#} = 2.8$.

8.3.2 Experimental results

Initially the nozzle of the vortex ring generator was placed collinearly with the light sheet to provide cross-sectional cuts through the vortex ring (figure 8.26). This provided reference data as well as information on the ring's circulation and stability. In the second configuration, that is, position 2 in figure 8.24, the generator was placed normal to the light sheet. Figure 8.27 shows a pair of two-component velocity fields prior to their combination into a three-component data set. The stereoscopic view is clearly visible. Stereoscopic reconstruction using equation (7.3), equation (7.5) and equation (7.8) then produces the desired three-component data set of figure 8.28.

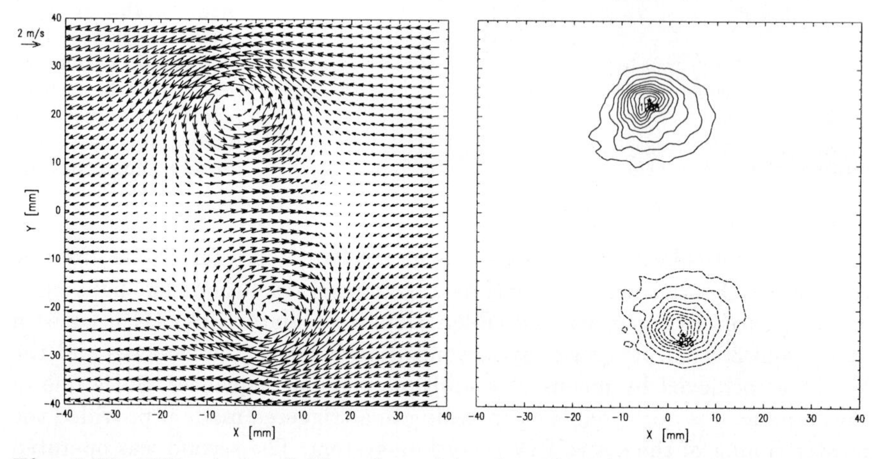

Fig. 8.26. PIV velocity (left) and vorticity data (right) in the symmetry plane of the vortex ring. A velocity of $U = 1.5\,\mathrm{m/s}, V = 0.25\,\mathrm{m/s}$ has been removed to enhance the visibility of the flow's features. The propagation of the ring is left to right and slightly upward (the nozzle was inclined with the horizontal). The vorticity contours are spaced in intervals of $100\,\mathrm{s}^{-1}$ excluding 0

In terms of processing, the image back projection was chosen such that the magnification factor was constant at $10\,\mathrm{pixel\,mm}^{-1}$ in all images after reconstruction. The final image size of 1450 horizontal by 1200 vertical pixel is about 70% larger than the original images. An interrogation area of 32×32 pixel with an overlap (oversampling) of 66% was chosen although only every fourth vector is shown in the plots (= 33% overlap). In physical space the interrogation window covers $3.2 \times 3.2\,\mathrm{mm}^2$ while the grid spacing is $1.0 \times 1.0\,\mathrm{mm}^2$. The particle image density was high enough to achieve valid data rates exceeding 99% over the entire field of view. It was also found that image preconditioning – adaptive background subtraction using a 7×7 pixel kernel highpass filter and subsequent binarization – significantly improved the data yield by bringing most particle images to the same intensity level.

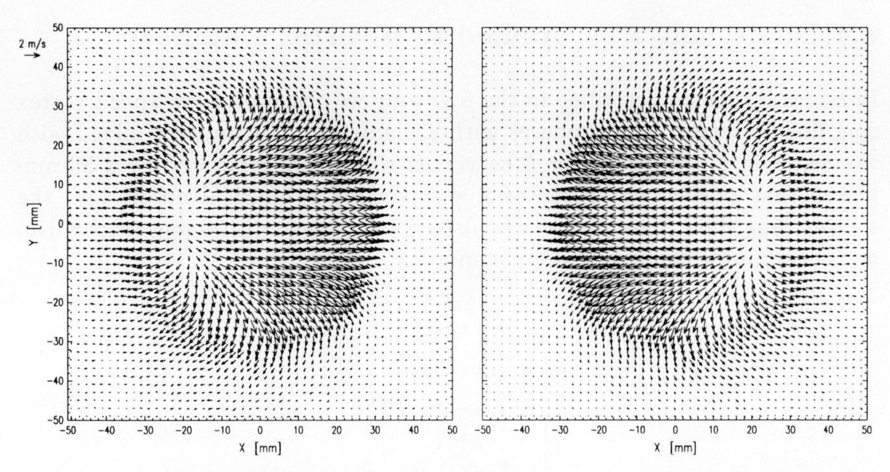

Fig. 8.27. Two-component PIV velocity data of the vortex ring propagating normal to the light sheet as viewed by camera 1 (left) and camera 2 (right)

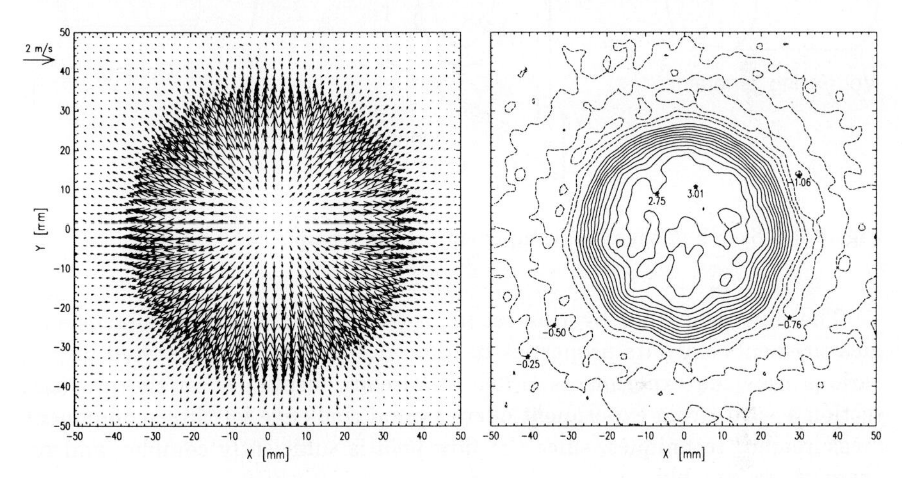

Fig. 8.28. Reconstructed three-component PIV velocity data obtained by combining the data sets shown in figure 8.27. The out-of-plane velocity component w is shown as a contour plot on the right (contour levels at 0.25 m/s)

After the displacement estimation automated outlier detection found on the order of 100 outliers per 16 600 vector data set. Most of these were found on the edges of the original image domain outside of which particle images do not exist. Only very few outliers (< 10) were detected in the central 95% of the field of view, especially in regions of high gradients. The detected outliers were then linearly re-interpolated to allow a subsequent three-dimensional reconstruction.

8.4 Dual-plane PIV applied to a vortex ring flow

In the experiments described in the following we observed a low-speed vortex ring flow in water. Glass spheres with a diameter of 10 μm were mixed with the water in a plexiglas tank. The vortex rings were generated by a 30 mm piston that pushes water out of a sharp-edged cylindrical nozzle into the surrounding fluid. The piston was driven by a linear traversing mechanism and a computer controlled stepper motor.

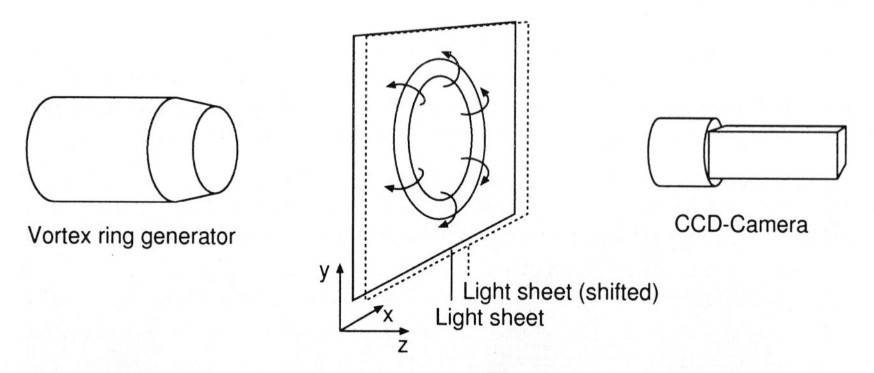

Fig. 8.29. Sketch of the main components of the set-up

The flow generated by this set-up is well suited for three-dimensional measurements since its properties have been documented and tested during various previous experiments [180]. As already mentioned in the previous section a vortex ring experiment offers a good challenge for three-component measurement techniques, since the flow field is sufficiently complex and reproduces reasonably.

8.4.1 Imaging configuration and hardware

Figure 8.29 shows the main components of the set-up except the light sheet shaping optics and the electronic equipment. The arrangement of the optical and the electro-mechanical components are shown in figure 8.30 and are described below.
An argon-ion laser produced a continuous beam of about 6 W output power. An electro-mechanical shutter controlled by a timer box generated light pulses with a pulse length of $t_e = 5\,\text{ms}$ and a pulse separation time of $\Delta t = 33\,\text{ms}$. The shutter was phase locked with the video camera which had a frame-transfer time of $t_f = 2\,\text{ms}$. The aperture of the shutter was of a size that cuts off the outer area of the laser beam of lower intensity. A computer

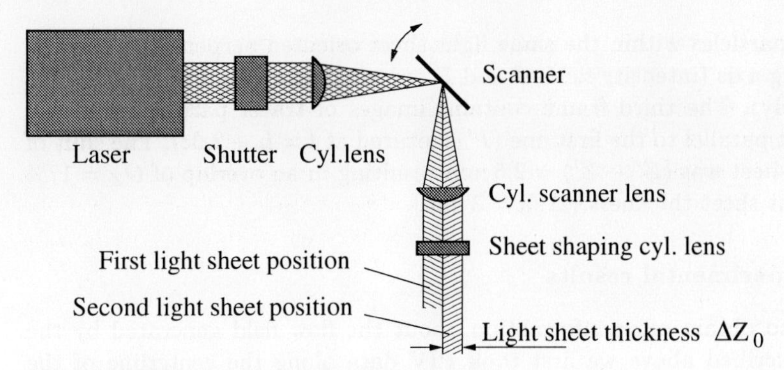

Fig. 8.30. Sketch of the optical components

controlled micro stepper motor with a mirror mounted to one end of the shaft was used as a scanner, which, together with the cylindrical scanner lens (see figure 8.30), generated a parallel displacement of the light sheet. An additional cylindrical lens in front of the scanner mirror focused the light on to the mirror and thus compensated for the focusing effect of the scanner lens on to the beam, which would result in a variation of the light sheet thickness. The light sheet shaping lens had a focal length small enough to generate a light sheet height that was twice as large as the height of the observation field. As a result, the variation of the light intensity was held small with respect to the observed field.

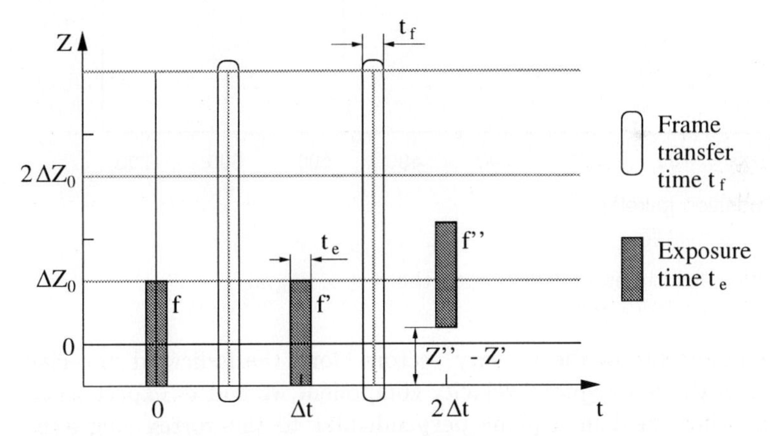

Fig. 8.31. Timing diagram of image capture and light sheet position

The scanner was phased locked to the video signal of the recording camera and alternated the light sheet location after each second capture of a complete video frame (see figure 8.31). Synchronized with the motion of the piston three subsequent video frames were captured. Two frames contain images of tracer particles within the same light sheet oriented perpendicular to the vortex ring axis (intensity fields I and I' captured at $t = t_0$ and $t' = t_0 + \Delta t$, respectively). The third frame contains images of tracer particles within a light sheet parallel to the first one (I'' captured at $t = t_0 + 2\Delta t$). The shift of the light sheet was $(Z'' - Z') = 2.5\,\text{mm}$ resulting in an overlap of $O_Z = 17\%$ of the light sheet thickness ($\Delta Z_0 = 3\,\text{mm}$).

8.4.2 Experimental results

In order to obtain more information about the flow field generated by the set-up described above we first took PIV data along the centerline of the vortex ring (see figure 8.32).

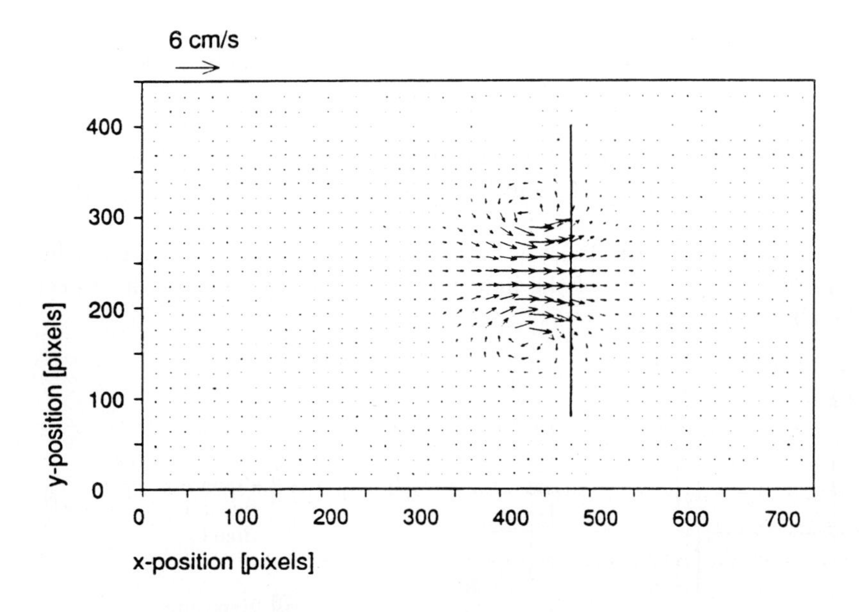

Fig. 8.32. Flow field in an intersection on the vortex ring axis

The axial components of the velocity vectors along the indicated line give information on the out-of-plane velocity component we had to expect when observing the flow field in a plane perpendicular to the vortex ring axis. The magnitude of this velocity component parallel to the axis is plotted in figure 8.33.

Following the described method, we then captured images of particles within two parallel light sheets on to three different frames. Both light sheet planes were orientated perpendicular to the vortex ring axis as shown in figure 8.29. The frames were evaluated by correlating interrogation windows I with I' and I' with I'', detecting the location of the stronger peak, and

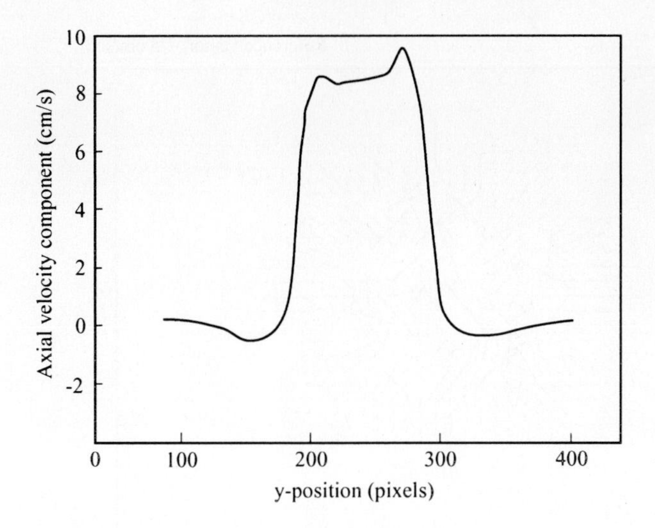

Fig. 8.33. Velocity component parallel to the vortex ring axis along the line shown in the above figure

storing the normalized intensities of both correlation planes at location $\boldsymbol{d}$ for each interrogation area. The size of the interrogation windows was 32×32 pixels and the interrogation stepwidth in both the x and y directions was 16 pixels. The results of the evaluation of the interrogation windows I and I' containing images of particles within the same light sheet show outliers in a ring close to the center of the flow field (see figure 8.34). This area of low detection probability is caused by the decreased seeding density near the center of the vortex ring and by the strong out of plane motion in the center of the observed field.

The normalized heights of the tallest peaks in the cross-correlation planes $R_{\mathrm{II'}}$ are shown in figure 8.35. They clearly show the influence of the out-of-plane velocity component (i.e. low correlation peak heights in the center of the vortex ring).

The results of the evaluation of the interrogation areas I'' and I' show outliers in a ring further outward (see figure 8.36). The normalized heights of the correlation planes $R_{\mathrm{I'I''}}$ are shown in figure 8.37. In this case out-of-plane velocity components increase the correlation peak heights.

The following evaluation procedure was used to take advantage of the images captured in different planes. The correlations $R_{\mathrm{II'}}$ between the interrogation windows I and I' and the correlations $R_{\mathrm{I'I''}}$ between the interrogation windows I' and I'' were computed and normalized in order to obtain $c_{\mathrm{II'}}$ and $c_{\mathrm{I'I''}}$. The distribution containing the highest peak of the interrogated cell was then used to determine the particle image displacement.

This procedure reduces the number of outliers (see figure 8.38) and therefore shows that compared to conventional PIV a larger out-of-plane component can be tolerated for identical Δt. The peak positions found by this procedure were used to find the correct and identical locations in both cross-correlation planes for intensity analysis. Figure 8.39 shows the plot of the

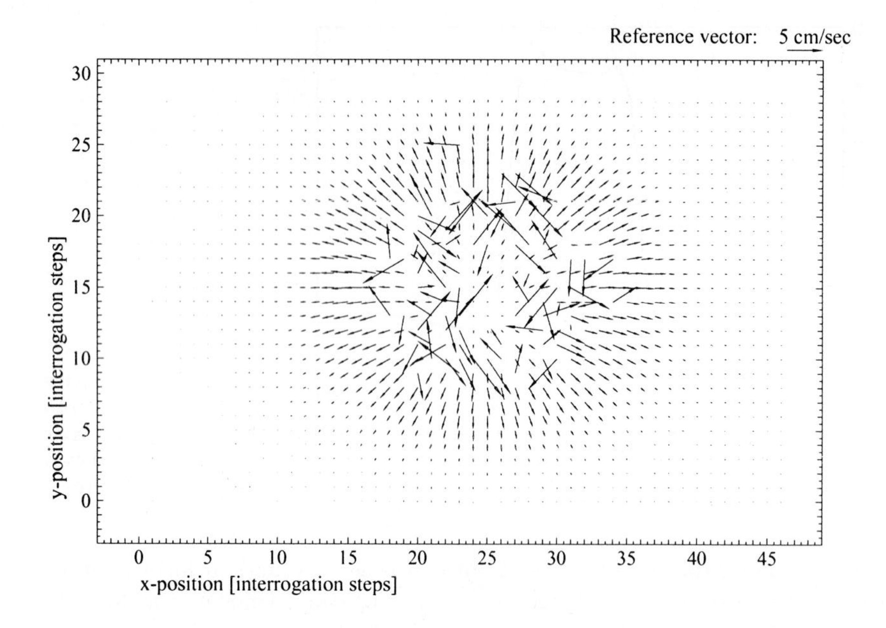

Fig. 8.34. Velocity vector map obtained by images of particles illuminated by the same light sheet

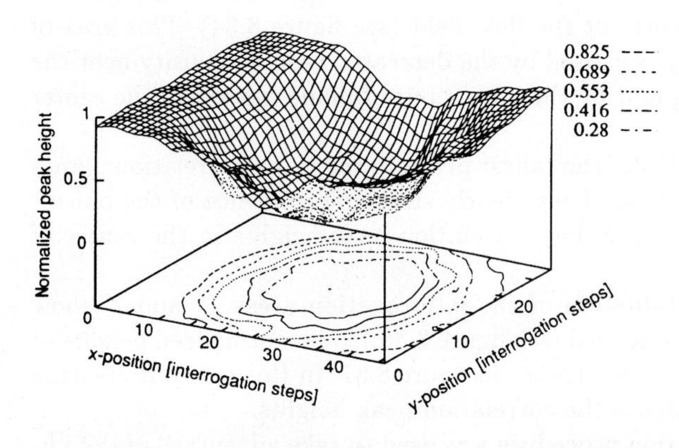

Fig. 8.35. Cross-correlation coefficients $c_{II'}$ for images of particles illuminated by the same light sheet (smoothed by a spatial averaging (3 × 3 kernel) for this representation)

out-of-plane velocity distribution computed from the intensities found in the procedure described in section 7.2 and according to equation (7.21). In contrast to the results obtained by evaluating only two frames (see figure 8.35 and figure 8.37) the expected structures of the flow can now be seen in figure 8.39.

The maximum value of the out-of-plane velocity obtained by the dual-plane correlation technique, of 2.9 mm/33 ms = 8.79 cm/s is in good correspondence with the maximum shown in figure 8.33. Minimum values of

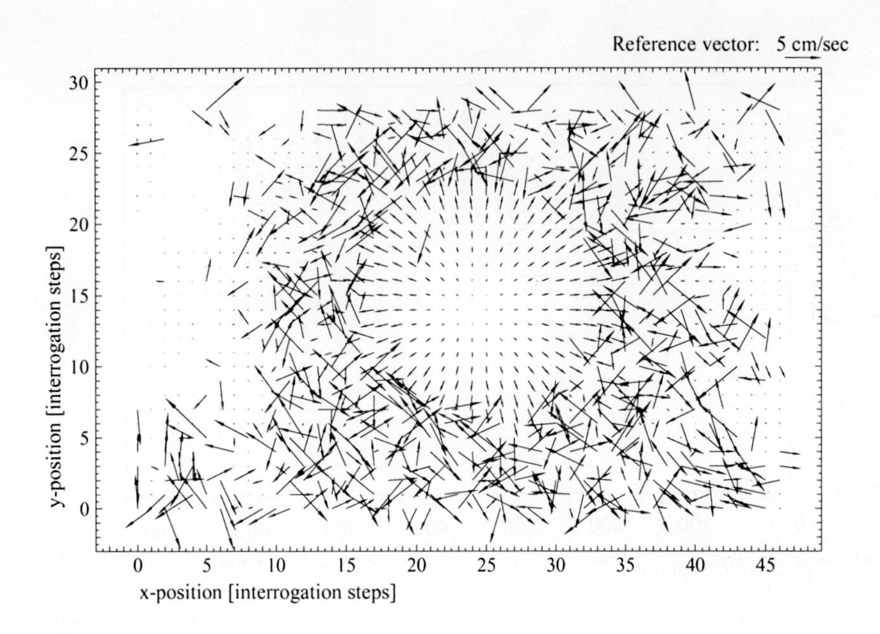

Fig. 8.36. Velocity vector map obtained by images of particles illuminated by different light sheets

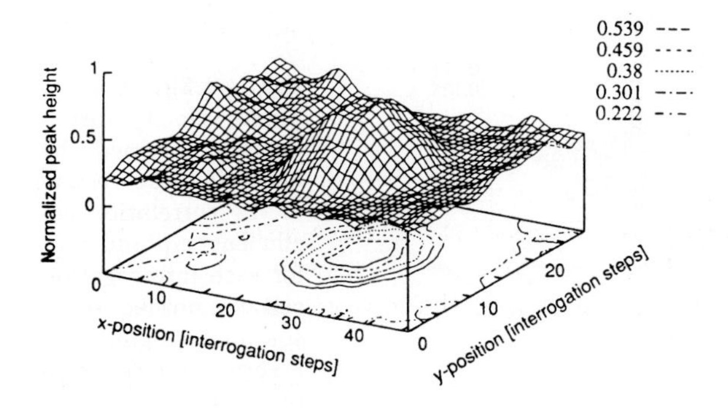

Fig. 8.37. Cross-correlation coefficients $c_{I'I''}$ for images of particles illuminated by different light sheets (smoothed by a spatial averaging (3×3 kernel) for this representation)

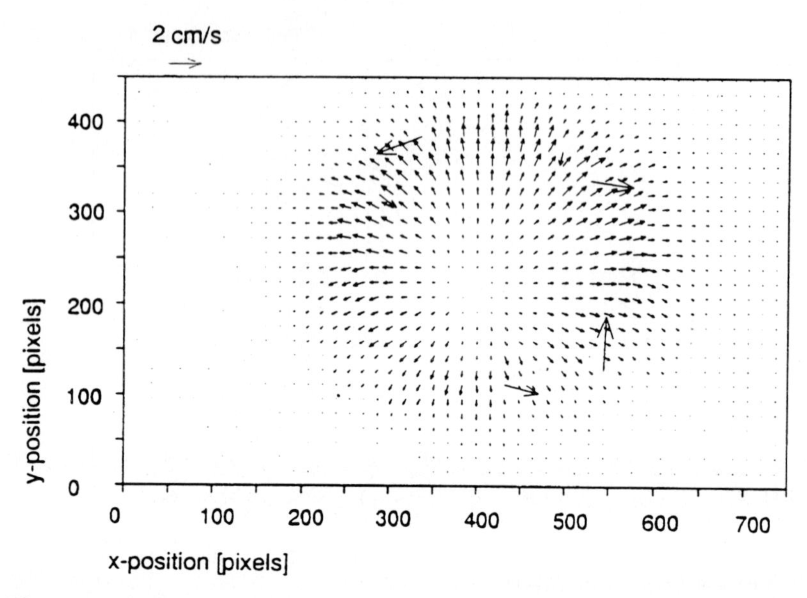

Fig. 8.38. In-plane velocity vector map obtained by considering the strongest peak of both correlations

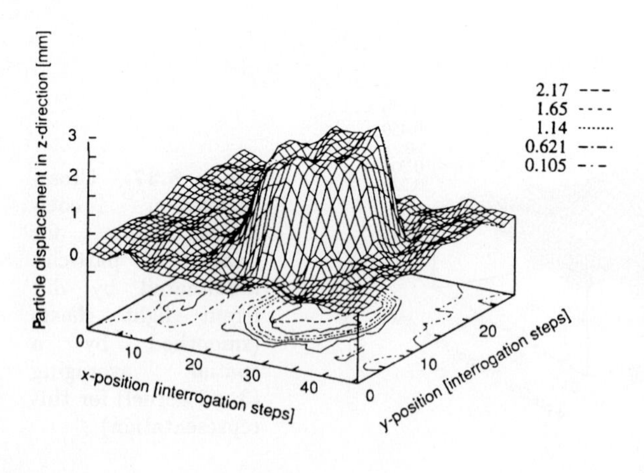

Fig. 8.39. Out-of-plane velocity distribution obtained by analyzing the results of the correlation coefficients $c_{II'}$ and $c_{I'I''}$ for each interrogation cell (smoothed by a spatial averaging (3×3 kernel) for this representation)

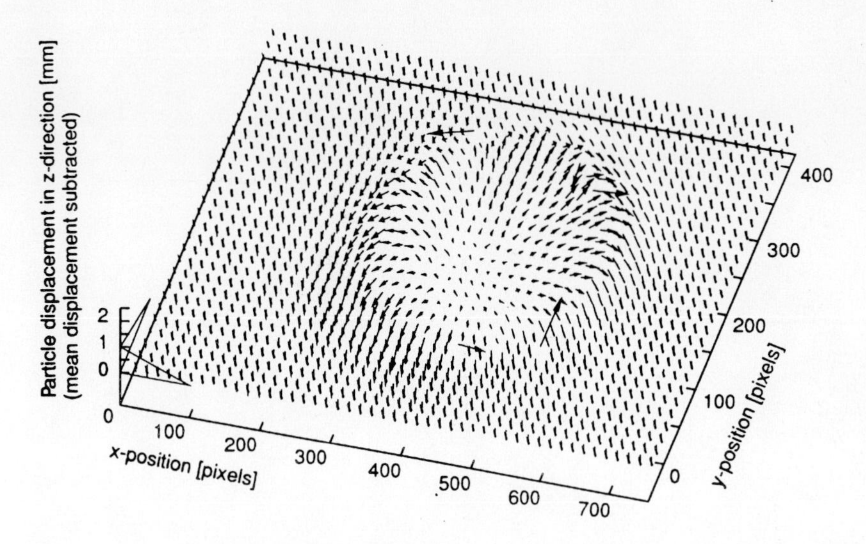

Fig. 8.40. Three-dimensional representation of the velocity vectors of the observed plane (raw data without any smoothing, data validation or interpolation)

the measured velocity distribution are approximately zero in both cases. The final result is shown in figure 8.40 in a three-dimensional representation.

Bibliography

References marked with ($\star$) can be found as reprints in [4].

Books on PIV or other Fluid Measurement Techniques

1. Adrian R.J. (1996): Bibliography of particle image velocimetry using imaging methods: 1917 – 1995, TAM Report 817, UILU-ENG-96-6004, University of Illinois (available from fluid@tsi.com).
2. Dracos Th., ed. (1996): *Three-Dimensional Velocity and Vorticity Measuring and Image Analysis Techniques*, Kluwer Academic Publishers, Dordrecht.
3. Goldstein R. J. (1996): *Fluid Mechanics Measurements, 2nd Edition*, Taylor & Francis, Washington, DC.
4. Grant I., ed. (1994): *Selected papers on particle image velocimetry* SPIE Milestone Series **MS 99**, SPIE Optical Engineering Press, Bellingham, Washington.
5. Merzkirch W. (1987): *Flow Visualization, 2nd Edition*, Academic Press, Orlando.
6. Riethmuller M.L., ed. (1996): Particle Image Velocimetry, *von Karman Institute for Fluid Dynamics, Lecture Series* 1996–03.
7. Westerweel J. (1993): *Digital particle image velocimetry – Theory and application* Ph.D. Dissertation, Delft University Press, Delft.

Books on optics, image processing, statistics and numerical methods

8. Bendat J.S., Piersol A.G. (1971): *Random data: Analysis and measurement procedures*, Wiley-Interscience, John Wiley & Sons, New York.
9. Bracewell R.N (1978): *The Fourier transform and its applications, second edition*, McGraw-Hill Kogakusha LTD., Tokyo.
10. Brigham E.O. (1974): *The fast Fourier transform*, Prentice-Hall, Englewood Cliffs, New Jersey.
11. Gonzalez R.C., Wintz P. (1987): *Digital image processing, 2nd Edition*, Addison-Wesley Publishing Company, Reading, Massachusetts.
12. Goodman J.W. (1996): *Introduction to Fourier optics*, McGraw-Hill Book Company, San Francisco.
13. Hecht E., Zajac A. (1974): *Optics*, Addison-Wesley Pub. Company, Massachusetts.
14. Horner J.L. (1987): *Optical Signal Processing*, Academic Press, Orlando.
15. van de Hulst H.C. (1957): *Light scattering by small particles*, John Wiley & Sons, Inc., New York (republished 1981 by Dover Publications, New York).
16. Jähne B. (1997): *Digitale Bildverarbeitung, 4th Edition*, Springer-Verlag, Berlin Heidelberg (also available in English: *Digital image processing*).
17. Jain A.K. (1989): *Fundamentals of digital image processing*, Prentice Hall, Englewood Cliffs, New Jersey.
18. Kneubühl F.K., Sigrist M.W. (1995): *Laser*, Teubner Studienbücher,

Stuttgart.
19. Lauterborn W., Kurz T., Wiesenfeldt M. (1995): *Coherent optics – Fundamentals and applications*, Springer Verlag, Berlin.
20. Pratt W.K. (1991): *Digital image processing, 2nd Edition*, Wiley-Interscience, John Wiley & Sons, New York.
21. Press W.H., Teukolsky S.A., Vettering W.T., Flannery B.P. (1992): *Numerical recipes in C, 2nd Edition*, Cambridge University Press, Cambridge.
22. Papoulis A. (1977): *Signal Analysis*, Mc Graw Hill, New York.
23. Papoulis A. (1991): *Probability, Random Variables and Stochastic Processes*, Mc Graw Hill, New York.
24. Reynolds G.O., DeVelis J.B., Parrent G.B., Thompson B.J. (1989): *The New Physical Optics Notebook: Tutorials in Fourier Optics*, SPIE Optical Engineering Press, Washington.
25. Rosenfeld A., Kak A.C. (1982): *Digital picture processing, 2nd Edition, Volumes 1 & 2*, Academic Press, San Orlando.
26. Solf K.D. (1986): *Fotografie: Grundlagen, Technik, Praxis*, Fischer Taschenbuch Verlag, Frankfurt am Main.
27. Yaroslavsky L.P. (1985): *Digital Picture Processing*, Springer Verlag, Berlin.

Papers of general interest to PIV including applications

28. Adrian R.J. (1986): Multi-point optical measurements of simultaneous vectors in unsteady flow – a review, *Int. Journal of Heat and Fluid Flow*, **7**, pp. 127–145 (⋆).
29. Adrian R.J. (1991): Particle-imaging techniques for experimental fluid mechanics, *Ann. Rev. Fluid Mech.*, **23**, pp. 261–304 (⋆).
30. Grant I. (1997): Particle image velocimetry: a review, *Proceedings Institute of Mechanical Engineers*, **211**, pp. 55-76.
31. Hinsch K.D. (1993): Particle image velocimetry, in *Speckle Metrology*, ed. R.S. Sirohi, Marcel Dekker, New York, pp. 235–323.
32. Hinsch K.D. (1995): Three-dimensional particle velocimetry, *Meas. Sci. Tech.*, **6**, pp. 742–753.
33. Kompenhans J., Raffel M., Willert C. (1996): PIV applied to aerodynamic investigations in wind tunnels, *von Karman Institute for Fluid Dynamics, Lecture Series* 1996-03, Particle Image Velocimetry.
34. Lai W.T. (1996): Particle Image Velocimetry: A new approach in experimental fluid research, in *Th. Dracos (ed.), Three-Dimensional Velocity and Vorticity Measuring and Image Analysis Techniques*, Kluwer Academic Publishers, Dordrecht, pp. 61-92.
35. Lauterborn W., Vogel A. (1984): Modern optical techniques in fluid mechanics, *Ann. Rev. Fluid Mech.*, **16**, pp. 223–244 (⋆).

Papers about aspects of two-dimensional PIV

36. Adrian R.J. (1986): Image shifting technique to resolve directional ambiguity in double-pulsed velocimetry, *Appl. Optics*, **25**, pp. 3855–3858 (⋆).
37. Adrian R.J. (1988): Statistical properties of particle image velocimetry measurements in turbulent flow, in *Laser Anemometry in Fluid Mechanics III*, Springer-Verlag, Berlin Heidelberg, pp. 115–129 (⋆).
38. Adrian R.J. (1995): Limiting resolution of particle image velocimetry for turbulent flow, in *Advances in Turbulence Research-1995*, Proc. 2nd Turbulence Research Assoc. Conf. Pohang Inst. Tech., pp. 1–19.
39. Adrian R.J., Yao C.S. (1985): Pulsed laser technique application to liquid and gaseous flows and the scattering power of seed materials, *Appl. Optics*, **24**, pp.

44-52 (⋆).

40. Anderson D.J., Greated C.A., Jones J.D.C., Nimmo G., Wiseall S. (1996): Fibre optic PIV studies in an industrial combustor, Proc. *8th Intl. Symp. on Appl. of Laser Techniques to Fluid Mechanics, 8-11July, Lisbon, Portugal.*
41. Bjorkquist D.C. (1990): Particle image velocimetry analysis system, Proc. *5th Intl. Symp. on Appl. of Laser Techniques to Fluid Mechanics, 9–12 July, Lisbon, Portugal.*
42. Bryanston-Cross P.J., Epstein A. (1990): The application of sub-micron particle visualisation for PIV (particle image velocimetry) at transonic and supersonic speeds, *Prog. Aerospace Sci.*, **27**, pp. 237–265.
43. Carasone F., Cenedese A., Querzoli G. (1995): Recognition of partially overlapped particle images using the Kohonen neural network, *Exp. Fluids*, **19**, pp. 225–232.
44. Coupland J.M., Pickering C.J.D., Halliwell N.A. (1987): Particle image velocimetry: theory of directional ambiguity removal using holographic image separation, *Appl. Optics*, **26**, pp. 1576–1578.
45. Cowen E.A., Monismith S.G. (1997): A hybrid digital particle tracking velocimetry technique, *Exp. Fluids*, **22**, pp. 199-211.
46. Echols W.H., Young J.A. (1963): Studies of portable air-operated aerosol generators, NRL Report 5929, Naval Research Laboratory, Washington D.C.
47. Gauthier V., Riethmuller M.L. (1988): Application of PIDV to complex flows: Resolution of the directional ambiguity, *von Karman Institute for Fluid Dynamics, Lecture Series* 1988-06, Particle Image Displacement Velocimetry.
48. Gogineni S., Trump D., Goss L., Rivir R., Pestian D. (1996): High resolution digital two-color PIV (D2CPIV) and its application to high free stream turbulent flows, Proc. *8th Intl. Symp. on Appl. of Laser Techniques to Fluid Mechanics, 8–11 July, Lisbon, Portugal.*
49. Goss L.P., Post M.E., Trump D.D., Sarka B. (1989): Two-color particle velocimetry, Proc. *ICALEO '89, L.I.A.,* **68**, pp. 101–111.
50. Grant I., Liu A. (1989): Method for the efficient incoherent analysis of particle image velocimetry images, *Appl. Optics*, **28**, pp. 1745–1748 (⋆).
51. Grant I., Liu A. (1990): Directional ambiguity resolution in particle image velocimetry by pulse tagging, *Exp. Fluids*, **10**, pp. 71–76 (⋆).
52. Grant I., Pan X. (1995): An investigation of the performance of multi layer neural networks applied to the analysis of PIV images, *Exp. Fluids*, **19**, pp. 159–166.
53. Grant I., Smith G.H., Owens E.H. (1988): A directionally sensitive particle image velocimeter, *J. Phys. E: Sci. Instrum.*, **21**, pp. 1190–1195.
54. Grant I., Smith G.H., Liu A., Owens E.H., Yan Y.Y. (1989): Measuring turbulence in reversing flows by particle image velocimeter, Proc. *ICALEO '89, L.I.A.,* **68**, pp. 92–100.
55. Guezennec Y.G., Kiritsis N. (1990): Statistical investigation of errors in particle image velocimetry, *Exp. Fluids*, **10**, pp. 138–146 (⋆).
56. Hart D.P. (1996): Sparse array image correlation, Proc. *8th Intl. Symp. on Appl. of Laser Techniques to Fluid Mechanics, 8–11 July, Lisbon, Portugal.*
57. Hinsch K., Arnold W., Platen W. (1987): Turbulence measurements by particle imaging velocimetry, Proc. *ICALEO '87 — Optical Methods in Flow and Particle Diagnostics, L.I.A.,* **63**, pp. 127–134 (⋆).
58. Höcker R., Kompenhans J. (1989): Some technical improvements of particle image velocimetry with regard to its application in wind tunnels, Proc. *Intl. Congr. on Instrumentation in Aerospace Facilities (ICIASF'89), Göttingen, Germany.*
59. Höcker R., Kompenhans J. (1991): Application of Particle Image Velocimetry

to Transonic Flows, in *Application of Laser Techniques to Fluid Mechanics,* ed. R.J. Adrian et al., Springer Verlag, pp. 416–434.
60. Huang H.T., Fiedler H.E. (1994): Reducing time interval between successive exposures in video PIV, *Exp. Fluids,* **17**, pp. 356–363.
61. Humphreys W.M. (1989): A histogram-based technique for rapid vector extraction from PIV photographs, Proc. *4th Intl. Conf. on Laser Anemometry, Advances and Applications, Cleveland, Ohio.*
62. Humphreys W.M., Bartram S.M., Blackshire J.L. (1993): A survey of particle image velocimetry applications in langley aerospace facilities, Proc. *31st Aerospace Sciences Meeting, 11–14 January, Reno, Nevada, (AIAA Paper 93-041).*
63. Hunter W.W., Nichols C.E. (1985): Wind tunnel seeding systems for laser velocimeters, Proc. *NASA Workshop, 19–20 March, NASA Langley Research Center (NASA Conference Publication 2393).*
64. Keane R.D., Adrian R.J. (1990): Optimization of particle image velocimeters. Part I: Double pulsed systems, *Meas. Sci. Tech.*, **1**, pp. 1202–1215.
65. Keane R.D., Adrian R.J. (1991): Optimization of particle image velocimeters. Part II: Multiple pulsed systems, *Meas. Sci. Tech.*, **2**, pp. 963–974.
66. Keane R.D., Adrian R.J. (1992): Theory of cross-correlation analysis of PIV images, *Appl. Sci. Res.*, **49**, pp. 191–215 ($\star$).
67. Kompenhans J., Reichmuth J. (1986): Particle imaging velocimetry in a low turbulent wind tunnel and other flow facilities, Proc. *AGARD Conference on Advanced Instrumentation for Aero Engine Components, 19–23 May, Philadelphia (AGARD-CP 399-35).*
68. Landreth C.C., Adrian R.J., Yao C.S. (1988): Double-pulsed particle image velocimeter with directional resolution for complex flows, *Exp. Fluids,* **6**, pp. 119–128 ($\star$).
69. Landreth C.C., Adrian R.J. (1988): Electrooptical image shifting for particle image velocimetry, *Appl. Optics,* **27**, pp. 4216–4220 ($\star$).
70. Landreth C.C., Adrian R.J. (1988): Measurement and refinement of velocity data using high image density analysis in particle image velocimetry, Proc. *4th Intl. Symp. on Appl. of Laser Techniques to Fluid Mechanics, 11–14 July, Lisbon, Portugal (6-14).*
71. Lecordier B., Mouquallid M., Vottier S., Rouland E., Allano D., Trinite (1994): CCD recording method for cross-correlation PIV development in unstationary high speed flow, *Exp. Fluids,* **17**, pp. 205–208.
72. Lourenço L.M., Krothapalli A., Buchlin J.M., Riethmuller M.L. (1986): A non-invasive experimental technique for the measurement of unsteady velocity and vorticity fields, *AIAA Journal,* **24**, pp. 1715–1717.
73. Lourenço L.M. (1988): Some comments on particle image displacement velocimetry, *von Karman Institute for Fluid Dynamics, Lecture Series* 1988-06, Particle Image Displacement Velocimetry.
74. Lourenço L.M. (1993): Velocity bias technique for particle image velocimetry measurements of high speed flows, *Appl. Optics,* **32**, pp. 2159–2162.
75. Meinhart C.D., Prasad A.K., Adrian R.J. (1993): A parallel digital processor system for particle image velocimetry, *Meas. Sci. Tech.*, **4**, pp. 619–626.
76. Melling A. (1986): Seeding gas flows for laser anemometry, Proc. *AGARD Conference on Advanced Instrumentation for Aero Engine Components, 19–23 May, Philadelphia (AGARD-CP 399-8).*
77. Meynart R. (1983): Mesure de champs de vitesse d' ecoulements fluides par analyse de suites d' images obtenues par diffusion d' un feuillet lumineux, Ph.D. dissertation, Faculté des Sciences Appliquées, Universite Libre de Bruxelles.
78. Meyers J.F. (1991): Generation of particles and seeding, *von Karman Institute*

for Fluid Dynamics, Lecture Series 1991-05, Laser Velocimetry.
79. Morck T., Andersen P.E., Westergaard C.H. (1992): Processing speed of photorefractive optical correlators in PIV-processing, Proc. *6th Intl. Symp. on Appl. of Laser Techniques to Fluid Mechanics, 20–23 July, Lisbon, Portugal (27-1).*
80. Oschwald M., Bechle S., Welke S. (1995): Systematic errors in PIV by realizing velocity offsets with the rotating mirror method, *Exp. Fluids*, **18**, pp. 329–334.
81. Pickering C.J.D., Halliwell N.A. (1984): Speckle photography in fluid flows: signal recovery with two-step processing, *Appl. Optics*, **23**, pp. 1128–1129 (⋆).
82. Pickering C.J.D., Halliwell N.A. (1984): Laser speckle photography and particle image velocimetry: photographic film noise, *Appl. Optics*, **23**, pp. 2961–2969.
83. Pierce W.F., Delisi D.P. (1995): Effects of interrogation window size on the measurement of vortical flows with digital particle image velocimetry, in *J. Crowder (ed.), Flow Visualization VII,* Begell House, New York, pp. 728–732.
84. Prasad A.K., Adrian R.J., Landreth C.C., Offutt P.W. (1992): Effect of resolution on the speed and accuracy of particle image velocimetry interrogation, *Exp. Fluids*, **13**, pp. 105–116.
85. Raffel M., Kompenhans J. (1994): Error analysis for PIV recording utilizing image shifting, Proc. *7th Intl. Symp. on Appl. of Laser Techniques to Fluid Mechanics, 11–14 July, Lisbon, Portugal (35-5).*
86. Raffel M., Kompenhans J., Stasicki B., Bretthauer B., Meier G.E.A. (1994): Velocity measurement of compressible air flows utilizing a high-speed video camera, *Exp. Fluids*, **18**, pp. 204–206.
87. Raffel M., Kompenhans J. (1995): Theoretical and experimental aspects of image shifting by means of a rotating mirror system for particle image velocimetry, *Meas. Sci. Tech.*, **6**, pp. 795–808.
88. Roesgen T., Totaro R. (1995): Two-dimensional on-line particle imaging velocimetry, *Exp. Fluids*, **19**, pp. 188-193.
89. Rouland E., Vottier S., Lecordier B., Trinité M. (1994): Cross-correlation PIV development for high speed flows with a standard CCD camera, Proc. *2nd Int. Seminar on Opt. Methods and Data Processing in Heat and Fluid Flow, London.*
90. Sinha S.K. (1988): Improving the accuracy and resolution of particle image or laser speckle velocimetry, *Exp. Fluids*, **6**, pp. 67–68 (⋆).
91. Somerscales E.F.C. (1980): Fluid velocity measurement by particle tracking, in *Flow, its Measurement and Control in Science and Industry, Vol. I,* ed. R.E. Wendt, , pp. 795–808.
92. Thomas P. (1991): Experimentelle und theoretische Untersuchungen zum Folgeverhalten von Teilchen in kompressibler Strömung, *Deutsche Forschungsanstalt für Luft- und Raumfahrt, Research Report* DLR-FB 91-25.
93. Towers C.E., Bryanston-Cross P.J., Judge T.R. (1991): Application of particle image velocimetry to large-scale transonic wind tunnels, *Optics and Laser Technology*, **23**, pp. 289–295.
94. Vogt A., Raffel M., Kompenhans J. (1992): Comparison of optical and digital evaluation of photographic PIV recordings, Proc. *6th Intl. Symp. on Appl. of Laser Techniques to Fluid Mechanics, 20–23 July, Lisbon, Portugal (27-4).*
95. Vogt A., Reichel F., Kompenhans J. (1996): A compact and simple all optical evaluation method for PIV recordings, in *Developments in Laser Techniques and Applications to Fluid Mechanics,* ed. R.J. Adrian et al., Springer Verlag, pp. 423–437.
96. Westerweel J., Dabiri D., Gharib M. (1997): The effect of a discrete window offset on the accuracy of cross-correlation analysis of PIV recordings, *Exp. Fluids*, **23**, pp. 20–28.
97. Willert C.E., Gharib M. (1991): Digital particle image velocimetry, *Exp. Flu-*

ids, **10**, pp. 181–193 (⋆).
98. Willert C., Stasicki, B., Raffel M., Kompenhans J. (1995): A digital video camera for application of particle image velocimetry in high-speed flows, Proc. SPIE 2546 *Intl. Symp. on Optical Science, Engineering and Instrumentation, 9–14 July, San Diego, USA,* pp.124–134.
99. Willert, C. (1996): The fully digital evaluation of photographic PIV recordings, *Appl. Sci. Res.*, **56**, pp. 79–102.
100. Wormell D.C., Sopchak J. (1993): A particle image velocimetry system using a high-resolution CCD camera, Proc. SPIE 2005, *Optical Diagnostics in Fluid and Thermal Flow,* ed. S S Cha, J D Trollinger, pp. 648–654.

Papers about advanced PIV methods

101. Arroyo M.P., Greated C.A. (1991): Stereoscopic particle image velocimetry, *Meas. Sci. Tech.*, **2**, pp. 1181–1186.
102. Barnhart D.H., Adrian R.J., Papen G.C. (1994): Phase-conjugate holographic system for high-resolution particle image velocimetry, *Appl. Optics*, **33**, pp. 7159–7170.
103. Brücker C. (1996): Spatial correlation analysis for 3-D scanning PIV: simulation and application of dual-color light-sheet scanning, Proc. *8th Intl. Symp. on Appl. of Laser Techniques to Fluid Mechanics, 8–11 July, Lisbon, Portugal.*
104. Brücker C. (1996): 3-D PIV via spatial correlation in a color-coded light sheet, *Exp. Fluids*, **21**, pp. 312–314.
105. Brücker C. (1996): 3-D scanning particle image velocimetry: technique and application to a spherical cap wake flow, *Appl. Sci. Res.*, **56**, pp. 157–179.
106. Coupland J.M., Halliwell N.A. (1992): Particle image velocimetry: three-dimensional fluid velocity measurements using holographic recording and optical correlation, *Appl. Optics*, **31**, pp. 1005-1007.
107. Gauthier V., Riethmuller M.L. (1988): Application of PIDV to complex flows: Measurement of the third component, *von Karman Institute for Fluid Dynamics, Lecture Series* 1988-06, Particle Image Displacement Velocimetry.
108. Gaydon M., Raffel M., Willert C., Rosengarten M., Kompenhans J. (1997): Hybrid stereoscopic particle image velocimetry, *Exp. Fluids*, **23**, pp. 331–334.
109. Heckmann W., Hilgers S., Merzkirch W., Wagner T. (1994): PIV-Messungen in einer Zweiphasenströmung unter Verwendung von zwei CCD-Kameras, Proc. *4. Fachtagung Lasermethoden in der Strömungsmesstechnik, 12–14 September, Aachen, Germany.*
110. Keane R.D., Adrian R.J., Zhang Y (1995): Super-resolution particle image velocimetry, *Meas. Sci. Tech.*, **6**, pp. 754–768.
111. van Oord J. (1997): The design of a stereoscopic DPIV-system, Report MEAH-161 Delft, the Netherlands: Delft University of Technology.
112. Prasad A.K., Adrian R.J. (1993): Stereoscopic particle image velocimetry applied to liquid flows, *Exp. Fluids*, **15**, pp. 49–60.
113. Prasad A.K., Jensen K. (1995): Scheimpflug stereocamera for particle image velocimetry to liquid flows, *Appl. Optics*, **34**, pp. 7092–7099.
114. Raffel M., Westerweel J., Willert C., Gharib M., Kompenhans J. (1996): Analytical and experimental investigations of dual-plane particle image velocimetry, *Optical Engineering*, **35**, pp. 2067–2074.
115. Raffel M., Gharib M., Ronneberger O., Kompenhans J. (1995): Feasibility study of three-dimensional PIV by correlating images of particles within parallel light sheet planes, *Exp. Fluids*, **19**, pp. 69–77.
116. Royer H., Stanislas M. (1996): Stereoscopic and holographic approaches to get the third velocity component in PIV, *von Karman Institute for Fluid Dynamics, Lecture Series* 1996–03, Particle Image Velocimetry.

117. Westerweel J., Nieuwstadt F.T.M. (1991): Performance tests on 3-dimensional velocity measurements with a two-camera digital particle-image velocimeter, in *Laser Anemometry Advances and Applications, Vol. 1* (ed. Dybbs A. and Ghorashi B.), ASME, New York, pp. 349–55.

Papers about post-processing PIV data

118. Abrahamson S., Lonnes S. (1995): Uncertainty in calculating vorticity from 2D velocity fields using circulation and least-squares approaches, *Exp. Fluids*, **20**, pp. 10–20.
119. Dieterle L. (1997): *Entwicklung eines abbildenden Messverfahrens (PIV) zur Untersuchung von Mikrostrukturen in turbulenten Strömungen*, PhD thesis, Deutscher Universitäts Verlag GmbH, Wiesbaden.
120. Lecuona A., Nogueira J.I., Rodriguez P.A. (1997): Flowfield vorticity calculation using PIV data, Proc. *2nd Intl. Workshop on PIV'97, 8–11 July, Fukui, Japan.*
121. Lourenço L., Krothapalli A. (1995): On the accuracy of velocity and vorticity measurements with PIV, *Exp. Fluids*, **18**, pp. 421-428.
122. Lourenço L.M. (1996): Particle image velocimetry: post-processing techniques, *von Karman Institute for Fluid Dynamics, Lecture Series* 1996-03, Particle Image Velocimetry.
123. Raffel M., Leitl B., Kompenhans J. (1993): Data validation for particle image velocimetry, in *Laser Techniques and Applications in Fluid Mechanics,* R.J. Adrian et al., Springer-Verlag, pp. 210–226.
124. Raffel M., Kompenhans J. (1996): Post-processing: data validation, *von Karman Institute for Fluid Dynamics, Lecture Series* 1996-03, Particle Image Velocimetry.
125. Westerweel J. (1994): Efficient detection of spurious vectors in particle image velocimetry data, *Exp. Fluids*, **16**, pp. 236–247.

Papers concerned mainly with the application of PIV

126. Böhm C., Wulf P., Egbers C., Rath H.J. (1997): LDV- and PIV-measurements on the dynamics in spherical Couette flow, Proc. *Int. Conf. on Laser Anemometry-Advances and Appl., 8–11 May, Karlsruhe.*
127. Cenedese A., Querzoli G. (1995): PIV for Lagrangian scale evaluation in a convective boundary layer, in *Flow Visualisation, vol. VI,* (eds. Tanida Y, Miyshiro H.), Springer Verlag, Berlin, pp. 863–867.
128. Kähler C.J. (1997): Ortsaufgelöste Geschwindigkeitsmessungen in einer turbulenten Grenzschicht, *Deutsche Forschungsanstalt für Luft- und Raumfahrt, Research Report* DLR-FB 97–32.
129. Kompenhans J., Reichmuth J. (1987): 2-D flow field measurments in wind tunnels by means of particle image velocimetry, Proc. *6th Intl. Congr. on Appl. of Lasers and Electro-Optics, 8–12 Nov., San Diego, USA.*
130. Kompenhans J., Höcker R. (1988): Application of particle image velocimetry to high speed flows, *von Karman Institute for Fluid Dynamics, Lecture Series* 1988-06, Particle Image Displacement Velocimetry pp. 67–84 (⋆).
131. Kompenhans J., Raffel M., Vogt A., Fischer M. (1993): Aerodynamic investigations in low- and high-speed wind tunnels by means of particle image velocimetry, Proc. *15th Intl. Congr. on Instrumentation in Aerospace Simulation Facilities, 20–23 Sept., St. Louis, France (46).*
132. Kompenhans J., Raffel M. (1993): Application of PIV technique to transonic flows in a blow-down wind tunnel, Proc. SPIE 2005, *Intl. Symp. on Optics, Imaging and Instrumentation, 11–16 July, San Diego, USA. Optical Diagnostics*

in Fluid and Thermal Flow, ed. S.S. Cha, J.D. Trollinger, pp. 425–436.

133. Kompenhans J., Raffel M. (1994): The importance of image shifting to the applicability of the PIV technique for aerodynamic investigations, Proc. *7th Intl. Symp. on Appl. of Laser Techniques to Fluid Mechanics, 11–14 July, Lisbon, Portugal (35-6).*
134. Kompenhans J., Raffel M., Wernert P., Schäfer H.J. (1994): Instantaneous flow field measurements on pitching airfoils by means of particle image velocimetry, Proc. *Optical Methods and Data Processing in Heat and Fluid Flow, 14–15 April, London* pp. 117–121.
135. Kompenhans J., Raffel M., Willert C. (1996): PIV applied to aerodynamic investigations in wind tunnels, *von Karman Institute for Fluid Dynamics, Lecture Series* 1996–03, Particle Image Velocimetry.
136. Kompenhans J., Raffel M., Willert C., Wiegel M., Kähler C., Schröder A., Bretthauer B., Vollmers H., Stasicki B. (1996): Investigation of unsteady flow fields in wind tunnels by means of particle image velocimetry, in *Three-Dimensional Velocity and Vorticity Measuring and Image Analysis Techniques* ed. Th. Dracos, Kluwer Academic Publishers, Dordrecht, pp. 113-127.
137. Lecordier B., Mouqallid M., Trinité M. (1994): Simultaneous 2D measurements of flame front propagation by high speed tomography and velocity field by cross correlation, Proc. *7th Intl. Symp. on Appl. of Laser Techniques to Fluid Mechanics, 11–14 July, Lisbon, Portugal.*
138. Lee J., Farrel P.V. (1992): Particle image velocimetry measurements of IC engine valve flows, Proc. *6th Intl. Symp. on Appl. of Laser Techniques to Fluid Mechanics, 20–23 July, Lisbon, Portugal.*
139. Liu Z.C., Landreth C.C., Adrian R.J., Hanratty T.J. (1991): High resolution measurement of turbulent structure in a channel with particle image velocimetry, *Exp. Fluids*, **10**, pp. 301–312 (⋆).
140. Liu Z.C., Adrian R.J., Hanratty T.J. (1996): A study of streaky structures in a turbulent channel flow with particle image velocimetry, Proc. *8th Intl. Symp. on Appl. of Laser Techniques to Fluid Mechanics, 8–11 July, Lisbon, Portugal.*
141. Meinhart C.D. (1994): Investigation of turbulent boundary-layer structure using particle-image velocimetry, Ph.D. dissertation, Department of Theoretical and Applied Mechanics, University of Illinois, Urbana, Illinois.
142. Molezzi M.J., Dutton J.C. (1993): Application of particle image velocimetry in high-speed separated flows, *AIAA Journal*, **31**, pp. 438–446.
143. Raffel M. (1993): PIV-Messungen instationärer Geschwindigkeitsfelder an einem schwingenden Rotorprofil, Ph.D. dissertation, Universität Hannover, DLR-FB 93-50.
144. Raffel M., Kompenhans J. (1993): PIV measurements of unsteady transonic flow fields above a NACA 0012 airfoil, Proc. *5th Intl. Conf. on Laser Anemometry, Veldhoven, Netherlands,* pp. 527–535.
145. Raffel M., Seelhorst U., Willert C., Vollmers H., Bütefisch K.A. Kompenhans J. (1996): Measurement of vortical structures on a helicopter rotor model in a wind tunnel by LDV and PIV, Proc. *8th Intl. Symp. on Appl. of Laser Techniques to Fluid Mechanics, 8–11 July, Lisbon, Portugal (14-3).*
146. Raffel M., Höfer H., Kost F., Willert C., Kompenhans J. (1996): Experimental aspects of PIV measurements of transonic flow fields at a trailing edge model of a turbine blade, Proc. *8th Intl. Symp. on Appl. of Laser Techniques to Fluid Mechanics, 8–11 July, Lisbon, Portugal (28-1).*
147. Reuss D.L. (1993): Two-dimensional particle-image velocimetry with electrooptical image shifting in an internal combustion engine, Proc. SPIE 2005, *Optical Diagnostics in Fluid and Thermal Flow,* ed. S.S. Cha, J.D. Trollinger, pp. 413–424.

148. Schröder A. (1996): Untersuchung der Struktur des laminaren Zylindernachlaufs mit Hilfe der Particle Image Velocimetry, Diplomarbeit, Universität Göttingen.
149. Sebastian B. (1995): Untersuchung einer Motorinnenströmung mit der Particle Image Velocimetry, Proc. *4. Fachtagung Lasermethoden in der Strömungsmesstechnik, 12–14 September, Rostock, Germany.*
150. Vogt A., Baumann P., Gharib M., Kompenhans J. (1996): Investigations of a wing tip vortex in air by means of DPIV, Proc. *19th AIAA Advanced Measurements and Ground Testing Technology Conference, June 17–20, New Orleans, AIAA 96-2254.*
151. Wernet P.W., Pline A.D. (1991): Particle image velocimetry for the surface tension driven convection experiment using a particle displacement tracking technique, Proc. *4th Intl. Conf. on Laser Anemometry, Advances and Applications, Cleveland, Ohio..*
152. Westerweel J., Draad A.A., Van der Hoeven J.G.Th., Van Oord J. (1996): Measurement of fully-developed turbulent pipe flow with digital particle image velocimetry, *Exp. Fluids*, **20**, pp. 165–177.
153. Wiegel M., Fischer M. (1995): Proper orthogonal decomposition applied to PIV data for the oblique transition in a Blasius boundary layer, Proc. SPIE 2546 *Intl. Symp. on Optical Science, Engineering and Instrumentation, 9–14 July, San Diego, USA*, pp. 87–97.
154. Willert C.E. (1992): The interaction of modulated vortex pairs with a free surface, Ph.D. dissertation, Department of Applied Mechanics and Engineering Sciences, University of California, San Diego.
155. Willert C., Gharib M. (1997): The interaction of spatially modulated vortex pairs with free surfaces, *J Fluid Mech.*, **345**, pp. 227–250.
156. Willert C., Raffel M., Stasicki B., Kompenhans J. (1996): High-speed digital video camera systems and related software for application of PIV in wind tunnel flows, Proc. *8th Intl. Symp. on Appl. of Laser Techniques to Fluid Mechanics, 8–11 July, Lisbon, Portugal (18-1).*
157. Willert C., Raffel M., Kompenhans J., Stasicki B., Kähler C. (1997): Recent applications of particle image velocimetry in aerodynamic research, *Flow. Meas. Instrum.*, **7**, pp. 247–56.

Papers describing other particle imaging techniques

158. Agüi J.C, Jiménez J. (1987): On the performance of particle tracking, *J Fluid Mech.*, **185**, pp. 447–468 (⋆).
159. Dadi M., Stanislas M., Rodriguez O., Dyment A. (1991): A study by holographic velocimetry of the behaviour of free particles in a flow, *Exp. Fluids*, **10**, pp. 285–294.
160. Dracos Th. (1996): Particle tracking velocimetry (PTV) – basic concepts, in *Three-Dimensional Velocity and Vorticity Measuring and Image Analysis Techniques* ed. Th. Dracos, Kluwer Academic Publishers, Dordrecht, pp. 155–160.
161. Dracos Th. (1996): Particle tracking in three-dimensional space, in *Three-Dimensional Velocity and Vorticity Measuring and Image Analysis Techniques* ed. Th. Dracos, Kluwer Academic Publishers, Dordrecht, pp. 209–227.
162. Gharib M., Willert C.E. (1990): Particle tracing – revisited, in *Lecture Notes in Engineering: Advances in Fluid Mechanics Measurements* **45**, ed. M. Gad-el-Hak, Springer-Verlag, New York, pp. 109–126.
163. Guezennec Y.G., Brodkey R.S., Trigui N., Kent J.C. (1994): Algorithms for fully automated three-dimensional particle tracking velocimetry, *Exp. Fluids*, **17**, pp. 209–219.

164. Imaichi K., Ohmi K. (1983): Numerical processing of flow-visualization pictures – measurement of two-dimensional vortex flow, *J Fluid Mech.*, **129**, pp. 283–311.
165. Kent J.C., Eaton A.R. (1982): Stereo photography of neutral density He-filled bubbles for 3-D fluid motion studies in an engine cylinder, *Appl. Optics*, **21**, pp. 904-912.
166. Prandtl, L. (1905): Über Flüssigkeitsbewegung bei sehr kleiner Reibung, Proc. *Verhandlungen des III. Internationalen Mathematiker-Kongresses, Heidelberg, 1904, Teubner, Leipzig, pp. 404–491.*
167. Siu Y.W., Taylor A.M.K.P., Whitelaw J.H. (1994): Lagrangian tracking of particles in regions of flow recirculation, Proc. *First International Conference on Flow Interaction, Hong Kong.* 330–333
168. Virant M. (1996): Anwendung des dreidimensionalen "Particle-Tracking-Velocimetry" auf die Untersuchung von Dispersionsvorgängen in Kanalströmungen, Ph.D. dissertation, Institut für Hydromechanik und Wasserwirtschaft, ETH Zürich.
169. Virant M., Dracos Th. (1996): Establishment of a videogrammetic PTV system, in *Three-Dimensional Velocity and Vorticity Measuring and Image Analysis Techniques* ed. Th. Dracos, Kluwer Academic Publishers, Dordrecht, pp. 229–254.

Papers regarding techniques utilized in PIV

170. Ashley P.R., Davis J.H. (1987): Amorphous silicon photoconductor in a liquid crystal spatial light modulator, *Appl. Optics*, **26**, pp. 241–246.
171. Efron U., Grinberg J., Braatz P.O., Little M.J., Reif P.G., Schwartz R.N. (1985): The silicon liquid-crystal light valve, *J. Appl. Phys.*, **57**, pp. 1356–1368.
172. Gabor A.M., Landreth B., Moddel G. (1993): Integrating mode for an optically addressed spatial light modulator, *Appl. Optics*, **37**, pp. 3064–3067.
173. Klein F. (1968): *Elementarmathematik vom höheren Standpunkte aus, Zweiter Band: Geometrie*, Springer Verlag, Berlin.

Papers not directly related to PIV

174. Garg V.K. (1992): Natural convection between concentric spheres, *Int. J. Heat Mass Transfer*, **35**, pp. 1938–1945.
175. König. M, Noack B.R., Eckelmann H. (1993): Discrete shedding modes in the von Kármán vortex street, *Phys. Fluids A*, **5**, pp. 1846–1848.
176. Mack L.R., Hardee H.C. (1968): Natural convection between concentric spheres at low Rayleigh numbers, *Int. J. Heat Mass Transfer*, **11**, pp. pages???.
177. Meyers J.F., Komine H. (1991): Doppler global velocimetry – a new way to look at velocity, Proc. *ASME Fourth International Conference on Laser Anemometry, Cleveland.*
178. Röhle I. (1997): Three-dimensional Doppler global velocimetry in the flow of a fuel spray nozzle and in the wake region of a car, *Flow Measurement and Instrumentation*, **7**, pp. 287–294.
179. Rotta J. (1990): *Die Aerodynamische Versuchsanstalt in Göttingen, ein Werk Ludwig Prandtls*, Vandenhoek & Ruprecht, Göttingen.
180. Weigand A., Gharib M. (1994): On the decay of a turbulent vortex ring, *Phys. Fluids*, **6**, pp. 3806–3808.

A. Matrices for perspective projection

The transformation D_H describes the displacement of a particle by D_X, D_Y, D_Z in figure 2.28 and figure 4.15:

$$\mathsf{D_H} = \begin{bmatrix} 1 & 0 & 0 & D_X \\ 0 & 1 & 0 & D_Y \\ 0 & 0 & 1 & D_Z \\ 0 & 0 & 0 & 1 \end{bmatrix}.$$

$\mathsf{P_H}$ is required for imaging a point on to the image plane:

$$\mathsf{P_H} = \begin{bmatrix} 1 & 0 & 0 & 0 \\ 0 & 1 & 0 & 0 \\ 0 & 0 & 0 & 0 \\ 0 & 0 & -1/z_0 & 1 \end{bmatrix}.$$

$\mathsf{P_H^{-1}}$ is the transformation of a single image point from the image plane on to the (virtual) light sheet plane:

$$\mathsf{P_H^{-1}} = \begin{bmatrix} 1 & 0 & 0 & 0 \\ 0 & 1 & 0 & 0 \\ 0 & 0 & 0 & -z_0(1+M) \\ 0 & 0 & 0 & -M \end{bmatrix}.$$

The movement of a virtual image of a point in the light sheet plane within the homogeneous mirror coordinate system x^*, y^*, z^* (figure 4.15) due to a mirror rotating with an angular velocity of ω_m is given by:

$$\mathsf{Rot}_{y,\mathrm{H}} = \begin{bmatrix} \cos(2\omega_m t) & 0 & \sin(2\omega_m t) & 0 \\ 0 & 1 & 0 & 0 \\ -\sin(2\omega_m t) & 0 & \cos(2\omega_m t) & 0 \\ 0 & 0 & 0 & 1 \end{bmatrix}.$$

The transformation T_H is required for changing from camera coordinates to mirror coordinates:

$$\mathsf{T_H} = \begin{bmatrix} 1 & 0 & 0 & -X_m \\ 0 & 1 & 0 & 0 \\ 0 & 0 & 1 & Z_m - (z_0 + Z_0) \\ 0 & 0 & 0 & 1 \end{bmatrix}.$$

The inverse transformation for changing from homogeneous mirror coordinates back to homogeneous camera coordinates is $\mathsf{T}_{\mathrm{H}}^{-1}$:

$$\mathsf{T}_{\mathrm{H}}^{-1} = \begin{bmatrix} 1 & 0 & 0 & X_{\mathrm{m}} \\ 0 & 1 & 0 & 0 \\ 0 & 0 & 1 & -Z_{\mathrm{m}} + (z_0 + Z_0) \\ 0 & 0 & 0 & 1 \end{bmatrix}.$$

B. Mathematical appendix

B.1 Convolution with the Dirac delta function

For a real function f of the variable $\boldsymbol{x}$, and a given vector $\boldsymbol{x_i}$, we have:

$$f(\boldsymbol{x} - \boldsymbol{x_i}) = f(\boldsymbol{x}) * \delta(\boldsymbol{x} - \boldsymbol{x_i})$$

where $*$ is a convolution product and δ the Dirac delta function.

B.2 Particle images

For infinite small geometric particle images the particle image intensity distribution (intensity profile) is given by the point spread function $\tau(\boldsymbol{x})$, which has been assumed to have a Gaussian shape [66]:

$$\tau(\boldsymbol{x}) = K \exp\left(-\frac{8\,|\boldsymbol{x}|^2}{d_\tau^2}\right)$$

with

$$K = \frac{8\,\tau_0}{\pi {d_\tau}^2}\,. \tag{B.1}$$

B.3 Convolution of Gaussian image intensity distributions

If we assume Gaussian image intensity distributions as given in equation (B.1) the product of two displaced images yields:

$$\tau(\boldsymbol{x} - \boldsymbol{x_i})\,\tau(\boldsymbol{x} - \boldsymbol{x_i} + \boldsymbol{s}) = K^2 \exp\left[-8\left(|\boldsymbol{x} - \boldsymbol{x_i}|^2 + |\boldsymbol{x} - \boldsymbol{x_i} + \boldsymbol{s}|^2\right)/{d_\tau}^2\right]\,.$$

For two vectors $\boldsymbol{a}$ and $\boldsymbol{b}$, it can be shown that:

$$|\boldsymbol{a}|^2 + |\boldsymbol{a} + \boldsymbol{b}|^2 = |\boldsymbol{b}|^2/2 + 2|\boldsymbol{a} + \boldsymbol{b}/2|^2\,.$$

Hence:

$$
\begin{aligned}
\int_{a_I} \tau(\boldsymbol{x}-\boldsymbol{x}_i)\,\tau(\boldsymbol{x}-\boldsymbol{x}_i+\boldsymbol{s})\,d\boldsymbol{x} &= \exp\left(-\frac{4\,|\boldsymbol{s}|^2}{d_\tau^2}\right) \\
&\times \int_{a_I} K^2 \exp(-16|\boldsymbol{x}-\boldsymbol{x}_i+\boldsymbol{s}/2|^2/{d_\tau}^2)\,d\boldsymbol{x} \\
&= \exp\left(-\frac{8\,|\boldsymbol{s}|^2}{\left(\sqrt{2}\,d_\tau\right)^2}\right) \\
&\times \int_{a_I} \tau^2(\boldsymbol{x}-\boldsymbol{x}_i+\boldsymbol{s}/2)\,d\boldsymbol{x}\;.
\end{aligned}
$$

B.4 Expected value

We defined $f_1(\boldsymbol{X}) = V_0(\boldsymbol{X})V_0(\boldsymbol{X}+\boldsymbol{D})$ in equation (3.8). Here we will determine

$$
E\left\{\sum_{i=1}^{N} f_1(\boldsymbol{X}_i)\right\}\;.
$$

The sum must be considered as a function of N random variables $\boldsymbol{X}_1, \boldsymbol{X}_2 ... \boldsymbol{X}_N$. Hence:

$$
E\left\{\sum_{i=1}^{N} f_1(\boldsymbol{X}_i)\right\} = \sum_{i=1}^{N} E\left\{f_1(\boldsymbol{X}_i)\right\} = \sum_{i=1}^{N} \frac{1}{V_F} \int f_1(\boldsymbol{X}_i)\,d\boldsymbol{X}_i
$$

$$
\Rightarrow \quad E\left\{\sum_{i=1}^{N} f_1(\boldsymbol{X}_i)\right\} = \frac{N}{V_F} \int_{V_F} f_1(\boldsymbol{X})\,d\boldsymbol{X}\;.
$$

C. List of Symbols

$\boldsymbol{a}$	local acceleration vector
a_{I}	interrogation area
C_{I}	spatial auto-covariance
C_{II}	spatial cross-covariance
C_{R}	constant factor of the correlation function
c_{II}	spatial correlation coefficient
c_{II}	cross-correlation coefficient
c_1, c_2	constant factors for outlier detection
$\boldsymbol{D}$	particle displacement within flow field
D_{a}	aperture diameter
D_{H}	particle displacement between the light pulses
D_{max}	maximum particle displacement
D_{Photo}	photographic emulsion density
$\boldsymbol{d}$	particle image displacement
$\overline{d}$	mean value of measured image displacement
d'	approximation of the image diameter
d_{diff}	diffraction limited imaging diameter = diameter of Airy disk
d_{max}	maximum particle image displacement
d_{min}	minimum particle image displacement
d_{opt}	optimum particle image displacement
d_{p}	particle diameter
d_{r}	difference between real and ideal particle image diameter
d_{s}	diameter of the Airy pattern
d_{shift}	particle image displacement due to the rotating mirror system
d_{τ}	particle image diameter
E	exposure during recording
$E\{\}$	expected value
F_{i}	in-plane loss of correlation
F_{o}	out-of-plane loss of correlation
f	lens focal length
$f_{\#}$	lens f-number
$\boldsymbol{g}$	acceleration due to gravity
H	index for homogeneous coordinate system
I	image intensity field of the first exposure

I'	image intensity field of the second exposure
$I_0(Z)$	light sheet intensity profile in the Z direction
I_{Inc}	light intensity for interrogation of photographic recordings
I_{trans}	light intensity locally transmitted through a photographic recording
I_z	maximum intensity of the light sheet
I^+	correlation of the intensity field with itself
$\hat{I}, \hat{I}'$	Fourier transforms of I and I'
M	magnification factor
$\tilde{M}_{\mathrm{TF}}(r')$	modulation transfer at a certain spatial freqency ν
M, N	interrogation window size in pixels
m_{p}	particle mass
N	total number of vectors in the data set
$\mathcal{N}$	particle image density (per unit area)
$\mathcal{N}_{\mathrm{I}}$	number of particle images per interrogation window
n_{exp}	number of exposures per recording
P	matrix of perspective projection
P_{H}	projection tensor
Pr	Prandtl number
QE	quantum efficiency of a sheet
QL	number of quantization levels
r'	spatial frequency
Ra	Rayleigh number
R_{C}	mean background correlation
R_{D}	displacement correlation peak
R_{D^+}	positive displacement correlation peak
R_{D^-}	negative displacement correlation peak
$\mathsf{Ref}(\theta)$	matrix for the reflexion in a plane tilted by θ with respect to the $y^\star$-$z^\star$ plane
R_{F}	noise term due to random particle correlations
R_{I}	spatial auto-correlation
R_{II}	spatial cross-correlation
Rot_y	matrix of mirror rotation
R_{P}	particle image self-correlation peak
R_τ	correlation of a particle image
$\boldsymbol{s}$	separation vector in the correlation plane
$\boldsymbol{s}_{\mathrm{D}}$	displacement vector in the correlation plane
T	local transmittance of a photographic emulsion
T	transformation matrix for rotation between camera and mirror coordinates
$T(x, y)$	local varying intensity transmittance of photographic emulsion
t	time of the first exposure
t'	time of the second exposure
t''	time of the third exposure
t_e	frame transfer time
t_f	pulse length
Δt	exposure time delay

$\Delta t_{\min}$	minimum time delay
$\Delta t_{\text{transfer}}$	charge transfer time in CCD sensor
$\boldsymbol{U}$	flow velocity vector
U, V	in-plane components of the velocity $\boldsymbol{U}$
$\boldsymbol{U}_{\text{g}}$	gravitationally induced velocity
$U_{\max}$	maximum flow velocity in streamwise direction
U_{mean}	mean flow velocity in streamwise direction
$U_{\min}$	minimum flow velocity in streamwise direction
$\boldsymbol{U}_{\text{p}}$	velocity of the particle
$\boldsymbol{U}_{\text{s}}$	velocity lag
U_{shift}	shift velocity
U_{τ}	friction velocity, $\sqrt{\tau_{\text{w}}/\rho}$
U_{∞}	free stream velocity
ΔU_I	fluctuation of the velocity within the interrogation volume
V_{I}	interrogation volume in the flow
V_{F}	fluid volume that has been seeded with particles
$V_0(\boldsymbol{x}_i)$	intensity transfer function for individual particle images
v_I	interrogation area (image plane)
W	out-of-plane component of the velocity $\boldsymbol{U}$
$W_0(X, Y)$	interrogation window function back projected into the light sheet
X, Y, Z	flow field coordinate system
X_{m}	distance between rotating mirror and optical axis
$\boldsymbol{X}_{\text{p}}$	particle position within flow field
$\boldsymbol{X}_{\text{v}}, \boldsymbol{X'}_{\text{v}}$	point in the virtual light sheet plane
x, y, z	image plane coordinate system
$x^{\star}, y^{\star}, z^{\star}$	mirror coordinate system
$\boldsymbol{x}$	point in the filme plane, $\boldsymbol{x} = \boldsymbol{x}(x, y)$
$\Delta x_0, \Delta y_0$	interrogation area dimensions
$\Delta X_0, \Delta Y_0$	horizontal, vertical interrogation area dimensions within light sheet
$\Delta x_{\text{step}}, \Delta y_{\text{step}}$	distance between two interrogation areas
$\Delta X, \Delta Y$	grid spacing in object plane
Z_0	distance between object plane and lens plane
Z_{m}	distance between object plane and mirror axis
z_0	distance between image plane and lens plane
ΔZ_0	light sheet thickness

Greek symbols

$\delta(\boldsymbol{x})$	Dirac delta function at position $\boldsymbol{x}$
δ_Z	depth of focus
ϵ_{tot}	total displacement error
ϵ_{bias}	displacement bias error
ϵ_{sys}	systematic error

$\epsilon_{\mathrm{resid}}$	residual (nonsystematic) error
$\epsilon_{\mathrm{thresh}}$	threshold for outlier detection
ϵ_U	velocity measurement uncertainty
γ	photographic gamma
$\boldsymbol{\Gamma}$	state of the ensemble
λ	wavelength of light
μ	dynamic viscosity
μ_{I}	spatial average of I
ν	kinematic viscosity, μ/ρ
ω_{m}	angular velocity of the rotating mirror
ρ	fluid density
ρ_{m}	spatial resolution limit during recording
ρ_{p}	particle density
σ	width parameter of Gaussian bell curve
σ_I	spatial variance of I
$\tau(\boldsymbol{x})$	point spread function of imaging lens
τ_{s}	relaxation time
$\boldsymbol{\omega}$	vorticity vector
$\omega_x, \omega_y, \omega_z$	vorticity components

Abbreviations

CCD	charge coupled device
CCIR	video transmission standard
CW	continuous wave
DLR	Deutsches Zentrum für Luft- und Raumfahrt (= German Aerospace Center)
DPIV	digital particle image velocimetry
DSPIV	digital stereo particle image velocimetry
DSR	dynamic spatial range [38] Ratio of the largest observable length scale to the smallest observable length scale (typically interrogation window size)
DVR	dynamic velocity range [38] Ratio of the maximum measurable velocity to the minimum resolvable velocity
FT	Fourier transformation
FFT	fast Fourier transformation
Mod	image modulation
MTF	modulation transfer function
NTSC	National Telivision System Committee
PAL	phase alternating line
PIDV	particle image displacement velocimetry
PIV	particle image velocimetry

PTF	phase transfer function
QE	quantum efficiency
SNR	signal-to-noise ratio
Tu	turbulence level in a flow
TUG	low turbulence wind tunnel at DLR Göttingen
pixel,px	picture element
rms	root mean square

Index

Druck: Mercedes-Druck, Berlin
Bindung: Buchbinderei Lüderitz & Bauer, Berlin

Printing: Mercedesdruck, Berlin
Binding: Buchbinderei Lüderitz & Bauer, Berlin